SpringerWienNewYork

Helga Stan-Lotter
Sergiu Fendrihan
Editors

Adaption of Microbial Life to Environmental Extremes

Novel Research Results and Application

SpringerWienNewYork

Prof. Dr. Helga Stan-Lotter
Department of Microbiology, University of Salzburg, Salzburg, Austria

Dr. Sergiu Fendrihan
Romanian Bioresource Center and Advanced Research Association,
Bucharest, Romania

This work is subject to copyright.

All rights are reserved, whether the whole or part of the material is concerned, specifically those of translation, reprinting, re-use of illustrations, broadcasting, reproduction by photocopying machines or similar means, and storage in data banks.

Product Liability: The publisher can give no guarantee for all the information contained in this book. The use of registered names, trademarks, etc. in this publication does not imply, even in the absence of a specific statement, that such names are exempt from the relevant protective laws and regulations and therefore free for general use.

© 2012 Springer-Verlag/Wien
Printed in Germany

SpringerWienNewYork is part of
Springer Science + Business Media
springer.at

Cover Illustrations: © H. Jóhannesson/Boiling pit in Vonarskard geothermal area in Iceland
Typesetting: Thomson Press (India) Ltd. Chennai
Printing: Strauss GmbH, 69509 Mörlenbach, Germany

Printed on acid-free and chlorine-free bleached paper
SPIN: 12738830

With 27 (partly coloured) Figures

Library of Congress Control Number: 2011936882

ISBN 978-3-211-99690-4 SpringerWienNewYork

Foreword

Life on the edge, life at the physico-chemical limits, weird and eccentric life – these are the descriptive terms assigned to extremophiles. The first European workshop entitled "Microbial adaptation to extreme environments" was held at the University of Nijmegen in the Netherlands in 1973 (Heinen et al. 1974). The first use of the term "extremophiles" is credited to Robert MacElroy, a NASA exobiologist, who participated in this workshop (MacElroy 1974).

Many more conferences on extremophiles followed, the journal "Extremophiles" was launched and the CAREX initiative (Coordination Action for Research Activities on Life in Extreme Environments) was started (www.carex-eu.org), providing an extensive data base and recently a new roadmap.

The often surprising ranges of physico-chemical factors, within which life is possible, stimulated many scientists to explore astrobiological perspectives, with the reasoning that findings on the evolution and mechanisms of adaptation of life at extremes would help to understand the environments of other planets or moons. These astrobiological aspects are dealt with in several chapters of this volume (*Billi; Mapelli et al.; Gomez and Parro; Moissl-Eichinger; Stan-Lotter*).

The biodiversity of extreme environments appears of unexpected and enormous size; its magnitude is still largely unknown. Classical and molecular approaches are used for its investigation and described by *Enache et al., Fendrihan and Negoiță, Mapelli et al., Heulin et al., Hreggvidsson et al.* and *Pearce*.

Several updated lists of taxonomic descriptions of extremophilic species and strains are provided: halophilic archaea and bacteria (*Enache et al.*), desiccation-resistant and thermotolerant bacteria (*Heulin et al.*), thermophilic bacteria (*Hreggvidsson et al.*), psychrophilic bacteria, archaea, algae and fungi (*Fendrihan and Negoiță; Pearce*).

Adaptations to extreme environmental conditions, especially on the molecular level, are a topic of uninterrupted fascination since the first investigations were made, and are described by *Billi, Heulin et al., Fendrihan and Negoiță,* and *Pearce*.

Numerous applications of extremophiles for biotechnological purposes have already been implemented, although the whole potential is certainly not realized yet – treatments of this subject are included in the chapters by *Billi, Enache et al., Fendrihan and Negoiță,* and *Pearce.*

Sadly, our dear colleague Teodor Negoiță passed away during the completion of this volume (on March 23, 2011 at the age of 64). Dr. Negoiță was the founder of the Romanian Institute of Polar Research. He had organized several international expeditions into the Arctic and in the year 2005 managed to establish the Romanian Law-Racoviță Research Station in the Antarctic.

Our thanks go to the initiators and organizers of the CAREX project – Cynan Ellis-Evans, Nicolas Walter, Petra Rettberg – and for the opportunity of participating in great meetings at wonderful places – Sant Feliu de Guixols, Sasbachwalden, – which facilitated the interactions between people who would otherwise not have met, and which was also the partial foundation for the contributions in this book.

We are also very grateful to Amrei Strehl and Eva-Maria Oberhauser from the Springer-Verlag for helpful suggestions and excellent cooperation.

Salzburg, Bucharest, May 2011   *Helga Stan-Lotter, Sergiu Fendrihan*

References

Heinen W, Skinner FA, Tempest WD, van den Ende G, Vogels GD, van den Drift C, Dundas I, Harder W, Weerkamp A, Gounot AM, van Beeumen J, MacElroy RD, Knoope P, Schwartz AW, Schlegel HG (1974) Proceedings of the first European workshop on microbial adaptation to extreme environments. Biosystems 6: 57–80

MacElroy RD (1974) Some comments on the evolution of extremophiles. Biosystems 6: 74–75

Contents

Physico-chemical boundaries of life
Helga Stan-Lotter 1

1 Introduction 1
2 Brief history of life on Earth 3
3 Prokaryotes, eukaryotes, the tree of life and viruses 4
4 Extreme environments and their inhabitants 7
    4.1 Temperature ranges of microorganisms 7
    4.2 Low water activity: halophiles, osmophiles, and xerophiles 8
    4.3 Extremes of pH and low nutrients 9
    4.4 High pressure 10
    4.5 Viruses 10
    4.6 Extremophilic multicellular organisms 11
5 Microbial survival of extreme conditions 12
6 Concluding remarks 14

Microbial diversity in deep hypersaline anoxic basins
Francesca Mapelli, Sara Borin, Daniele Daffonchio 21

1 Introduction 21
2 Localization and origin of DHABs 21
    2.1 The anoxic brine basins in the Red Sea: the Shaban and Kebrit Deeps 22
    2.2 The Orca basin and other seafloor anoxic brines in the Gulf of Mexico . 22
    2.3 Eastern Mediterranean Sea: DHABs and the Chefren mud volcano 23
3 Geochemical features of DHABs 23
    3.1 DHABs of the Red Sea 24
    3.2 DHABs of the Gulf of Mexico 24
    3.3 DHABs of the Eastern Mediterranean Sea 25
    3.4 DHABs of the Mediterranean Sea: the Chefren mud volcano (Nile Deep Sea Fan) ... 27
4 Microbial life in the DHABs: biodiversity and adaptation 27
    4.1 Prokaryotic biodiversity of the Kebrit and Shaban Deeps, Red Sea 28

    4.2 Phylogenetic analyses of the microbial communities colonizing different seafloor hypersaline brines of the Gulf of Mexico 29
    4.3 Investigation of the microbiota composition inhabiting the Chefren mud volcano (Nile Deep Sea Fan) 30
    4.4 Stratified microbial communities at the deep seawater-brine interface of hypersaline anoxic basins of the Eastern Mediterranean Sea................. 31
5 Concluding remarks... 33

Microbial speciation in the geothermal ecosystem
Gudmundur Oli Hreggvidsson, Solveig K. Petursdottir, Snaedis H. Björnsdottir, Olafur H. Fridjonsson 37

1 Introduction... 37
2 Geothermal areas 39
    2.1 High- and low-temperature fields 41
    2.2 Origin of hot spring water 42
    2.3 Surface characteristics of geothermal areas..................................... 43
3 Ecology of thermophiles 48
    3.1 Temperature.. 49
    3.2 pH... 49
    3.3 Energy sources as a selective pressure 50
4 Biogeography of thermophiles 52
    4.1 Dispersal of thermophiles 55
    4.2 Biogeography of *Thermus* 60
    4.3 Ecological adaptations of *Thermus* 61

Bacterial adaptation to hot and dry deserts
Thierry Heulin, Gilles De Luca, Mohamed Barakat, Arjan de Groot, Laurence Blanchard, Philippe Ortet, Wafa Achouak................................ 69

1 Introduction... 69
2 Characteristic of hot and dry deserts with emphasis on Sahara 70
3 Mechanisms for desiccation tolerance 71
4 Counting and describing bacterial populations in desert environments 73
5 *Ramlibacter* and its life cycle... 78
6 *Deinococcus* and DNA repair 80
7 Conclusions 81

Extremophiles in Antarctica: life at low temperatures
David A. Pearce.. 87

1 Introduction... 88
2 Environmental extremes associated with the Antarctic 90

viii

3 Extreme environments that also occur in the Antarctic	92
4 Extreme environments particular to the Antarctic	97
5 Key Antarctic extremophiles	99
6 Biodiversity	100
7 Methodology	102
8 Adaptations	104
8.1 *Pseudoalteromonas haloplanktis*	107
8.2 *M. burtonii*	108
8.3 *M. frigidum*	108
9 Discussion and future perspectives	109

Anhydrobiotic rock-inhabiting cyanobacteria: potential for astrobiology and biotechnology
Daniela Billi 119

1 Introduction	119
2 Cyanobacteria in hot and cold desert rocks	120
3 Cyanobacterial adaptation to Earth's deserts	123
4 Survival of desert cyanobacteria beyond Earth	125
5 Biotechnological exploitation of anhydrobiosis	127

Psychrophilic microorganisms as important source for biotechnological processes
Sergiu Fendrihan, Teodor G. Negoiță 133

1 Introduction	133
2 Diversity of cold-adapted microorganisms	134
3 Ecology and biology	135
4 Cold environments	136
5 Adaptation to cold environments	140
6 Applications of psychrophilic microorganisms	144
7 Conclusions	157

Halophilic microorganisms from man-made and natural hypersaline environments: physiology, ecology, and biotechnological potential
Madalin Enache, Gabriela Popescu, Takashi Itoh, Masahiro Kamekura 173

1 Introduction	173
2 Halophilic microorganisms	173
2.1 Haloarchaea	174
2.2 Halophilic Bacteria	181

2.3 Romanian hypersaline environments...	182
2.4 Halophiles from salty environments in Romania...............................	185
3 Biotechnological potential...	186
3.1 Bioremediation...	186
3.2 Nanobiotechnology..	187
3.3 S-layers..	187
3.4 Extracellular enzymes...	188
3.5 Halocins..	189
3.6 Exopolysaccharides..	189
3.7 Resistance to heavy metals...	190
3.8 Therapeutical value...	191
4 Concluding remarks..	192

Applications of extremophiles in astrobiology: habitability and life detection strategies
Felipe Gómez, Víctor Parro... 199

1 Introduction...	199
2 Extremophiles and astrobiology...	200
2.1 Extremophiles	200
2.2 Habitability..	203
3 Extremophiles as sources of biomarkers for the search for extraterrestrial life	204
3.1 Preservation of molecular biomarkers..	207
3.2 Stability of organic and living matter in planetary bodies: the case of Mars......	211
4 Life detection strategies..	212
4.1 Immunological systems for the detection of signs of life	213
4.2 Detection systems for extant life based on metabolic activity	217
4.3 Other systems for the detection of signs of life................................	219
4.4 Signs of life by remote sensing systems...	221

Extremophiles in spacecraft assembly clean rooms
Christine Moissl-Eichinger.. 231

1 Introduction...	231
1.1 Planetary protection..	231
1.2 Clean rooms..	233
1.3 Contamination control and examinations..	233
1.4 Clean room microbiology..	234
1.5 Extremophiles and extremotolerants – definition	237
2 Spore-forming microorganisms..	237
2.1 Background...	237
2.2 Results..	239

3 Oligotrophic microorganisms	240
3.1 Background	240
3.2 Results	241
4 Alkaliphiles and acidophiles	242
4.1 Background	242
4.2 Results	242
5 Autotrophic and nitrogen fixing microorganisms	246
5.1 Background	246
5.2 Results	246
6 Anaerobes	246
6.1 Background	246
6.2 Results	247
7 Thermophiles and psychrophiles	249
7.1 Background	249
7.2 Results	249
8 Halophiles	250
8.1 Background	250
8.2 Results	250
9 Archaea	251
9.1 Background	251
9.2 Results	251
10 Other extremophiles	252
11 Lessons learned from the Herschel campaign: extremophiles are everywhere	253
12 The bacterial diversity beyond cultivation, or cultivation vs. molecular analyses	255
Subject index	263
Organism index	271
List of contributors	281

Physico-chemical boundaries of life

Helga Stan-Lotter

Department of Molecular Biology, University of Salzburg, Salzburg, Austria

1 Introduction

> All life we know no matter how freaky in other respects, is still based on organic molecules dissolved in water, and we all use the same basic cellular machinery. Extremophiles haven't fundamentally changed the way we think about strategies to look for life, but they have bolstered the optimism with which we search. Right now anywhere with liquid water is considered a possible habitat, and this guides our quest. (Grinspoon 2003)

When attempting to define extremophiles – meaning organisms living in an extreme environment – it is obvious that a strongly anthropocentric component emerges. Extreme environments are considered "hostile to higher forms of life" and "uninhabitable by other organisms" (Richard Johnson, NASA Director). Thomas D. Brock, a microbiologist, who pioneered in the 1970s the study of life in thermal springs in Yellowstone National Park, USA, defined the characteristics of extreme habitats from the taxonomists' point of view: "environments with a restricted species diversity and the absence of some taxonomic groups" (Brock 1969).

Why there is a limited diversity and why some groups are missing will be comprehensible when the physico-chemical factors, which are characteristics of extreme environments (see Table 1), are examined. These parameters define niches, which allow occupancy by only certain groups of organisms; none can be expected to survive the whole range of conditions. An early ecologist, Victor Shelford, presented already in 1913 the observation that organisms will usually be limited by abiotic factors in his "Law of tolerance" (cited in Krebs 2008). Each particular factor that an organism responds to in an ecological system has what he called limiting effects. The factors function within a range, that is, a maximum value and a minimum value for the factors exist, which Shelford designated "limits of tolerance" (see Table 1). For an organism to succeed in a given environment, each of a complex set of conditions must remain within the tolerance range of that organism, and if any condition exceeds the minimum or maximum tolerance of that organism, it will fail to thrive.

Table 1. Classes and examples of extremophilic prokaryotes

Physicochemical factor	Descriptive term	Genus/species	Lineage	Habitat	Minimum	Optimum	Maximum	Reference
Temperature								
High	Hyperthermophile	*Pyrolobus fumarii*	Archaea	hydrothermal	90°C	106°C	113°C	Blöchl et al. (1997)
High	Hyperthermophile	Strain 121	Archaea	Black smoker	85°C		121°C	Kashefi and Lovley (2003)
High	Hyperthermophile	*Methanopyrus kandleri* strain 116	Archaea	Black smoker fluid	90°C	105°C	122°C	Takai et al. (2008)
Low	Psychrophile	*Psychromonas ingrahamii*	Bacteria	Sea ice	−12°C	5°C	10°C	Auman et al. (2006)
pH low	Acidophile	*Picrophilus oshimae*	Archaea	Acidic hot spring	pH −0.06	pH 0.7	pH 4	Schleper et al. (1995)
	Acidophile	*Ferroplasma acidarmanus*	Archaea	Acid mine drainage	pH 0	pH 1.2	pH 2.5	Edwards et al. (2000)
pH high	Alkaliphile	*Alkaliphilus transvaalensis*	Bacteria	Deep gold mine	pH 8.5	pH 10	pH 12.5	Takai et al. (2001)
Hydrostatic pressure	Piezophile	*Moritella yayanosii*	Bacteria	Ocean sediment	50 MPa	70 MPa	110 MPa	Nogi and Kato (1999)
	Piezophile	*Pyrococcus* CH1	Archaea	Black smoker	20 MPa	52 MPa	120 MPa	Zeng et al. (2009)
Salt (NaCl)	Halophile	*Halobacterium salinarum*	Archaea	Saltern	15% NaCl	25% NaCl	32% NaCl	Grant et al. (2001)
Water activity	Xerophile	*Xeromyces bispora*	Fungus	Mouldy fruit	a_w 0.61	a_w 0.82	a_w 0.92	Hocking and Pitt (1999)

The recent years have seen an unprecedented expansion of knowledge about the physico-chemical limits of life. Several books and reviews have been published on extremophilic microorganisms, occasionally including higher organisms (e.g., Horikoshi and Grant 1998; Seckbach 2000; Rothschild and Mancinelli 2001; Oren and Rainey 2006; Gerday and Glansdorff 2007; Seckbach and Walsh 2009), and viruses from extreme environments (Le Romancer et al. 2007). Dedicated scientific journals are available, e.g. *Extremophiles*, *Archaea*, and *Astrobiology*. More specific books and articles, pertaining to certain groups of extremophiles, are mentioned in the chapters of this volume.

The purpose of this chapter is to provide an overview of the extreme environmental factors which influence life on Earth, a short survey of the various types of extremophiles – including viruses – as well as an update of current records for the limits of growth on the one hand and survival of extreme conditions on the other hand.

2 Brief history of life on Earth

The Earth is about 4.6 billion years old; the most ancient rocks were dated to be 3.5–3.86 billion years old (Westall 2005). They are located in Southern Africa (Swaziland and Barberton Greenstone Belt), Western Australia (Warrawoona series and Pilbara formation) and Greenland (Isua rocks). Besides volcanic and carbonaceous rocks, the sedimentary rocks are of particular interest, since their formation required liquid water, probably in the form of oceans, and therefore, the conditions for life did exist.

The evidence for early microbial life is derived from fossilized remains of cells in ancient rocks and on an abundance of the lighter carbon isotope (^{12}C), due to its generally assumed preference over the heavier isotope (^{13}C) by microorganisms (see Westall 2005, for a discussion). Microfossils, which were found in old sedimentary rocks, are similar in shape, size and arrangements as modern bacteria; the earliest forms were probably cocci (spheres) and short straight or curved rods (Westall 2005).

The environments on the early Earth were different than today, especially temperatures were likely much higher than now. It is thought that surface temperatures during the first 200 million years exceeded 100°. The accumulation of water occurred only later, when the Earth was cooling down. How much time this process took is not known; if life originated at the end of the first half million years – as the microfossils suggest – Earth was still fairly hot, and therefore the first microorganisms were likely thermophilic, or at least thermotolerant.

The atmosphere of the early Earth did not contain oxygen; the main gases were carbon dioxide, water vapor, hydrogen sulfide, with probably smaller amounts of nitrogen, methane, and carbon monoxide (Westall 2005). Only much later, about 2 billion years ago, oxygen was produced by phototrophic bacteria and its concentration in the atmosphere was slowly rising to reach the present value of 20.95% (Nealson and Conrad 1999). Early life forms were thus anaerobes; even up to now, there are still many niches where anaerobic microorganisms are thriving. The lack of oxygen meant that there was no protective layer of ozone in the upper atmosphere and therefore, a high flux of ultraviolet radiation reached the surface of Earth (Westall 2005). Larger unicellular organisms did not exist before about 1 billion years ago, as was deduced from microfossils. Multicellular organisms probably did not appear before 600–700 million years ago.

The early Earth presented extreme conditions compared to the general situation today. But life was present on Earth throughout most of its history, and that life was solely microbial for very long periods of geological time, developing a high diversity and adaptations to various environmental extremes.

3 Prokaryotes, eukaryotes, the tree of life and viruses

Living organisms consist of cells and much of what is known about cell structures is due to information from microscopy. Microscopic examination revealed early the presence of two fundamental types of cells – eukaryotic and prokaryotic cells, to denote the presence or absence of a cell nucleus, a feature which was easily identifiable. Figure 1 contains a general scheme of a eukaryotic and a prokaryotic cell. The nucleus encloses the genetic material of the eukaryotic cell, which is present in the form of chromosomes; prokaryotic cells contain their genetic material as an aggregated mass of DNA, which is called nucleoid. Other differences between the two types of cells are the sizes (eukaryotic cells are on average 10–20 times larger than the typical prokaryotic cell) and the presence of organelles in the eukaryotes. Organelles are small membrane-enclosed structures within cells, which were traced back to bacteria-like precursors; they were incorporated into early eukaryotic cells and became the mitochondria and chloroplasts of modern cells.

Prokaryotic cells are simpler; besides lack of a nucleus and lack of organelles, their morphology, protein synthesis, cell proliferation, and genetic apparatus are less diverse and more streamlined than comparative features of eukaryotic cells. On the other hand, prokaryotes display a stunning diversity of metabolic functions; they can exist in a wide range of inhospitable environments, where eukaryotes would not be able to survive. Their influence on all processes of the biosphere is probably still severely underestimated; their biomass alone is thought to be equal to or even superseed the total plant mass on Earth (Whitman et al. 1998; Pedersen 2000).

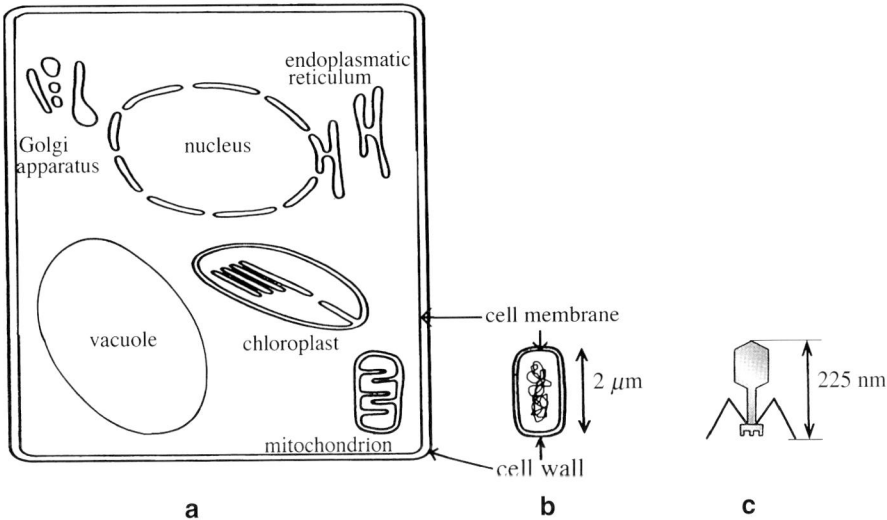

Fig. 1. Schemes of a eukaryotic cell from a plant (**a**), a prokaryotic cell (**b**) and a bacterial virus (**c**). The presence of organelles, nucleus and inner compartments (vacuole, Golgi apparatus, and endoplasmatic reticulum) in the eukaryotic cell is indicated. Mitochondria and chloroplasts (**a**) are of similar sizes as prokaryotic cells (**b**) and possess inner membranes. The DNA of prokaryotes is a coiled structure inside the cell (**c**). Viruses are much smaller than prokaryotes (note the difference of scale) and of simple composition

Viruses are separate entities – they need a host for replication; without one they are inert particles, and therefore the question if they are alive or not has been debated for years. Newer insights have prompted suggestions (see also below) to include them into the living world (Le Romancer et al. 2007). Viruses are generally much smaller than even most prokaryotic cells (Fig. 1c) and contain often only a nucleic acid (DNA or RNA) and some proteins.

Relationships between organisms and knowledge about their evolution can be obtained by comparing the sequences of their nucleic acid sequences. Figure 2 shows a phylogenetic tree, which is based on the sequences of a molecule which is present in similar form in all organisms, the ribonucleic acid (RNA) of the small subunit of ribosomes. The greater the differences in the sequences between the molecules from two or more organisms, the greater is their evolutionary distance. This is reflected in the length of the branches of the tree. Three main lineages of organisms have emerged from these analyses – the Eukarya (or eukaryotes), which include all animals, plants, fungi, and many unicellular microorganisms, and two lineages of prokaryotes, the Bacteria (or eubacteria) and the Archaea (or archaebacteria). The length of the archaeal branch is shorter than the bacterial and eukaryal branches (Fig. 2), which is generally interpreted as a sign of closeness to the common ancestor of all life. The Archaea consist of three main groups, which were

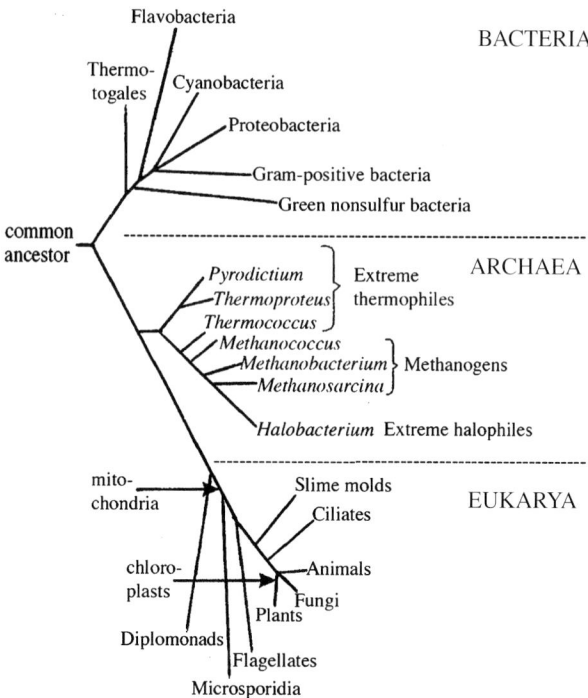

Fig. 2. Phylogenetic tree of all organisms, based on sequences of small ribosomal RNA genes. Prokaryotes form two lineages (Archaea and Bacteria); eukaryotes (Eukarya) form one lineage. The length of the branches represents the approximate evolutionary distances between groups, based on the number and positions of different bases in their small rRNAs

termed extreme halophiles, extreme thermophiles, and methanogens, hinting at the preference of extreme habitats by many archaea. Although groups of Archaea, which exist in moderate environments, are being increasingly identified (DeLong 1998), many record holders with respect to living conditions are found only within the Archaea (see below).

Nucleic acid sequences showed also clearly the similarity between bacteria and the organelles – mitochondria and chloroplasts – in Eukarya (Fig. 2). The uptake of the forerunners of mitochondria is assumed to have occurred about 1.9 billion years ago, the uptake of photosynthetic endosymbionts, which became modern chloroplasts, occurred somewhat later in Earth's history.

Viruses are not included in this general tree of life, since they do not have the genes for small subunit rRNAs on which the tree is based. In fact, there is no common informational molecule which could be the basis for a phylogenetic tree of viruses (Forterre and Prangishvili 2009a). It is becoming increasingly clear that modern viruses are not fragments of genetic materials that escaped from mother cells, but descendants of an ancient "virosphere" that possibly even preceded the origin of modern cells (Le Romancer et al. 2007; Forterre and Prangishvili 2009b). Like all other organisms, extremophiles serve as hosts for viral replication and many extremophilic viruses have become known (Le Romancer et al. 2007). It is thus reasonable to study their characteristics and to view life on early Earth considering the existence of a possible virosphere.

4 Extreme environments and their inhabitants

4.1 Temperature ranges of microorganisms

Microorganisms can be grouped into broad, although not very precise, categories, according to their temperature ranges for growth (Rothschild and Mancinelli 2001; Burgess et al. 2007):

- Psychrophiles (cold-loving) can grow at 0°C, and some even as low as −10°C or maybe −16°C; their upper limit is often about 20–25°C.
- Mesophiles grow in the moderate temperature range, from about 20°C (or lower) to 45°C.
- Thermophiles are heat-loving, with an optimum growth temperature of 50°C or more, a maximum of up to 70°C or more, and a minimum of about 20°C.
- Hyperthermophiles have an optimum above 75–106°C and thus can grow at the highest temperatures tolerated by any organism (Stetter 2002). Some will not grow below 80°C.

Temperatures lower than 55°C are widespread on Earth and are associated with sun-heated habitats, but temperatures higher than 55–60°C are much rarer and are almost exclusively associated with geothermal habitats (Brock 1986). Therefore, a "thermophile boundary" of 55–60°C has sometimes been suggested for prokaryotes, beyond which thermophiles are growing (Brock 1986).

Hot springs and geothermal vents are found in several parts of the world, such as in Yellowstone National Park (Wyoming, USA), Iceland, Southern Italy, New Zealand, Japan, and on the floors of the oceans, where spreading zones occur and superheated fluids are emitted (see also Hreggvidsson et al. 2011, this volume).

Life beyond the boiling point of water was unthinkable well into the 1970s, until Stetter (1982) transcended this boundary by reporting on a hyperthermophile which grew at 105°C. Subsequently, numerous microorganisms have been described, which flourish at temperatures up to 113°C and possibly even at 122°C (Table 1). The upper boundary of thermophilic life is still a matter of debate (Cowan 2004). Several isolated enzymes from hyperthermophiles (thermozymes) were shown to be active and stable up to 130°C (Daniel 1996; Lévêque et al. 2000), but microbial growth at that temperature has not been identified so far.

With respect to low temperature limits of life, psychrophiles (see Fendrihan and Negoiță 2011, this volume) were reported to perform metabolic activities (uptake of ^{14}C-labeled acetate) at −20°C under defined conditions (Rivkina et al. 2000; Gilichinsky 2002; D'Amico et al. 2006). The temperature limit for microbial reproduction is considered to be −12°C (see Table 1) and that for metabolism −20°C (Bakermans 2008).

4.2 Low water activity: halophiles, osmophiles, and xerophiles

High concentrations of salt and sugar have traditionally been used to prevent spoilage of food by microbial growth. The availability of water is greatly reduced by such additions, yet several types of halophilic bacteria, osmophilic yeasts, and xerophilic fungi are known to grow under these circumstances (Gilmour 1990; Grant 2004a). Inhibition of many fungi and yeasts occurs at water activities (a_w) between 0.8 and 0.75, but the mould *Xeromyces bisporus* is exceptional in being able to grow at a lower water activity (a_w 0.61) than any other organism described (Grant 2004a; Table 1). Other environments of low water activity are hypersaline sites, e.g. the Dead Sea in Israel; the Great Salt Lake in Utah, USA; natural and man-made salterns, where water availability is limited by a high concentration of salts (usually NaCl). These hypersaline waters are populated by halophilic microorganisms, mainly haloarchaea, which require substantial amounts of salt for growth (at least 12–15% NaCl), and are able to grow up to saturation (Table 1; see also Enache et al. 2011, this volume). Numerous strains were isolated from such sites (Grant 2004a); a regularly updated listing of halophilic genera and species is provided on the website of the ICSP (International Committee on Systematics of Prokaryotes; http://www.the-icsp.org).

When hypersaline waters are evaporating, halophilic microorganisms become entrapped within fluid inclusions in halite crystals (Norton and Grant 1988; Fendrihan et al. 2009). Haloarchaea remain viable within the fluid inclusions for considerable periods of time; laboratory cultures kept in this way have been recovered after more than 10 years (Grant 2004a). Ancient halite deposits, which are widely distributed on Earth (see Grant 2004a) range in age up to the Precambrian, while the most massive salt sediments – in the order of an estimated 1.3 million km^3 – stem from the late Permian and early Triassic period (ca. 245–290 million years old; Zharkov 1981). There have been several reports of viable halophilic prokaryotes, being recovered from such geologically old halite deposits (e.g. Grant et al. 1998; Stan-Lotter et al. 1999, 2002; McGenity et al. 2000; Gruber et al. 2004), suggesting that the microorganisms might be of the same age as the sediments and were originally entrapped when the ancient brines dried down. These extreme halophiles can be considered desiccation-resistant, oligotrophic and long-term survivors.

4.3 Extremes of pH and low nutrients

Many familiar biological processes are taking place at pH values around neutral, but organisms from all three domains contain representatives, which thrive at pH values on both ends of the scale.

Low pH: Acidophiles are extremophiles living in conditions with a pH of 2.0 or below. Many species of acidobacteria have been found in poor or polluted acidic soils and in acid mine drainages. Numerous acidophilic archaeal species are known, which are found in geothermal sulfurous sites, e.g., acidic springs, which contain sulfuric acid (Hreggvidsson et al. 2011, this volume). *Ferroplasma acidarmanus* has been described growing at pH 0 in acid mine drainage (Table 1). The genome of one isolated strain of the species was sequenced and compared with sequence data from an environmental population of the same species, revealing genomic heterogeneity within the population (Allen et al. 2007). Acidophiles have evolved efficient mechanisms that pump protons out of the intracellular space to keep the cytoplasm at or near neutral pH, but there are some exceptions: the recently described acidophilic archaean, *Picrophilus torridus*, grows optimally at pH of 0.7 and apparently maintains an intracellular pH of 4.6 (Ciaramella et al. 2004).

High pH: Alkaliphiles are extremophiles thriving in conditions with a pH of 9.0 or above. Very stable alkaline environments on Earth are soda lakes and soda deserts, which occur on all continents (Grant 2004b). In particular, numerous alkaline lakes are situated in the East African Rift Valley, which, due to their high content of NaCl and Na_2CO_3, have pH values of 10.5–11.5 (Grant 2004b). Haloalkaliphilic bacteria-like cyanobacteria, proteobacteria, and halomonads were isolated from these waters as well as archaea of the genera *Natronobacterium*, *Natronomonas* and *Natronococcus* (Grant 2004b). Another highly alkaline ecosystem is the Lost City hydrothermal field, which is located on the seafloor mountain Atlantis massif, near the Mid-Atlantic Ridge (Kelley et al. 2005). Reactions between seawater and upper mantle peridotite produce methane- and hydrogen-rich fluids whose pH values range from 9 to 11, with temperatures ranging from 40°C to 90°C. A variety of microorganisms live in, on, and around the vents. Archaeal populations include *Methanosarcinales* which form thick biofilms inside the vents, subsisting on methane and hydrogen. Bacteria related to *Firmicutes* live also inside the vents; on the outside, methane- and sulfur-oxidizing proteobacteria are present (Kelley et al. 2005).

Low nutrients: Microorganisms have evolved certain physiological and metabolic strategies for growing in nutrient-poor environments. In the open ocean, particularly in the deep-water column, in deep sediments, and in deep crustal environments, carbon and energy sources are extremely scarce, but active microorganisms are present. Heterotrophic prokaryotes that can reproduce at very low levels of organic carbon concentrations are known as oligotrophs (their counterparts, which would reproduce at high organic carbon concentrations, are called copiotrophs). There is no generally accepted definition of these categories, but typically the value for distinction is 1–10 mg of C per liter for oligotrophic environments (Giovannoni and Rappé 2000). Recently, *Pelagibacter ubique*, a representative of one of the most cosmopolitan microorganisms in oligotrophic oceans, was found to grow only in the in situ micromolar concentrations of organic carbon. This bacterium has one of the smallest genomes known from free-living organism. *P. ubique* and related marine oligotrophs are setting the lower limits of concentration of organic compounds that can support the growth of heterotrophs (Giovannoni et al. 2005).

4.4 High pressure

"Barophilic" and "piezophilic" are sometimes used interchangably, but barophilic should refer to an organism that lives and thrives under high barometric (atmospheric) pressure, and organisms, which live in the deep oceans, are called piezophilic following an agreement in 1995 (Simonato et al. 2006). The hydrostatic pressure in the oceans increases by 1 atm for every 10 m of depth. The maximum depth is in the Mariana Trench; with its 10,900 m depth, the pressure is about 1100 atm (ca. 110 MPa). Even from this site viable microorganisms have been isolated. The extremely piezophilic bacterium *Moritella yayanosii* is unable to grow at pressures below 500 atm; its optimum growth occurs at 700–800 atm, and it can grow up to 1035 atm, which is the pressure of its natural habitat (Table 1). Other isolates from the ocean floors and sediments of about 4000–6000 m depth grow optimally at pressures of 400–600 atm, but have usually retained the capacity for growth at normal pressure (1 atm). In laboratory experiments, model organisms such as *Saccharomyces cerevisiae* and *Escherichia coli* were subjected for short times to pressures of 200 and 275 MPa, respectively, and their survival and responses were tested (reviewed in Simonato et al. 2006).

4.5 Viruses

A review on viruses in extreme environments was recently published by Le Romancer et al. (2007), who described viruses from the following sites: hypersaline (Dead Sea, solar salterns), alkaline and hypersaline (soda lakes; Mono Lake), deserts (Sahara), polar regions (lakes in the Taylor Valley of Antarctica; arctic and antarctic sea ice), high temperature (terrestrial hot springs in Yellowstone National Park, Iceland and Japan; deep sea hydrothermal vents), and a deep subsurface sediment (drilling hole near the West Canadian coast).

It is perhaps noteworthy that so far only double-stranded DNA viruses were isolated from extreme environments, but no RNA viruses, and it has been suggested that this very stable form of genome may be necessary to face the harsh constraints of extreme habitats (Le Romancer et al. 2007).

So far, no viral fossils have been detected, but interestingly, silicification of viruses (bacteriophage T4) under conditions of silica-depositing hot spring environments has been achieved recently in the laboratory (Laidler and Stedman 2010). Viral morphology and elemental biosignatures were preserved, thus raising the exciting possibility of finding viruses in the geological record (Laidler and Stedman 2010).

4.6 Extremophilic multicellular organisms

Halophilic representatives occur in all three domains of life – besides prokaryotes, there are halophilic or halotolerant fungi, algae, and animals (Oren 2007). Similarly, thermophilic members are found in all three branches, although their maximum temperatures of growth vary from 38°C for vertebrates (fish) to 50°C (insects and mosses) to >100°C (prokaryotes), depending on the taxonomic group as outlined by Brock (1986).

There is no comprehensive review yet on extremophilic higher organisms, but there are books in the COLE series (Cellular origin and life in extreme habitats), edited by Seckbach, which contain much information on this subject (e.g., Seckbach and Chapman 2010; Seckbach and Grube 2010). Some further sources and a few examples of special interest are given here.

Weber et al. (2007) addressed in their review the metabolism of extremophilic algae and plants, with some coverage of metazoa and fungi. A website from 1998 "Eukaryotes in extreme environments" (http://www.nhm.ac.uk/research-curation/research/projects/euk-extreme/) provides some older literature and still timely discussion on the subject.

Alvinella pompejana, the Pompeii worm, was detected by Desbruyeres and Laubier (1986) near hydrothermal vents. It is living at 80°C, possesses chemolithotrophic thermophilic epi-bacteria as symbionts (Jeanthon and Prieur 1990) and is probably the most thermotolerant animal known.

Tardigrades (meaning "slow walkers"), known as water bears, are microscopic, water-dwelling, segmented animals with eight legs. The adults may reach a body length of 1.5 mm. They are found on lichens and mosses, on dunes, beaches, soil, and in marine or freshwater sediments. The survival capacities of tardigrades are stunning – several species are able to survive temperatures of −180°C or −196°C, respectively (Jönsson 2007, and references therein), about 1000 times more radiation than other animals (Horikawa et al. 2006), and almost a decade without water (Guidetti and Jönsson 2002). In 2007, tardigrades were taken into low Earth orbit on the FOTON-M3 mission and were exposed to the vacuum of space and solar radiation for 10 days. Back on Earth, it was discovered that many of them survived and laid eggs that hatched normally (Jönsson et al. 2008). The review by Jönsson (2007) deals, besides tardigrades, also with the radiation and desiccation tolerance of a few more animals, such as larvae from *Polypedium vanderplanki* (an insect), eggs from the brine shrimps *Artemia salina* and *Artemia franciscana*, larvae and adults of nematodes, and rotifers (genera *Mniobia* and *Macrotrachela*).

5 Microbial survival of extreme conditions

Table 2 shows several examples of microorganisms, which survived exposure to extreme physico-chemical factors. The time of exposure varied greatly, for example, *Deinococcus radiodurans* will, depending on the source of gamma radiation, survive a dose of about 20,000 Gy, obtained by a pulse lasting a few hours (Ito et al. 1983). These levels of radiation are not found naturally on present-day Earth. Several more

Table 2. Microbial survival of extreme conditions

Microorganism	Physico-chemical parameter	Time of exposure	Other information	Reference
Deinococcus radiodurans	Ionizing radiation	5–20 h	ca. 20,000 Gy	Ito et al. (1983)
Streptococcus mitis	Surface of the Moon	2.5 years	In a camera	Website[a]
Numerous microorganisms	$-20°C$	ca. 10^6 years	Permafrost	Gilichinsky 2002
Numerous microorganisms[b]	$-193°C$	>10 years	Liquid N_2	Gherna (1994)
Numerous microorganisms[b]	$a_w < 0.75$	>10 years	Vacuum	Gherna (1994)
Halococcus salifodinae	NaCl >30%	>10^6 years (?)	In salt crystals	Denner et al. (1994) and McGenity et al. (2000)
Endospores (*Bacillus, Clostridium*)	Heat; chemicals	>3000 years	Sediments	Madigan et al. (2009)
Endospores (*Bacillus*)	Outer space	6 years	Surface of space probe	Horneck et al. (1994)

[a] http://science.nasa.gov/science-news/science-at-nasa/1998/ast01sep98_1/
[b] With protective substances (e.g., 25–40% glycerol)

species of *Deinococcus* are known, which possess high resistance against radiation and desiccation, and several genomes have been sequenced (see Heulin et al. 2011, this volume). Interestingly, irradiation at a temperature of $-79°C$ – simulating Martian surface conditions – increased the resistance of *D. radiodurans* markedly, compared to irradiation at room temperature or on ice (Dartnell et al. 2010).

The survival of the bacterium *Streptococcus mitis*, which has traveled to the Moon during the Apollo missions and apparently survived there for 2.5 years in a camera, has, unfortunately, not been rigorously documented – in some descriptions, survival of 20–500 cells is mentioned, however, without reference to the method of determination. Nevertheless, the fact of its apparent resistance to space conditions over several years is of profound importance.

Careful documentation of survival was carried out with endospores from *Bacillus subtilis*, which were fully exposed to space conditions in the LDEF (Long Duration Exposure Facility) mission for 6 years (from 1984 to 1990), from which a significant fraction was recovered in a viable state (Horneck et al. 1994). Samples of the same mission, which were shielded against UV light with metal foils, exhibited about 70% survivors, following the 6 years in space (Horneck et al. 1994; reviewed by Horneck et al. 2010).

Permafrost is an environment, where many viable microorganisms have been found; some stem from sediments which are several million years old (Gilichinsky 2002). Their metabolism is, due to the frozen state, extremely slow, or perhaps in some cases, nonexistent. Some microorganisms in permafrost soils are apparently forming resting stages with thick walls, but these are not endospores (Soina et al. 2004). Freezing will prevent microbial growth, but it does not necessarily kill microbes. Culture collections are using freezing in the presence of protective substances for long-term preservation of microbial strains and other biological materials (Gherna 1994). Prerequisites are the addition of water-miscible liquids such as glycerol or dimethylsulfoxide in concentrations of 20% or more; they will penetrate the cells and protect them, mainly by hindering crystal formation, upon freezing. Temperatures around that of liquid nitrogen ($-196°C$) are survived by strains for up to 30 years (Gherna 1994).

Endospores are a survival form of certain bacteria (mostly from the genera *Bacillus* and *Clostridium*), which are formed when they find themselves in an unfavorable environment, such as low nutrients or high concentrations of salt. Endospores are perhaps the most resilient life forms on Earth (Madigan et al. 2009). They are resistant to extreme temperatures (an autoclave operating at $121°C$ will kill endospores of most species, but exceptional endospores survive up to $150°C$), to most disinfectants, radiation, and drying. They can survive for thousands of years in this dormant state, since they have been found in Egyptian mummies and 8000-year-old lake sediments (Madigan et al. 2009). They may survive even longer, since it was reported that *Bacillus* species were revived from bees encased in 25–30-million-year-old amber (Cano and Borucki 1995). However, since phylogenetic analysis showed that the isolates from ancient amber were very similar to extant microbes, some caution is advised, and an independent confirmation of these results would be desirable.

Still older microbial isolates were reported from salt sediments and other deep subterranean sources (bore cores and mines), which are of geological ages up to 290 million years (McGenity et al. 2000). If any metabolic maintenance activities of these microorganisms, including potential repair of damaged DNA, are occurring, awaits further research.

The survival of endospores in adverse conditions, including the space environment (Nicholson et al. 2000) raises the possibility that bacterial endospores could travel to Mars on the surface of spacecraft and survive on or in Martian soil. This could seriously compromise future efforts to establish whether there is, or has been, life on Mars, as it would be difficult for researchers to know whether any endospores found originated from Earth or Mars. Similarly, survival of nonspore forming microorganisms (*S. mitis*, permafrost bacteria, various subterranean bacteria, and haloarchaea) under Martian or other space conditions should be explored in detail.

There are regulations already agreed upon by all space-faring nations, notably by COSPAR (the International Commitee of Space Research), with the goal of preventing contamination by spacecrafts (see Moissl-Eichinger 2011, this volume); the rules will have to be updated and revised, taking into account the potential of extreme survival of microorganisms.

6 Concluding remarks

Organisms in all three domains of life have adapted to many terrestrial extremes. High-temperature, low-pH, and high-salinity environments represent probably very ancient sites, as may be frozen environments. Extreme environments are not rare at all: most of the ocean is cold and deep, and a vast portion of the subsurface of Earth is hot. Extreme environments are characterized by one or several physico-chemical parameters. Species diversity in extreme environments may be lower than in moderate environments; however, improved methods have led to the identification of numerous novel phylogenetic groups, including viruses, and these data can be expected to increase substantially in the future.

Extremophilic microorganisms survive exposure to space conditions, desiccation, high and low temperatures, and low nutrient environments, perhaps for millions of years. They may therefore also survive space travel and be capable of living in extraterrestrial environments, which suggests that the old concept of panspermia is an acceptable idea for the distribution of life in the Universe.

Acknowledgments

Thanks go to former and present coworkers and students for their contributions. Support to my laboratory work by the Austrian Science Foundation (FWF), grants P16260 and P18256, and the Austrian Research Promotion Agency (FFG), grant ASAP 815141, is gratefully acknowledged.

References

Allen EE, Tyson GW, Whitaker RJ, Detter JC, Richardson PM, Banfield JF (2007) Genome dynamics in a natural archaeal population. Proc Natl Acad Sci USA 104:1883–1888

Auman AJ, Breezee JL, Gosink JJ, Kämpfer P, Staley JT (2006) *Psychromonas ingrahamii* sp. nov., a novel gas vacuolate, psychrophilic bacterium isolated from Arctic polar sea ice. Int J Syst Evol Microbiol 56:1001–1007

Bakermans C (2008) Limits for microbial life at subzero temperatures. In: Margesin R, Schinner R, Marx JC, Gerday C (eds) Psychrophiles: from biodiversity to biotechnology. Springer, Berlin, Heidelberg, pp 17–28

Blöchl E, Rachel R, Burggraf S, Hafenbradl D, Jannasch HW, Stetter KO (1997) *Pyrolobus fumarii*, gen. and sp. nov., represents a novel group of archaea, extending the upper temperature limit for life to 113°C. Extremophiles 1:14–21

Brock TD (1969) Microbial growth under extreme conditions. Symp Soc Gen Microbiol 19:15–42

Brock TD (1986) Introduction: an overview of the thermophiles. In: Brock TD (ed) Thermophiles. General, molecular and applied microbiology. John Wiley & Sons, New York, pp 1–16

Burgess EA, Wagner ID, Wiegel J (2007) Thermal environments and biodiversity. In: Gerday C, Glansdorff N (eds) Physiology and biochemistry of extremophiles. ASM Press, Washington, DC, USA, pp 13–29

Cano RJ, Borucki MK (1995) Revival and identification of bacterial spores in 25- to 40-million-year-old Dominican amber. Science 268:1060–1064

Ciaramella M, Napoli A, Rossi M (2004) Another extreme genome: how to live at pH 0. Trends Microbiol 13:49–51

Cowan DA (2004) The upper temperature for life – where do we draw the line? Trends Microbiol 12:58–60

D'Amico S, Collins T, Marx JC, Feller G, Gerday C (2006) Psychrophilic microorganisms: challenges for life. EMBO Rep 7:385–389

Daniel RM (1996) The upper limits of enzyme thermal stability. Enzyme Microb Technol 19:74–79

Dartnell LR, Hunter SJ, Lovell KV, Coates AJ, Ward JM (2010) Low-temperature ionizing radiation resistance of *Deinococcus radiodurans* and Antarctic Dry Valley bacteria. Astrobiology 10:717–732

DeLong EF (1998) Everything in moderation: archaea as 'non-extremophiles'. Curr Opin Genet Dev 8:649–654

Denner EBM, McGenity TJ, Busse H-J, Wanner G, Grant WD, Stan-Lotter H (1994) *Halococcus salifodinae* sp. nov., an archaeal isolate from an Austrian salt mine. Int J System Bacteriol 44:–774–780

Desbruyeres D, Laubier L (1986) Les Alvinellidae, une famille nouvelle d'annelides polychetes infeodees aux sources hydrothermales sous-marines: systematique, biologie et ecologie. Can J Zool 64:2227–2245

Edwards KJ, Bond PL, Gihring TM, Banfield JF (2000) An archaeal iron-oxidizing extreme acidophile important in acid mine drainage. Science 287:1796–1799

Enache M, Popescu G, Itoh T, Masahiro Kamekura M (2011) Halophilic microorganisms from man-made and natural hypersaline environments: physiology, ecology and biotechnological potential, this volume

Fendrihan S, Negoiţă TG (2011) Psychrophilic microorganisms as important source for biotechnological processes, this volume

Fendrihan S, Berces A, Lammer H, Musso M, Ronto G, Polacsek TK, Holzinger A, Kolb C, Stan-Lotter H (2009) Investigating the effects of simulated Martian ultraviolet radiation on *Halococcus dombrowskii* and other extremely halophilic archaebacteria. Astrobiology 9: 104–112

Forterre P, Prangishvili D (2009a) The origin of viruses. Res Microbiol 160:466–472

Forterre P, Prangishvili D (2009b) The great billion-year war between ribosome- and capsid-encoding organisms (cells and viruses) as the major source of evolutionary novelties. In: Natural Genetic Engineering and Natural Genome Editing: Ann NY Acad Sci 1178:65–77

Gerday C, Glansdorff N (eds) (2007) Physiology and biochemistry of extremophiles. ASM Press, Washington, DC, USA

Gherna RL (1994) Culture preservation. In: Gerhardt P, Murray RGE, Wood WA, Krieg NR (eds) Manual of methods for general microbiology. American Society for Microbiology, Washington, DC, pp 278–292

Gilichinsky DA (2002) Permafrost. In: Bitton G (ed) Encyclopedia of environmental microbiology. John Wiley & Sons, New York, pp 2367–2385

Gilmour D (1990) Halotolerant and halophilic microorganisms. In: Edwards C (ed) Microbiologogy of extreme environments. Open University Press, Milton Keynes, pp 147–177

Giovannoni S, Rappé M (2000) Evolution, diversity and molecular ecology of marine prokaryotes. In: Kirchman DL (ed) Microbial ecology of the oceans. John Wiley & Sons, New York, pp 47–84

Giovannoni SJ, Tripp HJ, Givan S, Podar M, Vergin KL, Baptista D, Bibbs L, Eads J, Richardson TH, Noordewier M, Rappé MS, Short JM, Carrington JC, Mathur EJ (2005) Genome streamlining in a cosmopolitan oceanic bacterium. Science 309:1242–1245

Grant WD (2004a) Life at low water activity. Philos Trans R Soc Lond B 359:1249–1267

Grant WD (2004b) Introductory chapter: half a lifetime in soda lakes. In: Ventosa A (ed) Halophilic microorganisms. Springer, Berlin, Heidelberg, pp 17–32

Grant WD, Gemmell RT, McGenity TJ (1998) Halobacteria: the evidence for longevity. Extremophiles 2:279–287

Grant WD, Kamekura M, McGenity TJ, Ventosa A (2001) Order I. Halobacteriales Grant and Larsen 1989b, 495VP (effective publication: Grant and Larsen 1989a, 2216). In: Boone DR, Castenholz RW, Garrity GM (eds) Bergey's manual of systematic bacteriology, vol 1, 2nd edn. Springer, Berlin, Heidelberg, New York, pp 294–299

Grinspoon D (2003) Lonely planets. The natural philosophy of alien life. HarperCollins Publishers, New York, p 139

Gruber C, Legat A, Pfaffenhuemer M, Radax C, Weidler G, Busse H-J, Stan-Lotter H (2004) *Halobacterium noricense* sp. nov., an archaeal isolate from a bore core of an alpine Permian salt deposit, classification of *Halobacterium* sp. NRC-1 as a strain of *H. salinarum* and emended description of *H. salinarum*. Extremophiles 8:431–439

Guidetti R, Jönsson KI (2002) Long-term anhydrobiotic survival in semi-terrestrial micrometazoans. J Zool 257:181–187

Heulin T, De Luca G, Barakat M, de Groot A, Blanchard L, Ortet P, Achouak W (2011) Bacterial adaptation to hot and dry deserts, this volume

Hocking AD, Pitt JI (1999) *Xeromyces bisporus* Frazer. In: Robinson RK, Batt CA, Patel PD (eds) Encyclopaedia of food microbiology, vol 3. Academic Press, London, pp 2329–2333

Horikawa DD, Sakashita T, Katagiri C, Watanabe M, Kikawada T, Nakahara Y, Hamada N, Wada S, Funayama T, Higashi S, Kobayashi Y, Okuda T, Kuwabara M (2006) Radiation tolerance in the tardigrade *Milnersium tardigradum*. Int J Rad Biol 82:843–848

Horikoshi K, Grant WD (eds) (1998) Extremophiles: microbial life in extreme environments. Wiley series in ecological and applied microbiology. John Wiley & Sons Inc., New York

Horneck G, Bucker H, Reitz G (1994) Long-term survival of bacterial spores in space. Adv Space Res 14:41–45

Horneck G, Klaus DM, Mancinelli RL (2010) Space microbiology. Microbiol Mol Biol Rev 74:121–156

Hreggvidsson GO, Petursdottir SK, Björnsdottir SH, Fridjonsson OH (2011) Microbial speciation in the geothermal ecosystem, this volume

Ito H, Watanabe H, Takeshita M, Iizuka H (1983) Isolation and identification of radiation-resistant cocci belonging to the genus *Deinococcus* from sewage sludges and animal feeds. Agric Biol Chem 47:1239–1247

Jeanthon C, Prieur D (1990) Susceptibility to heavy metals and characterization of heterotrophic bacteria isolated from two hydrothermal vent polychaete annelids, *Alvinella pompejana* and *Alvinella caudata*. Appl Environ Microbiol 56:3308–3314

Jönsson KI (2007) Tardigrades as a potential model organism in space research. Astrobiology 7:757–766

Jönsson KI, Rabbow E, Schill RO, Harms-Ringdahl M, Petra Rettberg P (2008) Tardigrades survive exposure to space in low Earth orbit. Curr Biol 18:R729–R731

Kashefi K, Lovley DR (2003) Extending the upper temperature limit for life. Science 301:934

Kelley DS, Karson JA, Früh-Green GL, Yoerger DR, Shank TM, Butterfield DA, Hayes JM, Schrenk MO, Olson EJ, Proskurowski G, Jakuba M, Bradley A, Larson B, Ludwig K, Glickson D, Buckman K, Bradley AS, Brazelton WJ, Roe K, Elend MJ, Delacour A, Bernasconi SM, Lilley MD, Baross JA, Summons RE, Sylva SP (2005) A serpentinite-hosted ecosystem: the Lost City hydrothermal field. Science 307:1428–1434

Krebs CJ (2008) The ecological world view. CSIRO Publishing, Collinwood, Australia, pp 19–39

Laidler JR, Stedman KM (2010) Virus silicification under simulated hot spring conditions. Astrobiology 10:569–576

Le Romancer M, Gaillard M, Geslin C, Prieur D (2007) Viruses in extreme environments. Rev Environ Sci Biotechnol 6:17–31

Lévêque E, Janecek S, Haye B, Belarbi A (2000) Thermophilic archaeal amylolytic enzymes. Enzyme Microb Technol 26:3–14

Madigan MT, Martinko JM, Dunlap PV, Clark DP (2009) Chapter one: microorganisms and microbiology. In: Brock biology of microorganisms, 12th edn. Pearson Benjamin Cummings, San Francisco, pp 1–24

McGenity TJ, Gemmell RT, Grant WD, Stan-Lotter H (2000) Origins of halophilic microorganisms in ancient salt deposits. Environ Microbiol 2:243–250

Moissl-Eichinger C (2011) Extremophiles in spacecraft assembly clean rooms, this volume

Nealson KH, Conrad PG (1999) Life: past, present and future. Philos Trans R Soc Lond B 354:1923–1939

Nicholson WL, Mukenata N, Horneck G, Melosh HJ, Setlow P (2000) Resistance of *Bacillus* endospores to extreme terrestrial and extraterrestrial environments. Microbiol Mol Biol Rev 64:548–563

Nogi Y, Kato C (1999) Taxonomic studies of extremely barophilic bacteria isolated from the Mariana Trench and description of *Moritella yayanosii* sp. nov., a new barophilic bacterial isolate. Extremophiles 3:71–77

Norton CF, Grant WD (1988) Survival of halobacteria within fluid inclusions in salt crystals. J Gen Microbiol 134:1365–1373

Oren A (2007) Biodiversity in highly saline environments. In: Gerday C, Glansdorff N (eds) Physiology and biochemistry of extremophiles. ASM Press, Washington, DC, USA, pp 223–231

Oren A, Rainey F (eds) (2006) Methods in microbiology. Extremophiles, vol 35. Elsevier, Oxford

Pedersen K (2000) MiniReview. Exploration of deep intraterrestrial microbial life: current perspectives. FEMS Microbiol Lett 185:9–16

Rivkina EM, Friedmann EI, McKay CP, Gilichinsky DA (2000) Metabolic activity of permafrost bacteria below the freezing point. Appl Environ Microbiol 66:3230–3233

Rothschild LJ, Mancinelli RL (2001) Life in extreme environments. Nature 409:1092–1101

Schleper C, Puehler G, Holz I, Gambacorta A, Janekovic D, Santarius U, Klenk HP, Zillig W (1995) *Picrophilus* gen. nov., fam. nov.: a novel aerobic, heterotrophic, thermoacidophilic genus and family comprising archaea capable of growth around pH 0. J Bacteriol 177:7050–7059

Seckbach J (ed) (2000) Journey to diverse microbial worlds. Adaptation to exotic environments. Series: Cellular origins and life in extreme habitats, vol 2. Kluwer, Dordrecht, The Netherlands

Seckbach J, Chapman DJ (eds) (2010) Red algae in the genomic age. Series: Cellular origin and life in extreme habitats and astrobiology, vol 13. Springer, Heidelberg, Germany

Seckbach J, Grube M (eds) (2010) Symbioses and stress. Series: Cellular origin and life in extreme habitats and astrobiology, vol 17. Springer, Heidelberg, Germany

Seckbach J, Walsh M (eds) (2009) From fossils to astrobiology. Series: Cellular origin and life in extreme habitats and astrobiology, vol 12. Springer, Heidelberg, Germany

Simonato F, Campanaro S, Lauro FM, Vezzi A, D'Angelo M, Vitulo N, Valle G, Bartlett DH (2006) Review. Piezophilic adaptation: a genomic point of view. J Biotechnol 126:11–25

Soina VS, Mulyukin AL, Demkina EV, Vorobyova EA, El-Registan GI (2004) The structure of resting bacterial populations in soil and subsoil permafrost. Astrobiology 4:345–358

Stan-Lotter H, McGenity TJ, Legat A, Denner EBM, Glaser K, Stetter KO, Wanner G (1999) Very similar strains of *Halococcus salifodinae* are found in geographically separated Permo-Triassic salt deposits. Microbiology 145:3565–3574

Stan-Lotter H, Pfaffenhuemer M, Legat A, Busse H-J, Radax C, Gruber C (2002) *Halococcus dombrowskii* sp. nov., an Archaeal isolate from a Permian alpine salt deposit. Int J System Evol Microbiol 52:1807–1814

Stetter KO (1982) Ultrathin mycelia-forming organisms from submarine volcanic areas having an optimum growth temperature of 105°C. Nature 300:258–260

Stetter KO (2002) Hyperthermophilic microorganisms. In: Horneck G, Baumstark-Khan C (eds) Astrobiology. The quest for the conditions of life. Springer Verlag, Berlin, New York, pp 169–184

Takai K, Moser DP, Onstott TC, Spoelstra N, Pfiffner SM, Dohnalkova A, Frederickson JK (2001) *Alkaliphilus transvaalensis* gen. nov., sp. nov., an extremely alkaliphilic bacterium isolated from a deep South African gold mine. Int J Sys Evol Microbiol 51:1245–1256

Takai K, Nakamura K, Toki T, Tsunogai U, Miyazaki M, Miyazaki J, Hirayama H, Nakagawa S, Nunoura T, Horikoshi K (2008) Cell proliferation at 122°C and isotopically heavy CH_4

production by a hyperthermophilic methanogen under high-pressure cultivation. Proc Natl Acad Sci USA 105:10949–10954

Weber APM, Horst RJ, Barbier GG, Oesterhelt C (2007) Metabolism and metabolomics of eukaryotes living under extreme conditions. Int Rev Cytol 256:1–34

Westall F (2005) Early life on Earth and analogies to Mars. In: Tokano T (ed) Water on Mars and life. Advances in Astrobiology and Biogeophysics. Springer Verlag, Berlin, Heidelberg, pp 45–64

Whitman WB, Coleman DC, Wiebe WJ (1998) Prokaryotes: the unseen majority. Proc Nat Acad Sci USA 95:6578–6583

Zeng X, Birrien JL, Fouquet Y, Cherkashov G, Jebbar M, Querellou J, Oger P, Cambon-Bonavita MA, Xiao X, Prieur D (2009) *Pyrococcus* CH1, an obligate piezophilic hyperthermophile: extending the upper pressure–temperature limits for life. ISME J 3:873–876

Zharkov MA (1981) History of paleozoic salt accumulation. Springer Verlag, Berlin

Microbial diversity in deep hypersaline anoxic basins

Francesca Mapelli, Sara Borin and Daniele Daffonchio

Università degli Studi di Milano, DiSTAM, Milan, Italy

1 Introduction

Deep hypersaline anoxic basins (DHABs) have been discovered on the seafloor in different oceanic regions, such as the Gulf of Mexico (Shokes et al. 1977), the Red Sea (Pautot et al. 1984) and the Eastern Mediterranean Sea (Jongsma et al. 1983; Dupré et al. 2007). DHABs contain brines that due to their temperature and/or salinity do not mix with the upper seawater layers and lead to the stratification of the water column. Gradients of oxygen, sulphide, ammonium and other electron acceptors/donors can occur, quite often contributed by microbial activity (Sass et al. 2001). The chemoclines at the interface between the hypersaline waters of the brines and the overlaying seawater have been recognized as hot spot for microbial abundance and activity (van der Wielen et al. 2005; Daffonchio et al. 2006). Environmental gradients represent high turnover spots in the biosphere and have to be taken into account to model global biogeochemical cycles and their response to environmental changes (Brune et al. 2000).

All the DHABs studied so far have been found to be colonized by microbial communities active and rich in biodiversity. Microorganisms living in these environments have to cope with multiple stresses that limit the activity of organisms thriving in conventional environments. Hypersalinity, 5–10 times higher than seawater, lacks of oxygen and highly reducing conditions, high pressure and absence of light make microorganisms of DHABs true extremophiles.

2 Localization and origin of DHABs

DHABs have been discovered in different geographical regions, generated in different geological conditions. Brine pools are described on the caldera of active and ancient mud volcanoes, whilst brine basins (or lakes) originated by the dissolution of buried evaporitic rocks.

2.1 The anoxic brine basins in the Red Sea: the Shaban and Kebrit Deeps

During the past 50 years several deep-sea brine pools have been identified in the Red Sea, that can be defined as an ocean *in statu nascendi*, forming during the past 20 million years due to the divergent movement of the Arabian and African continental plates (Degens and Ross 1969; Backer and Schoell 1972; Girdler and Styles 1974; Pautot et al. 1984). This movement is associated with the genesis of new oceanic crust, which can be observed in the southern Red Sea as a 1500–2000 m deep axial graben. Towards the north the graben narrows up to 21°N, where isolated deeps can be found, approximately 25 of which are filled with extreme saline water, formed by the leaching of subbottom Miocene evaporite. Recent volcanic intrusions in proximity of some of the brines are responsible for their high temperatures (Hartmann et al. 1998). Two of the most studied deep-sea brine pools in the Red Sea are the Shaban and the Kebrit Deep. The Shaban Deep consists of four depressions where the brine-seawater interface can be found at a depth of 1325 ± 3 m and has remained constant since its discovery. The Kebrit Deep is a roundish basin of approximately 1 km in diameter and a maximum depth of 1549 m, filled by 84-m-thick anaerobic brine (Eder et al. 1999).

2.2 The Orca basin and other seafloor anoxic brines in the Gulf of Mexico

In the Gulf of Mexico several mud volcanoes, brine pools and brine basins have been described. This marine area is an economically relevant hydrocarbon basin containing late Jurassic age oil and gas (Macgregor 1983). The bathymetry of the Gulf of Mexico was shaped by the tectonics of an early Jurassic salt deposit which underlies the northwestern margin of the basin (Joye et al. 2005). The discovery of the Orca basin, a bathymetric depression on the northern slope of the Gulf of Mexico was reported in the late 1970s by Shokes et al. (1977). The Orca basin is elbow-shaped; it covers about 400 km^2 and it has a maximum depth of 2400 m. Several lines of evidence suggest that the extreme saline brine contained in the Orca basin originated from the dissolution of nearby salt deposits (LaRock et al. 1979; van Cappellen et al. 1998).

More recently shallower hypersaline habitats in the northern Gulf of Mexico have attracted the attention of many researchers and their ecosystems have been investigated. A quiescent brine pool (site GC233, 27°43.4′N, 91°16.8′W) covering 190 m^2 is located at a depth of 650 m and is considered the remaining of an ancient mud volcano. A second larger brine lake is constituted by an active mud volcano (site GB425, 27°33.2′N, 92°32.4′W) with a strong fluid flow through its crater emitting high rates of gas and fluidized hypersaline mud formed by a fine clay suspension (MacDonald et al. 2000; Joye et al. 2005).

2.3 Eastern Mediterranean Sea: DHABs and the Chefren mud volcano

Five DHAB lakes have been discovered and studied since the 1980s in the Eastern Mediterranean Sea. The basins Bannock, Tyro, Urania, L'Atalante and Discovery represent extreme and largely unexplored habitats. These lakes are the deepest known DHABs, far below the photic zone (3200–3500 m deep) and are located in the Mediterranean Ridge in the Eastern Mediterranean Sea, an accretionary complex subjected to continental collision. The origin of their brines has been attributed to the dissolution of the 5–8-million-year-old Messinian evaporites or to diagenetic relics of evaporated seawater entrapped as interstitial water in Messinian sediments released by tectonic deformation of the sediments on the Mediterranean Ridge. These waters have been subsequently modified from the original composition by geochemical events such dolomization and gypsum precipitation, as well as biological activities such as sulphate reduction and decomposition of sediment organic matter (Camerlenghi 1990; Vengosh et al. 1998). The downward flow of the dense brines into local depressions was followed by the progressive development of anoxia in the brine lakes (Wallmann et al. 1997). The haloclines established at the seawater-brine interfaces act as density barriers and limit the mixing of the upper oxygenated water columns with the underlying brines. This implies that DHABs have been physically isolated from other habitats on Earth for thousand of years (Ferrer et al. 2005), thus potentially leading to the selection of new very peculiar extremophiles.

In the Eastern Mediterranean Sea, one of the most oligotrophic marine water bodies, active mud volcanism associated with diverse ecosystems has recently been detected on the Nile Deep Sea Fan (Dupré et al. 2007; Omoregie et al. 2008). In 2000 the bathymetry surveys of this area allowed the discovery of the Chefren mud volcano of the Menes Caldera (Mascle et al. 2006). The Menes Caldera is a circular depression with a diameter of 8 km that is approximately 50–100 m deep, at a water depth of about 3000 m in the western area of the Nile Deep Sea Fan. The caldera contains three mud volcanoes named Chefren, Cheops and Mykerinos. Chefren is about 500 m in diameter and rises about 60 m above the bottom of the caldera. Omoregie et al. (2008) discovered that the centre of the Chefren volcano is filled by a brine and mud lake.

3 Geochemical features of DHABs

The different origins and locations make each DHABs a specific environment with a peculiar geochemical setting. Common traits are hypersalinity, anoxia, high pressure and absence of light.

All DHABs present a transition zone at the interface between upper seawater and the underlying brines. This layer is a few to ten metres thick, contains an oxic/anoxic boundary and constitutes a halocline/chemocline where chemical species are distributed in a gradient of concentrations. The profiles of concentration of the different chemical species allow evaluating if and how they are produced and/or consumed. Changes of concentrations can reflect the occurrence of chemical or biological processes, assuming that the concentration profiles are not significantly affected by lateral transport or changes of the transport properties over the considered range of depth. The strategy commonly used to eliminate the effects of variable vertical eddy diffusivity on the concentration distribution of solutes in the transition zone consists in plotting the concentration versus salinity. The concentration of a conservative dissolved species (e.g. NaCl) in such a diagram will appear as a straight line between the seawater and brine concentration values. In contrast, when the plot shows a curvature it indicates production (concave-up) or consumption (convex-down).

3.1 DHABs of the Red Sea

The Shaban Deep is formed by four depressions filled with anaerobic hydrogen sulphide-free brines. The brines in the north and west basins have a temperature of 22.7°C and a pH of 5.7 whereas in the south and east basins both the temperature (24.0–25.2°C) and pH value (6.0–6.2) of the brines are higher. The brines collected from the south and east basins are enriched in Na and Cl (4.0–4.4 M and 4.4–4.5 M, respectively) compared with the upper seawater. Suboxic condition within the brine and the lower part of the interface make possible an increase of the concentration of dissolved Fe (122–130 µM) and Mn (35.5–37.6 µM) with depth (Eder et al. 2002). In addition, many other elements show a relevant increase moving from the seawater to the brine (i.e. Mg, K, Ca, Li and Ba).

Similarly to the Shaban Deep, the Kebrit Deep exhibits a steep increase of NaCl concentration (from 4 to 26%, w/v) within only 3 m at the seawater-brine interface, but the brine chemistry is different. Within about 7 m the temperature increases from 21.6 up to 23.4°C, the CH_4 concentration increases from 50 nl/l to 22 ml/l and H_2S content rises up to 12–14 mg of S/l. In the same depth range, pH values decrease from 8.1 to 5.5 and O_2 concentration drops from 3.2 ml/l to zero. The high gas content of the brine is constituted mainly of CO_2 and H_2S and small amounts of N_2, CH_4 and C_2H_6 (Eder et al. 1999, 2001).

3.2 DHABs of the Gulf of Mexico

In correspondence with the oxic–anoxic interface of the Orca basin, the oxygen is depleted, the temperature increases slightly from 4.45 up to 5.35°C, the phosphorous concentration raises and the water salinity reaches saturation for NaCl (Shokes et al. 1977; LaRock et al. 1979) in 13 m, a larger interval than in DHABs of the Red

Sea and the Eastern Mediterranean (van Cappellen et al. 1998). The ammonium concentration follows a near-linear trend in correspondence to the 6–24.5% salinity range, indicating a conservative mixing of the ammonium-enriched brine with seawater along with depth (van Cappellen et al. 1998). Below 2200 m manganese and iron oxides have been indicated as the energetically most favourable terminal electron acceptors for heterotrophic bacteria (Stumm and Morgan 1996). Reductive dissolution of manganese oxides in the intermediate salinity range (6–18%) can be gathered from the concave-up curve of the diagram reporting dissolved Mn^{2+} concentration versus salinity (van Cappellen et al. 1998). Both alkalinity, which can be used as a tracer of pathways of organic matter breakdown, and Mn^{2+} concentrations show peaks at salinities around 15%. The data are in agreement with the stoichiometric oxidation of organic matter by MnO_2, according to the following reaction:

$$CH_2O + 2MnO_2 + 4H^+ \rightarrow 2Mn^{2+} + H_2CO_3 + 2H_2O$$

Below a depth of 2225 m, the presence of detectable dissolved sulphide that can react with Fe(II) and Fe(III) is resulting in the formation of iron sulphide in the water column. This reaction is consistent with the high concentration of particulate iron detected in the brine and the sediments rich in iron sulphides observed in the Orca basin (van Cappellen et al. 1998).

The brines recovered from the active mud volcano (site GB425) and the brine pool (site GC233) were constituted via halite dissolution. In both sites the brines showed the same Na^+ and Cl^- concentrations (1800 and 2100 mM, respectively), the absence of sulphate and a high concentration of dissolved hydrogen (Joyc et al. 2005). Hydrogen concentrations range from hundreds of nanomolar in the mud volcano brines up to about 6 µM in the brine pool. The pH of the brine collected from the brine pool is about 6.4–6.8 whereas at the mud volcano the pH value is higher (7.4). In both brines the most abundant alkane is methane and the dissolved organic carbon (DOC) increases along with depth, indicating the presence of a subsurface DOC sink (Joye et al. 2009).

3.3 DHABs of the Eastern Mediterranean Sea

The chemical composition of Urania, L'Atalante, Bannock and Discovery basins has been investigated extensively. All the hypersaline lakes show a distinct geochemistry, though Bannock, L'Atalante and Urania basins are more similar and significantly different from Discovery (van der Wielen et al. 2005).

The global salinity of Urania is lower than that of the other Mediterranean DHABs, with NaCl concentration 5.4–7 times higher than normal seawater.

Methane concentration in Urania reaches 5.56 mM, which is 10–100 times higher than in the other DHABs. Very high concentrations of sulphide (up to 16 mM in the brines) place Urania amongst the most sulphidic marine water bodies on Earth (Borin et al. 2009). Sulphide levels range between 0.7 and 3.0 mM in the other Mediterranean DHABs, whereas the seawater shows a concentration of 2.6×10^{-6} mM. In this lake two discontinuities in salinity and temperature have been found along the water column. The first constitutes a halocline of 2 m depth (interface I) where salinity strongly increases from 3.7% to 16.1% and temperature values change from 14°C to 16.5°C. After 67 m (brine I) a second chemocline is present (interface II) between body brines of different salinities and temperatures (Borin et al. 2009).

The main difference between the geochemistry of Discovery brine compared with the other three basins is an extremely high concentration of Mg^{2+} and the absence of Na^+. The Discovery lake contains a brine that shows the highest concentration of $MgCl_2$ (around 5 M) that has ever been measured in a marine environment (Wallmann et al. 2002). The highest $MgCl_2$ concentration previously reported (up to 2 M) in a natural environment is in the Dead Sea and Lake Bonney in Antarctica (Matsubaya et al. 1979; Oren 1999).

L'Atalante basin has a sharp 1.5-m-deep halocline showing a steep salinity gradient (from 3.9% at its upper part to 36.5% at the lower part). The decrease of the redox potential to negative values in the initial part of the halocline indicates a complete depletion of oxygen in the underlying brine lake. Along the halocline a progressive increase of ammonia from 5.5 up to 3000 µM has been described (Yakimov et al. 2007), making the Eastern Mediterranean DHABs the marine environments with the highest ammonia concentrations. The brine of the L'Atalante basin is characterized by anoxic, high sulphidic (2.9 mM) conditions and salt concentration to almost saturation (411 g/l NaCl).

In the Bannock basin the halocline is about 2.5 m deep. Down the halocline the nitrate concentration decreases from 6.6 to 0.3 µM and the ammonium concentration increases from 5 to 3450 µM (De Lange et al. 1990), a concentration that is 1–5 order of magnitudes greater than those observed in the open ocean and coastal waters (Könneke et al. 2005). The concentration of sulphate, the most abundant electron acceptor along the Bannock chemocline, increases from 31 to 84 mM. Manganese increases from 0.4 to 8.3 µM and redox potential drops from 210 mV to very negative values, as a result of total oxygen depletion in the lower part of the halocline. Dissolved manganese, nitrate, ammonium and sulphate concentrations have non-linear slopes, indicating a non-conservative behaviour, possibly due to biologically mediated processes (Daffonchio et al. 2006).

3.4 DHABs of the Mediterranean Sea: the Chefren mud volcano (Nile Deep Sea Fan)

During a dive with the submersible *Nautile* at the bottom of the Chefren mud volcano two kinds of closely associated microbial mats were discovered on top of the bottom sediments. The mats showed different colours (white and orange) and covered about 25 m². The surface of the cores collected from the white mat included thick white precipitates constituted by filamentous sulphur aggregates. On the surface of the cores recorded from orange mat a thick layer of fluffy yellow material and flaky, orange particles resembling Fe^{3+} (hydr)oxides were observed. Fluids from the brine pool at Chefren mud volcano are characterized by a salinity value of around 15%, relatively high methane (2.4 mM) and sulphide (7.2 mM) concentrations and sulphate concentration of 50 mM (Omoregie et al. 2008). In both the white and orange mats, the concentrations of Na^+ and Cl^- are higher than those of the bottom water, indicating the presence of upward brine flow through the sediments. About 6 cm below the white mat it is possible to observe a black, reduced sediment layer characterized by high iron and sulphur concentrations of the solid phase, consistent with the high content of FeS and pyrite. The sulphate concentration decreases from the surface layer (28 mM) moving towards the bottom of the core (5 mM at 15 cm depth), whereas the methane concentration increases by one order of magnitude from 0.01 up to 0.1 mM.

4 Microbial life in the DHABs: biodiversity and adaptation

More than 70% of the Earth's surface is covered by oceans and about 60% of this area is represented by water more than 2000 m deep. Nevertheless the oceans are the last environment to be investigated for their microbiology on our planet (Bull et al. 2000). Microscopy and biochemical investigations (i.e. lipid analysis, measurements of ATP and CH_4) indicated the existence of microbes in hypersaline brines (LaRock et al. 1979; Dickins and Van Vleet 1992). However only in the past two decades, thanks to the development of powerful cultivation-independent molecular surveying techniques, a more careful analysis of the prokaryotes became possible and a surprisingly high degree of biodiversity has been discovered in marine microorganisms representing all domains and viruses. Bacteria and Archaea belonging to new taxonomic lineages were discovered in high abundance in DHABs by the screening of 16 S rRNA libraries and fluorescent in situ hybridization (FISH). In some cases, they were considered salt-adapted and responsible for key microbial activities that were detected in the brines, without parallel detection of microbial taxa of known physiology.

4.1 Prokaryotic biodiversity of the Kebrit and Shaban Deeps, Red Sea

Bacterial and archaeal 16 S rRNA gene libraries were obtained from an extreme saline brine sediment of the Kebrit Deep. The phylogenetic investigation of these sequences showed that most of them had only low sequence similarity with cultivated Bacteria or Archaea.

The bacterial sequences grouped together forming a novel branch named KB1 in the 16 S rRNA tree, distantly related to *Aquificales* and *Thermotogales*. All the archaeal 16 S rRNA sequences retrieved from the Kebrit Deep belonged to the phylum *Euryarchaeota* and were only distantly related to sequences detected in other marine habitats (Eder et al. 1999). Many clone sequences belonging to the KB1 cluster were obtained also from the higher salinity (24%) part of the seawater-brine interface of the Shaban Deep and in the Mediterranean DHABs Bannock and Urania, showing that the KB1 group is ubiquitous within the brines of the Red Sea and the Mediterranean Sea. These lines of evidence confirm that the KB1 lineage comprises bacteria specifically adapted to anoxic highly hypersaline conditions, unfortunately still not obtained in culture. Only from the lower part of the halocline at the Shaban Deep a second well separated group of sequences, named SB1, was detected that did not exhibit high similarity with any of the cultivated Bacteria. The screening of 16 S rRNA gene libraries revealed that the Shaban basin differs from the Kebrit Deep also in the archaeal community, represented not only by *Euryarchaeota*, but also by *Crenarchaeota*.

The assessment of the microbiota diversity at the deep basins of the Red Sea has been carried out also through the application of cultivation methods. The first halophilic bacteria cultivated from DHABs were obtained from the Kebrit Deep. These strains are Gram-negative, motile rods with rounded ends that grow singly or in pairs. They are strongly related to members of the genus *Halanaerobium* comprising anaerobic, halophilic and fermentative bacteria. The same species was also detected in 16 S rRNA libraries of the seawater-brine interface of the Kebrit Deep, indicating its adaptation to the interface habitat (Eder et al. 2001). Recently Antunes et al. (2008) were able to obtain in pure culture a novel obligate anaerobic bacterium from the brine-sediment interface of the Shaban Deep. The isolate SSD-17B^T is Gram-negative, able to grow in a medium containing NaCl concentration that ranges from 1.5 to 18%; it exhibits denitrifying and fermentative metabolisms and it is characterized by pleomorphic cells without cell walls. It has 'tentacle-like' protrusions with motility function. Phylogenetic studies indicated that strain SSB-17B^T is a representative of a new lineage between the phyla Firmicutes and Mollicutes. SSB-17B^T isolate belongs to the novel species *Haloplasma contractile*, order *Haloplasmatales* (Antunes et al. 2008).

4.2 Phylogenetic analyses of the microbial communities colonizing different seafloor hypersaline brines of the Gulf of Mexico

The presence of an active microbial population at the seawater-brine interface of the Orca basin is suggested by several lines of evidence. Direct cell counts showed a peak of total cells in correspondence with the seawater-brine interface, the only portion of the basin in which the uptake of [5-^{3}H]-uridine was detected (LaRock et al. 1988). The highest ATP level (20 ng/l) was measured in the upper interface and the presence of Archaea has been demonstrated through the isolation of specific isoprenyl ether-linked lipids (Dickins and Van Vleet 1992). The seawater-brine interface of the Orca basin hosts a redoxcline essentially dominated by manganese. Iron- and manganese-reducing bacteria were isolated on selective media and the plate counts indicated that the cell number of these two functional groups decreases along seawater up to a depth of 2170 m, whereas they significantly increase between 2170 and 2225 m (van Cappellen et al. 1998). Despite the detection of high numbers of iron- and manganese-reducing bacteria at the intermediate salinity region of the interface, only manganese oxide resulted to be depleted at these depths, whereas the iron oxides showed a conservative profile. To explain the same pattern of distribution of iron- and manganese-reducing bacteria, van Cappellen et al. (1998) hypothesized the presence of bacteria able to use both iron and manganese oxides as electron acceptors.

Cell plate counts from samples of the mud volcano (site GB425) and the brine pool (site GC233) showed that prokaryotes are two orders of magnitude more abundant in the brines than in the seawater. The microbial diversity was evaluated through metabolic activity assays, 16 S rRNA and functional gene libraries. According to the different distribution of chemical species and activity assays, different metabolic processes seemed to prevail in the brine of sites GB425 and GC233. In the brine pool sulphate reduction was the dominating metabolic pathway, acetoclastic methanogenesis rates were lower and both hydrogenoclastic and anaerobic methane oxidation (AOM) were below the detection limit (Joye et al. 2009). In contrast, the mud volcano showed higher rates of acetoclastic methanogenesis and also hydrogenoclastic methanogenesis was detected. The absence of AOM was registered also at the mud volcano site, though high methane fluxes were observed at these two sites. In both brines, 16 S rRNA gene libraries showed the presence of sulphate-reducing bacteria (SRB). SRB diversity, described by dissimilatory sulphite reductase (*dsrAB*) gene libraries, was higher in the quiescent pool than in the mud volcano brines, in accordance with activity data. Moreover, the presence of *Epsilonproteobacteria*, able to oxidize sulphide and hydrogen, in the brine pool suggested that an active sulphur cycle occurs in this deep hypersaline habitat (Joye et al. 2009). Despite the low rates of acetoclastic and hydrogenoclastic methanogenesis measured at the brine pool, methyl coenzyme M reductase (*mcrA*) sequences taxonomically linked to acetoclastic, methylotrophic, hydrogenotrophic and methane-consuming microorganisms were retrieved at this site. The *mcrA* gene sequences obtained from the mud volcano were mainly represented by acetoclastic species whereas hydrogenoclastic methanogens were not identified, possibly due to the presence of novel unknown salt-adapted hydrogenotrophs (Joye et al. 2009). The diverse microbiota composition observed at the brine pool and mud volcano sites can be correlated with their different fluid flows and organic matter inputs.

4.3 Investigation of the microbiota composition inhabiting the Chefren mud volcano (Nile Deep Sea Fan)

The composition of the prokaryotic community that colonizes morphologically different microbial mats on the sediment surface at the Chefren mud volcano was analyzed by Omoregie et al. (2008) by the application of FISH and 16 S rRNA gene libraries. Significant differences were found comparing bacterial 16SrRNA gene libraries obtained from white and orange mats collected in this habitat, as expected from their different geochemistry. In both mats many clones were identified as SRB belonging to the *Deltaproteobacteria* group, similar to genera commonly found in cold seeps, such as *Desulfobacter, Desulfosarcina, Desulfocapsa* and *Desulfobulbus* (Knittel et al. 2005). *Gammaproteobacteria* were present in both the mats and prevalent in the orange one, where they represent up to 74% of the total bacterial community. Most of the *Gammaproteobacteria* sequences were correlated to bacteria capable of iron, sulphide or methane oxidation. A high number of sequences of sulphur-oxidizing *Epsilonproteobacteria* were also detected in the white mats, according to their chemical composition (Omoregie et al. 2008). FISH and 16 S rRNA libraries showed that up to 24% of the total cells within the white mat were targeted by an oligonucleotide probe specific for 'Candidatus Arcobacter sulfidicus'. This species produces long sulphur filaments, resulting from sulphide oxidation (Sievert et al. 2006).

The orange mat is composed of iron (hydr)oxide encrusted sheaths. Known iron-oxidizing bacteria were not recovered in 16 S rRNA gene libraries from this mat, but unexpectedly many 16 S rRNA sequences of species involved in the reductive portion of the iron cycle were retrieved. These sequences cluster close to *Sulfurospirillum deleyianum*, known to perform reduction of iron via sulphur cycling, and they could likely enhance sulphur and iron cycling by oxidizing sulphur compounds using Fe^{3+}.

Archaeal 16 S rRNA gene clones retrieved from the mats were mainly represented by anaerobic methane oxidizers (ANMEs) belonging to the ANME-2 and ANME-3 groups, constituting respectively 55% and 74% of the archaeal sequences in the white and orange mats, respectively. ANMEs are responsible of AOM, a process that occurs at the sediments underneath of both of the mats, as demonstrated by activity measurements and the distribution of methane concentration (Omoregie et al. 2008). The known stoichiometry of AOM to sulphate reduction (1:1) is very different from what was observed at the Chefren mud volcano (>28:1). The specific geochemical parameters that dominate this mud volcano possibly resulted in the selection of AOM and SRB consortia capable to utilize, in addition to methane, higher hydrocarbon compounds that have been measured within the pore waters of the Chefren volcano.

4.4 Stratified microbial communities at the deep seawater-brine interface of hypersaline anoxic basins of the Eastern Mediterranean Sea

The microbial diversity of brine-water interfaces and water columns of the DHABs of the Eastern Mediterranean are the most carefully studied cases by taking advantage of a simple sampling strategy that allowed to sample 5–10 cm wide water layers in the chemoclines (Daffonchio et al. 2006; Borin et al. 2009). Such a high resolution sampling allowed describing the stratification of the microbial groups within the chemoclines in response to changes of the concentration of electron donors and acceptors.

Total cell counts showed that the chemoclines of the DHABs are strongly enriched in comparison to overlying seawater and underlying brine. FISH allowed to determine the Bacteria/Archaea ratio and revealed that only the Urania basin is dominated by Archaea, whereas Bannock, L'Atalante and Discovery are dominated by Bacteria. Overall the bacterial diversity is higher than the archaeal diversity. Cluster analysis of the archaeal and bacterial 16 S rRNA sequences recovered from the four basins showed that they are colonized by different microbial communities, according to the differences in geochemistry. The prokaryotic composition of the Discovery lake is rather diverse from the others, whilst Bannock and L'Atalante show the highest similarity (van der Wielen et al. 2005). These data are in accordance with the cluster analysis performed on the geochemical parameters of the four basins. The distribution of the operational taxonomic units (OTUs; groups of sequences sharing >97% 16 S rRNA gene sequence similarity) indicated that the bacterial community of the Discovery interface is significantly different from those inhabiting the seawater and brine. Many OTUs were specific for the Discovery brine, belonging to microbial species not present in the other DHABs. In particular, sequences belonging to the archaeal species *Halorhabdus utahensis* were not detected in seawater but strongly increased along the chemocline moving towards the brine, where they constituted up to the 33% of the total archaeal community. This heterotrophic archaeon, able to grow performing a fermentative metabolism, tolerates up to 0.8 M $MgCl_2$ (Waino et al. 2000). The abundance of *H. utahensis* together with the high ectoenzymatic activities and the low sulphate reduction and methane production occurring in Discovery brines, indicated that heterotrophic prokaryotes might play a major role than just that of being responsible for methanogenesis and sulphate reduction. MSBL-1 is a new archaeal candidate division retrieved only in all of the Eastern Mediterranean DHABs, belonging to the archaeal phylum *Euryarchaeota*, which is phylogenetically distantly related to methanogens. Since only few sequences related to known methanogens were detected in the DHABs, the methane production measured in the brines is likely due to the activity of this new archaeal halophilic taxon (van der Wielen et al. 2005).

Differently from what was observed in Discovery, sulphur cycling and methanogenesis shaped the prokaryotic communities colonizing the Urania DHAB. In Urania, methanogenesis and methane concentration were higher than in the other DHABs (Borin et al. 2009), along with dominance of the group MSBL-1 in the archaeal clone libraries of brines and the most saline layers of the chemocline. Sulphate reduction rates and sulphide concentrations were greater in Urania than in the other DHABs in agreement with the high sulphide concentration. The distribution of bacterial phylogenetic groups observed along the Urania water-brine interface showed bacteria of taxonomic groups having a sulphur-based metabolism. Whereas the bacterial community of seawater was mainly represented by *Alphaproteobacteria* and *Fibrobacteres* (van der Wielen et al. 2005), the Urania chemoclines and brines were dominated by *Deltaproteobacteria* and *Epsilonproteobacteria*. *Deltaproteobacteria*, mostly represented by *Desulfobulbaceae* and *Desulfobacteraceae*, formed up to 22.2–30.8% of the total bacterial community along the Urania chemoclines (Borin et al. 2009) and included most of the known SRB (Castro et al. 2000). *Epsilonproteobacteria* (11.7–47.8% of the bacterial clones) comprised different species capable of sulphide and sulphur oxidation (Campbell et al. 2006).

16 S rRNA gene libraries were constructed and isolation of different bacterial strains was performed also with samples from the Bannock basin. Eighty-four isolates were obtained from the Bannock interface belonging to the *Firmicutes, Bacteroidetes, Alpha-, Gamma-* and *Epsilon-proteobacteria* taxa. The 16 S rRNA gene libraries showed nevertheless a higher bacterial taxonomic richness and permitted the identification of new candidate divisions (MSBL 2–6). The relatedness of MSBL-2 to the candidate division SB1, detected for the first time at the interface of Shaban Deep (Red Sea) and its absence from seawater and brine samples point out to the specific adaptation of MSBL-2 division to the seawater-brine interfaces (Daffonchio et al. 2006). The detection of the MSBL 3–6 candidate divisions only in the deeper part of the seawater-brine interface of Bannock basin suggested that they are specifically adapted to high NaCl concentrations. In the deeper part of Bannock and L'Atalante halocline sequences belonging to KB1 group, previously identified in

the Kebrit Deep, were found (Daffonchio et al. 2006). The presence of KB1 and SB1 taxa demonstrated analogies between the microbiota colonizing DHABs of the Eastern Mediterranean and the Red Sea. All the metabolically active bacteria detected in the seawater overlying the L'Atalante brine belong to the class *Gammaproteobacteria*, 20% of which were represented by the genus *Thiomicrospira*. This sulphur-oxidizing bacterium dominated the upper part of the halocline and represented 100% of *Gammaproteobacteria* detectable at salinity values in the range 13–20% (Yakimov et al. 2007). Salinity affected also the distribution of *Epsilonproteobacteria* in the deepest part of the haloclines, where the *Deltaproteobacteria* seemed to be the most dominant class.

Autotrophic bacteria, responsible for energy production in the dark and contributing to the primary production of the DHAB ecosystems, were found along the water-brine chemoclines. Chemoautotrophic activity and the detection of the expression of functional enzymes have been documented for L'Atalante and Discovery (Yakimov et al. 2007). At the oxic–anoxic boundary of Urania and L'Atalante 16S rRNA signature sequences of aerobic sulphur oxidizers were retrieved (Yakimov et al. 2007; Borin et al. 2009) and along the Urania chemocline, signatures of anaerobic ammonium oxidizers (Borin et al. 2009).

5 Concluding remarks

DHABs were shown to be colonized by highly diverse microbial communities, whose diversity is shaped by the specific geochemical context. Studying the diversity of the microbial communities inhabiting hypersaline isolated environments can facilitate the understanding of adaptation mechanisms and the evolution of biogeochemical cycles and life on the early Earth or on extraterrestrial bodies. Evolution and adaptation of microbes in hypersaline habitats can be advantageous for the horizontal transfer of genetic information. In all the brines from DHABs Urania, Bannock, L'Atalante and Discovery, dissolved plasmid DNA was demonstrated to be substantially preserved for a period of 32 days in axenic conditions, maintaining the ability to transform naturally competent bacteria (Borin et al. 2008). These results indicate the potential role of dissolved extracellular DNA in hypersaline environments as a reservoir of genetic information that can be spread by HGT mediated by natural transformation.

Hypersaline habitats represent a huge and still largely unknown source of biodiversity exploitable for biotechnological applications. Halophilic microorganisms produce a wide set of enzymes stable under conditions which would lead to denaturation of most proteins and which are commonly used during many industrial processes (Ventosa et al. 1998; DasSarma and Priya 2001). Esterases which exhibit halophilic and additional innovative properties and are active in hypersaline conditions, have been expressed from the metagenome of Urania DHAB brines (Ferrer et al. 2005), confirming that a wide biochemical potential is available for biotechnological exploitation.

References

Antunes A, Rainey FA, Wanner G, Taborda M, Patzold J, Nobre MF, da Costa MS, Huber R (2008) A new lineage of halophilic, wall-less, contractile bacteria from a brine-filled deep of the Red Sea. J Bacteriol 190:3580–3587

Backer H, Schoell M (1972) New deeps with brine and metalliferous sediments in the Red Sea. Nat Phys Sci 240:153–158

Borin S, Crotti E, Mapelli F, Tamagnini I, Corselli C, Daffonchio D (2008) DNA is preserved and maintains transforming potential after contact with brines of the deep anoxic hypersaline lakes of the Eastern Mediterranean Sea. Saline Syst 4:10

Borin S, Brusetti L, Mapelli F, D'Auria G, Brusa T, Marzorati M, Rizzi A, Yakimov M, Marty D, De Lange GJ, Van der Wielen PWJJ, Bolhuis H, McGenity TJ, Polymenakou PN, Malinverno E, Giuliano L, Corselli C, Daffonchio D (2009) Sulfur cycling and methanogenesis primarily drive microbial colonization of the highly sulfidic Urania deep hypersaline basin. Proc Natl Acad Sci USA 106:9151–9156

Brune A, Frenzel P, Cypionka H (2000) Life at the oxic–anoxic interface: microbial activities and adaptations. FEMS Microbiol Rev 24:691–710

Bull AT, Ward AC, Goodfellow M (2000) Search and discovery strategies for biotechnology: the paradigm shift. Microbiol Mol Biol Rev 64:573–606

Camerlenghi A (1990) Anoxic basins of the eastern Mediterranean: geological framework. Mar Chem 31:1–19

Campbell BJ, Engel AS, Porter ML, Takai K (2006) The versatile epsilon-proteobacteria: key players in sulfidic habitats. Nat Rev Microbiol 4:458–468

Castro HF, Williams NH, Ogram A (2000) A phylogeny of sulfate-reducing bacteria. FEMS Microbiol Ecol 31:1–9

Daffonchio D, Borin S, Brusa T, Brusetti L, van der Wielen PWJJ, Bolhuis H, Yakimov M, D'Auria G, Giuliano L, Marty D, Tamburini C, McGenity T, Hallsworth J, Sass A, Timmis KN, Tselepides A, de Lange G, Hübner H, Thomson J, Varnavas S, Gasparoni F, Gerber H, Malinverno E, Corselli C, Biodeep Scientific Party (2006) Stratified prokaryote network in the oxic–anoxic transition of a deep sea halocline. Nature 440:203–207

DasSarma S, Priya A (2001) Halophiles. Encyclopedia of life sciences. Nature Publishing Group, London, Vol 8, pp 458–466

De Lange GJ, Middelburg JJ, Van Der Weijden CH, Catalano G, Luther GW III, Hydes DJ, Woittiez JRW, Klinkhammer GP (1990) Composition of anoxic hypersaline brines in the Tyro and Bannock Basins, eastern Mediterranean. Mar Chem 31:63–88

Degens E, Ross DA (1969) Hot brines and recent heavy metal deposits in the Red Sea. Springer-Verlag, New York, NY

Dickins HD, Van Vleet ES (1992) Archaebacterial activity in the Orca basin determined by the isolation of characteristic isopranyl ether-linked lipids. Deep Sea Res 39:521–536

Dupré S, Woodside JM, Foucher JP, de Lange GJ, Mascle J, Boetius A, Mastalerz V, Stadnitskaia A, Ondreas H, Huguen C, Harmégnies F, Gontharet S, Loncke L, Deville E, Niemann H, Omoregie

E, Olu-Le Roy K, Fiala-Médioni A, Dahlmann A, Caprais JC, Prinzhofer A, Sibuet M, Pierre C, Sinninghe Damsté J and the Nautinil Scientific Party (2007) Seafloor geological studies above active gas chimneys off Egypt (Central Nile Deep Sea Fan). Deep Sea Res 54:1146–1172

Eder W, Ludwig W, Huber R (1999) Novel 16 S rRNA gene sequences retrieved from highly saline brine sediments of kebrit Deep, Red Sea. Arch Microbiol 172:213–218

Eder W, Jahnke LL, Schmidt M, Huber R (2001) Microbial diversity of the brine – seawater interface of the Kebrit Deep, Red Sea, studied via 16 S rRNA gene sequences and cultivation methods. Appl Environ Microbiol 67:3077–3085

Eder W, Schmidt M, Koch M, Garbe-Shonberg D, Huber R (2002) Prokaryotic phylogenetic diversity and corresponding geochemical data of the brine – seawater interface of the Shaban Deep, Red Sea. Environ Microbiol 4:758–763

Ferrer M, Golyshina OV, Chernikova TN, Khachane AN, Martins dos Santos VAP, Yakimov MM, Timmis KN, Golyshin PN (2005) Microbial enzymes mined from the Urania deep-sea hypersaline anoxic basin. Chem Biol 12:895–904

Girdler RW, Styles P (1974) Two stage Red Sea floor spreading. Nature 247:7–11

Hartmann M, Scholten JC, Stoffers P, Wehner F (1998) Hydrographic structure of brine-filled deeps in the Red Sea – new results from the Shaban Kebrit, Atlantis II, and Discovery Deep. Mar Geol 144:311–330

Jongsma D, Fortuin AR, Huson W, Troelstra SR, Klaver GT, Peters JM, van Harten D, de Lange GJ, ten Haven L (1983) Discovery of an anoxic basin within the Strabo trench, eastern Mediterranean. Nature 305:795–797

Joye SB, MacDonald IR, Montoya JP, Piccini M (2005) Geophysical and geochemical signatures of Gulf of Mexico seafloor brines. Biogeosciences 2:295–309

Joye SB, Samarkin VA, Orcutt BN, MacDonald IR, Hinrichs KU, Elvert M, Teske AP, Lloyd KG, Lever MA, Montoya JP, Meile CD (2009) Metabolic variability in seafloor brines revealed by carbon and sulphur dynamics. Nat Geosci 2:349–354

Knittel K, Losekann T, Boetius A, Kort R, Amann R (2005) Diversity and distribution of methanotrophic archaea at cold seeps. Appl Environ Microbiol 71:467–479

Könneke M, Bernhard AE, de la Torre JR, Walzer CB, Waterbury JB, Stahl DA (2005) Isolation of an autotrophic ammonia-oxidizing marine archaeon. Nature 437:543–546

LaRock PA, Lauer RD, Schwartz JR, Watanabe KK, Wiesenburg DA (1979) Microbial biomass and activity distribution in an anoxic, hypersaline basin. Appl Environ Microbiol 37:466–470

LaRock PA, Schwartz JR, Hofer KG (1988) Pulse labelling: a method for measuring microbial growth rates in the ocean. J Microbiol Methods 8:281–297

MacDonald IR, Buthman D, Sager WW, Peccini MB, Guinasso NL (2000) Pulsed oil discharged from a mud volcano. Geology 28:907–910

Macgregor DS (1983) Relationship between seepage, tectonics and subsurface petroleum reserves. Mar Petrol Geol 10:606–619

Mascle J, Sardou O, Loncke L, Migeon S, Camera L, Gaullier V (2006) Morphostructure of the Egyptian continental margin: insights from swath bathymetry survey. Mar Geophys Res 27:49–59

Matsubaya O, Sakai H, Torii T, Burton H, Kerry K (1979) Antarctic saline lakes – stable isotopic ratios, chemical compositions, and evolution. Geochim Cosmochim Acta 43:7–25

Omoregie EO, Mastalerz V, de Lange G, Straub KL, Kappler A, Roy H, Stadnitskaia A, Foucher JP, Boetius A (2008) Biogeochemistry and community composition of iron- and sulfur-precipitating microbial mats at the Chefren mud volcano (Nile Deep Sea Fan, Eastern Mediterranean). Appl Environ Microbiol 74:3198–3215

Oren A (1999) Microbiological studies in the Dead Sea: future challenges toward the understanding of life at the limit of salt concentrations. Hydrobiologia 205:1–9

Pautot G, Guennoc P, Contelle A, Lyberis N (1984) Discovery of a large brine deep in the northern Red Sea. Nature 310:133–136

Sass AM, Sass H, Coolen MJL, Cypionka H, Overmann J (2001) Microbial communities in the chemocline of a hypersaline deep-sea basin (Urania Basin, Mediterranean Sea). Appl Environ Microbiol 67:5392–5402

Shokes RF, Trabant PK, Presley BJ, Reid DF (1977) Anoxic, hypersaline basin in the Northern Gulf of Mexico. Science 196:1443–1446

Sievert SM, Wieringa EBA, Wirsen CO, Taylor CD (2006) Growth and mechanism of filamentous-sulfur formation by Candidatus *Arcobacter sulfidicus* in opposing oxygen-sulfide gradients. Environ Microbiol 9:271–276

Stumm W, Morgan JJ (1996) Acquatic chemistry. Chemical equilibria and rates in natural waters, 3rd edn. John Wiley and Sons, New York, NY

van Cappellen P, Viollier E, Roychoudhury A, Clark L, Ingall E, Lowe K, DiChristina T (1998) Biogeochemical cycles of manganese and iron at the oxic–anoxic transition of a stratified marine basin (Orca Basin, Gulf of Mexico). Environ Sci Technol 32:2931–2939

van der Wielen PWJJ, Bolhuis H, Borin S, Daffonchio D, Corselli C, Giuliano L, de Lange GJ, Varnavas SP, Thompson J, Tamburini C, Marty D, McGenity TJ, Timmis KN, BioDeep Scientific Party (2005) The enigma of prokaryotic life in deep hypersaline anoxic basins. Science 307:121–123

Vengosh A, de Lange GJ, Starinsky A (1998) Boron isotope and geochemical evidence for the origin of Urania and Bannock brines at the eastern Mediterranean: effect of water – rock interactions. Geochim Cosmochim Acta 62:3221–3228

Ventosa A, Nieto JJ, Oren A (1998) Biology of moderately halophilic aerobic bacteria. Microbiol Mol Biol Rev 62:504–544

Waino M, Tindall BJ, Ingvorsen K (2000) *Halorhabdus utahensis* gen. nov., sp. nov., an aerobic, extremely halophilic member of the Archaea from Great Salt Lake, Utah. Int J Syst Evol Microbiol 50:183–190

Wallmann K, Suess E, Westbrook GH, Winckler G, Cita MB, the Medriff consortium (1997) Salty brines on the Mediterranean sea floor. Nature 387:31–32

Wallmann K, Aghib FS, Castradori D, Cita MB, Suess E, Greinert J, Rickert D (2002) Sedimentation and formation of secondary minerals in the hypersaline Discovery basin, eastern Mediterranean. Mar Geol 186:9–28

Yakimov MM, La Cono V, Denaro R, D'Auria G, Decembrini F, Timmis KN, Golyshin PN, Giuliano L (2007) Primary producing prokaryotic communities of brine, interface and seawater above the halocline of deep anoxic lake L'Atalante, Eastern Mediterranean Sea. ISME J 1:1–13

Microbial speciation in the geothermal ecosystem

Gudmundur Oli Hreggvidsson[1,2], Solveig K. Petursdottir[1], Snaedis H. Björnsdottir[1] and Olafur H. Fridjonsson[1]

[1]Matís, Reykjavík, Iceland
[2]Department of Life and Environmental Sciences, University of Iceland, Reykjavík, Iceland

1 Introduction

Geothermal areas are unique in many aspects as microbial habitats. They are rare on a global scale and geographically confined. They can be regarded as islands, ecologically separated by large distances and physicochemical dispersal barriers. In a sense the global geothermal ecosystem can be considered to be a world of widely dispersed, often very different "archipelagos" with no mainland. These and other features make geothermal sites an attractive and perhaps ideal model system for studies of microbial divergence and speciation. Microbial speciation may even be more easily observable in geothermal habitats than in other ecosystems.

Geothermal ecosystems are the province of prokaryotes – of unicellular organism. In addition to the high temperature, conditions may range from one extreme to another in pH and other environmental variables. Chemical composition of water varies enormously, even between adjacent sites, and steep environmental gradients are common. The geothermal gases provide abundant sources of chemical energy, in many instances leading to the formation of unique microbial mat communities. The geothermal biosphere is a very dynamic environment where many different forces are at play shaping intricate relationships in a rich tapestry of life.

It is commonly accepted that bacteria are clonal but it is also recognized that the population is the subject or unit of evolution. This presupposes a cohesive force keeping and reinforcing population boundaries. What is the nature of such a force and how universal is it? Are there more than one? One view maintains that prokaryotic evolution can be understood primarily in terms of clonal divergences and intraspecies competition and that periodic selection acts as the cohesive force by repeatedly purging a population – the species – of genetic diversity (Levin 1981; Cohan 2002). The proponents of this view advocate that a microbial species is an ecologically distinct population (Ward 1998). Another view embraces lateral gene transfer as a causal force, responsible for patterns of both similarities and differences between populations and species of bacteria (Gogarten et al. 2002). According to this view lateral gene transfer acts not only as a cohesive force by genetic exchange and homologous recombination between related strains, but also as a major force of divergence by mediating rarer acquisitions of foreign genes and a subsequent phenotypic or ecological divergence. It follows that a species may consist of groups of ecologically heterogenous bacteria and that the overall species diversity is constrained by genetic exchange and homologous recombination (Gogarten et al. 2002; Lawrence 2002). This latter view is, essentially, the more practical view and conforms better to accepted but arbitrary species boundaries as they are demarcated by 16S rRNA similarity and DNA hybridization values.

The idea of bacterial species may remain elusive, but evolutionary studies of bacteria may not need a rigid concept of species. Perhaps it can be "dismissed" with altogether (Lawrence 2002), or the operational unit of evolution may be defined differently as "ecologically distinct populations" that are "species-like fundamental units of microbial communities" (Ward et al. 2008). In addition it may be important to recognize that "different microbial species may have evolved in different ways" (Cohan 2006; Ward et al. 2008).

The metabolic and physiological similarity between many thermophilic species is an issue that bears on the concept of species. The taxonomy of thermophiles has almost from the beginning been based on molecular methods. 16S rRNA sequencing and DNA:DNA hybridization have been fundamental for delineating taxa and establishing species status. Interestingly, this revealed the apparent phenotypic conformity not only within many widespread genera, such as *Thermus*, *Hydrogenobacter*, and *Thermococcus*, but also between members of higher taxa, such as within the orders of *Sulfolobales* and *Aquificales* (Hreggvidsson and Kristjansson 2003; Kristjansson et al. 2000).

Since so many thermophilic genera appear to be collections of genospecies with few observed phenotypically distinguishing characteristics it indicates that adaptive traits of thermophilic genospecies within a genus could be very few or involving subtle physiological or metabolic differences difficult to identify. This also raises related questions e.g. what dictates the distribution of the different thermophilic species? Do geographical, topological, or other physical dispersal barriers play the major role or is the distribution determined largely by physicochemical factors?

Culture-independent molecular methods are invaluable for analyzing community structures, estimating diversity, evenness and abundance of taxa in geothermal areas as well as for studying their distribution and phenomena, such as dispersal mechanisms and barriers, endemism, migration, and colonization. These methods have been helpful in revealing the geographical distribution of thermophilic bacteria and the relationships between environmental physicochemical conditions and species composition. They may also become important for answering questions about specific ecological adaptations, the roles of microorganisms in mineralization and cycling of nutrients as well as about microbial interactions in hot springs (Amann et al. 1995; Skirnisdottir et al. 2000).

Microbial evolutionary studies should involve an evaluation of the genetic structure of bacterial taxa, of the occurrence and the extent of gene flow, and of the relevance or importance of different cell apparatus in directing the adaptive evolution of a species. Genomic studies play an ever-increasing role in this context. Recent genomic studies have introduced important evolutionary concepts into microbiology, such as the pan genome, the core genome, and the dispensable genome. The results indicate that there is a certain fluidity of genetic information across species boundaries and they have put to test and renewed theories about the concept of microbial species.

Studies on more dynamic aspects of thermophilic microbial populations in an ecological context have lagged behind studies with more taxonomic or systematic emphasis, often in the pursuit of the novel exotic species. Now, we have many, perhaps most, of the major pieces, but too little of the context. Until now there has been scant information on species boundaries, genetic diversity, and the distribution of thermophilic microorganisms. However, descriptions of entire microbial communities, their interactions with the physical environment, and roles of constituent species are now becoming abundant as well as reflections on how geography and different evolutionary pressures may influence the population structure and divergence of thermophiles (Kristjansson and Hreggvidsson 1995; Ward et al. 1998; Petursdottir et al. 2000; Whitaker et al. 2003; Reysenbach et al. 2005; Ward and Cohan 2005; Hreggvidsson et al. 2006).

In this chapter we describe various features of the geothermal biosphere, and environmental factors that may influence speciation in these habitats. This includes physicochemical, geographical, and topological aspects and other factors and forces that may influence the distribution and dispersal of thermophilic organisms. Lastly we will give an example of the biogeography and ecology of *Thermus* to illustrate some of the questions and issues that have received attention in this particular field of study.

2 Geothermal areas

Geothermal areas on Earth are mainly connected with tectonic activity caused by plate movements on the fluid-like asthenosphere, where plates are moving apart, colliding or transforming (Fig. 1). The plate boundaries are commonly associated with geological events such as earthquakes and the creation of topographical features

Earth's Tectonic Plates

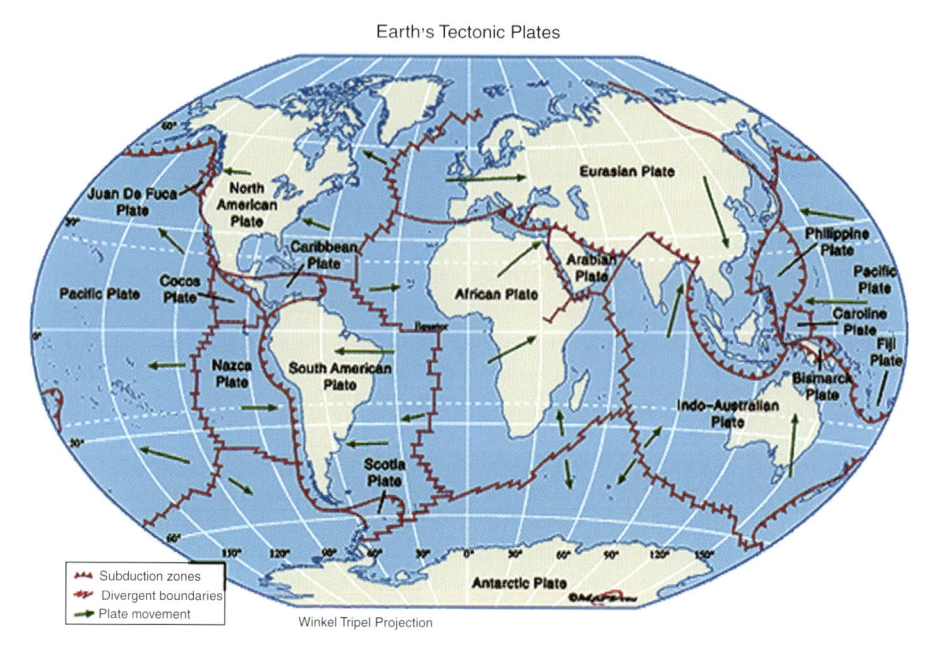

Fig. 1. Boundaries of tectonic plates on Earth (credit: http://media.maps.com/magellan/Images/tectonic.gif)

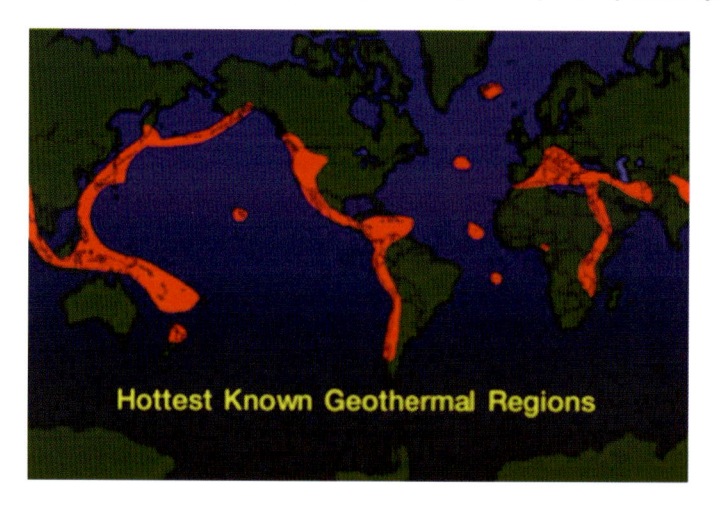

Fig. 2. The geothermal activities occur mainly along the plate boundaries (Slide 15 © 2000 Geothermal Education Office; http://geothermal.marin.org/geopresentation/)

such as mountains, volcanoes, midocean ridges, and oceanic trenches. The hottest known geothermal regions and the majority of the world's active volcanoes occur along plate boundaries (Fig. 2). The best known and biologically most studied geothermal areas are in North America (Yellowstone National Park), New Zealand, Japan, Italy, the Kamchatka Peninsula, and Iceland.

Geothermal areas in different parts of the world vary greatly in geology and chemistry, but belong mainly to two categories: Firstly, the solfataric type characterized by acidic soils, sulfur, mud pots, and fumaroles, and secondly the neutral–alkaline type, characterized by freshwater hot springs and geysers, which are neutral to alkaline in pH (Kristjansson and Hreggvidsson 1995). The chemistry may vary significantly in many other aspects. For example, high arsenic concentration in geothermal waters is typically associated with acidic volcanic systems in continental settings, particularly argillaceous sediments where it is known to be preferentially partitioned. In contrast, geothermal springs in volcanic regions associated with magmas of predominantly basaltic composition, such as in Iceland, have much lower levels of arsenic compounds (Arnórsson 2003). Geothermal systems known to contain high arsenic concentrations include the Yellowstone Park USA where arsenic concentrations have been reported to be as high as 150 mg/l and 50 mg/l, and the El Tatio geothermal field, Chile. Arsenite, As(III), can be the sole or primary arsenic species in hot anaerobic source waters where it is rapidly converted to arsenate, As(V), due to microbial oxidations (Langner et al. 2001; Romero et al. 2003).

2.1 High- and low-temperature fields

The two types of geothermal areas are the result of geological differences of the heat source. The high-temperature fields are located within active volcanic zones and have magma chamber as the heat source. They are defined by temperatures above 200°C at the depth of 1000 m and characterized by emissions of steam and volcanic gases on the surface. The gas is primarily N_2 and CO_2 but H_2S and H_2 can be up to 10% each of the total gas fraction. Traces of ammonia, methane, and carbon monoxide are also found. The pH of the subsurface steam is near neutrality because of the weak acids, CO_2 ($pK=6.3$) and H_2S ($pK=7.2$). Closer to the surface the sulfide of the steam is oxidized chemically and biologically to sulfur ($H_2S + \frac{1}{2} O_2 \leftrightarrow S + H_2O$). The oxidation can go further resulting in the formation of sulfuric acid ($H_2S + 2O_2 \leftrightarrow H_2SO_4$), thus lowering the pH and causing corrosion of the surrounding rocks and formation of the typical acidic mud of solfatara fields. As the temperature is high, there is little water coming to the surface and the hot springs are mostly mud pools or fumaroles. These areas are generally unstable and openings emerge and disappear quite rapidly. As there is no outflow, the water is static and becomes saturated by gases from the geothermal steam (Bödvarsson 1961; Kristjansson and Hreggvidsson 1995; Palmason 2005).

Low-temperature areas are located at the flank of the volcanically active zones. These are defined by temperature lower than 150°C at 1000 m depth. They are heated by deep lava flows or by dead magma chambers. Groundwater percolating into these hot areas is heated and returns up to the surface containing dissolved minerals such as silica and some dissolved gases, mainly carbon dioxide. The concentration of sulfide in the water is low. The subsurface pH is near neutrality and maintains at or above neutral pH at the surface. There is little or no H_2S to be oxidized and as the CO_2 escapes and silica precipitates this results in increased pH. The thermal manifestations at the surface are warm or hot springs, under or above 50°C, respectively, and individual hot springs are very constant in temperature and water flow. These areas are relatively stable as they are located outside the active volcanic zone. However, they may disappear or new ones are created during periods of seismic activity (Bödvarsson 1961; Kristjansson and Hreggvidsson 1995; Palmason 2005).

Both types of geothermal fields are also found on the seafloor, adding salinity and even sharper temperature gradients to the other factors. Also, the sulfide is oxidized to sulfur and sulfuric acid, but it cannot affect the pH to the same extent as in the terrestrial fields due to the huge water mass. Hot geothermal water originating inland can well up from the sea bottom especially in coastal areas. A remarkable example of such submarine freshwater hot springs is found in the fjord Eyjafjördur on the north coast of Iceland. Up to 60 m high cones of smectite rise from the sea bottom formed by the mixing of the hot SiO_2-rich geothermal fluid with the cold Mg-rich seawater (Marteinsson et al. 2001).

2.2 Origin of hot spring water

The origin of the hot spring water is groundwater, rainwater, or melting snow. Some of the rainwater that seeps deep into the Earth will eventually surface in various hot springs. The water of hot springs can be ancient, flowing from the high grounds toward the sea for centuries.

In high-temperature fields the geothermal steam originating from deep hot strata heats up the surface water from recently fallen rain and/or water seeping from glaciers or snow. The water volume of these hot springs can change rapidly with weather.

In low-temperature areas the groundwater is heated up by dead magma chambers or lava inserts. In both cases the water is enriched in minerals and dissolved chemicals from below. Final pH and temperature are variable according to the nature of the heat source.

All geothermal water contains silica (SiO_2). The level of dissolved silica in the water depends partly on temperature. As the water approaches the relatively cold surface the silica precipitates and forms the typical silica sinters common in geothermal areas. On the other hand, calcium carbonate ($CaCO_3$)

precipitates as the temperature gets higher and forms travertine. Sometimes precipitates of magnesium silicate are formed, but these are relatively rare (Palmason 2005).

2.3 Surface characteristics of geothermal areas

The surfaces of high-temperature geothermal areas can be very colorful (Fig. 3a–c). The different colors originate from the dissolved minerals and chemicals in the water and steam from deep below reacting with the oxygen dissolved in the aquifer closer to the surface. This is reflected in different colors at the surface of the hot spring areas. Elemental sulfur ($S°$) and sulfur compounds are characteristic for geothermal areas but these are much higher in the acidic environments. The evident yellow color is sulfur, the red is hematite (Fe_2O_3) and the dark gray color of the clay is a compound of crystallized iron and sulfur (FeS).

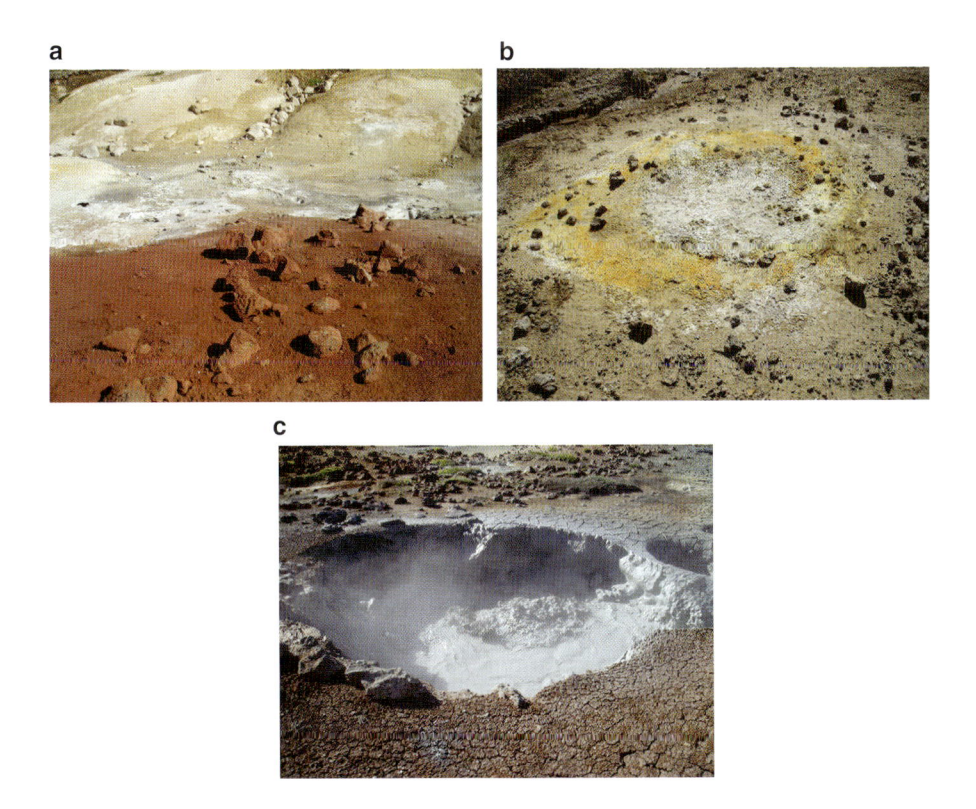

Fig. 3. (**a–c**) High-temperature geothermal areas are colorful; (**a**) a mud spot with fumaroles and red precipitates; (**b**) a fumarole with yellow sulfur precipitations; (**c**) a mud pool. The pictures were taken at Þeistareykir geothermal area in NE Iceland (photos by G. O. Hreggvidsson, 2008)

White clay can be high in caolinite, while dark clay is often high in smectite. Heavy metals are generally in high concentrations in the acidic environments. Silica sinters are common in geothermal acidic areas.

On the other hand, the travertines of calcium carbonate are generally found at pH around or slightly below neutral. Salts and minerals can be up to 2000 mg l^{-1} in water-rich alkaline hot springs. Nitrogen and phosphorus compounds are usually in high concentrations in alkaline hot springs (Palmason 2005).

The *water poor acidic hot springs* i.e. the boiling mud pools, steaming fumaroles, and regions of hot humid, and greatly transformed soil are common in the high-temperature fields (Fig. 4). Geothermal steam from deep hot strata heats up the surface water, which is mainly recently fallen rain and/or water seeping from glaciers or snow. The water volume of these hot springs can change rapidly with weather. As there is no outflow, the water is static and becomes saturated by gases from the geothermal steam. The high-sulfide content of the geothermal steam (5–15% H_2S) is oxidized chemically and biologically, first to sulfur and then to sulfuric acid. This lowers the pH, causing

Fig. 4. A colorful solfatara field in Þeistareykir geothermal area in NE Iceland (photo by G.O. Hreggvidsson, 2008)

Fig. 5. (a) A mud pool in the geothermal area in Vonarskard in Iceland (photo by H. Jóhannesson); (b) mud pools in Þeistareykir geothermal area (photo by Pétursdóttir 2008)

corrosion of the surrounding rocks (Fig. 5b) and the formation of the typical acidic mud of solfatara fields with pH around 2 (Kristjansson and Hreggvidsson 1995). This is the domain of strictly anaerobic archaea and they mainly utilize reduced sulfur compounds and hydrogen as energy sources. Also, ubiquitous aerobic hyperthermophilic *Sulfolobus* spp. proliferate in mud pits at and in the hot turbid acidic waters of characteristic stagnant pools. At lower temperatures, heterotrophic archaeal species *Thermoplasma* and *Picrophilus* can be found at pH around 3 and 0–1, respectively. Chemolithoautophic bacteria that belong to the genera *Thiomonas* and *Thiobacillus* and heterotrophic *Geobacillus* and *Deinococcus* species are found at the lowest temperatures.

Mud pools form in high-temperature fields where water is in relatively short supply. The available water rises to the surface and forms mud with the soil particles. The thickness of the mud depends on the water content (Fig. 5a, b).

Fumaroles are small openings often located in hills high above the groundwater levels (Fig. 6). These emit steam and volcanic gases such as CO_2, SO_2, and H_2S.

Boiling pits are still another hot springs forming shallow depressions in the Earth often with gravel in the bottom and with clear boiling water. They can be found in both high and low-temperature areas. In the high-temperature areas, the bubbles are caused by volcanic gases steaming through the water (Fig. 7), but in the low-temperature areas the water is actually boiling.

Sulfide-rich hot springs with pH values of 5.5–6.5 are relatively rare. These are hot mineral springs with very high concentrations of sulfides in the source water. As the water flow is high and these hot springs have outlets, the acid does not accumulate as in the mud pools, so the pH is maintained below neutral. The temperature at the source is often around 80–85°C but gradually lowers in the affluent where thick white sulfide utilizing microbial mats are formed by *Sulfurihydrogenibium* species. The growth seems to be limited at temperatures between 50 and 70°C (Fig. 8a, b)

Fig. 6. A fumarole in the geothermal area in Hveravellir in Iceland (photo by H. Jóhannesson)

Fig. 7. Boiling pit in Vonarskard geothermal area in Iceland (photo by H. Jóhannesson)

(Kristjansson and Hreggvidsson 1995; Skirnisdottir et al. 2000; Reysenbach et al. 2005).

Alkaline hot springs (>50°C) and *warm springs (<50°C)* are common in low-temperature fields, often located at the borders of the high-temperature fields, where the groundwater level is high. These are generally water rich and have outlets (Fig. 9, left).

The chemical compounds directing the pH of the alkaline hot springs are HCO_3^-/CO_3^{2-} and SiO_2. The pH becomes rather high or in the range of pH $\geq$ 7–10. The sulfide concentration of the alkaline hot springs can be up to 1 mM. The abundant water in these springs runs off in small streams with marked temperature gradients. Large fluctuations in temperature can occur in the springs depending on weather conditions. The water level and water flow from the alkaline

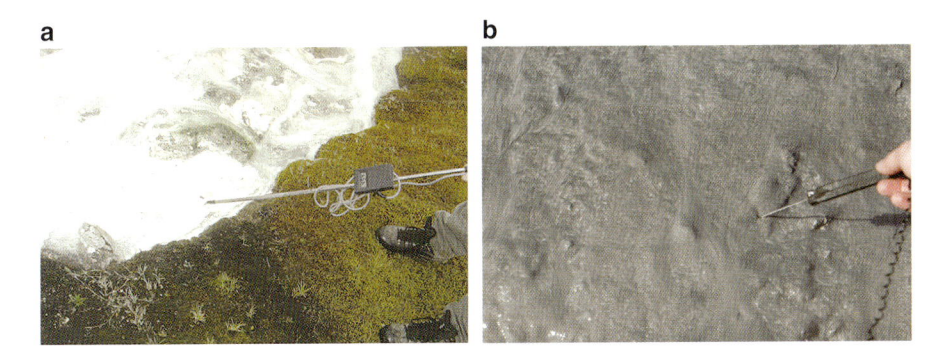

Fig. 8. (a) A thick, white microbial mat in a sulfide-rich hot spring in Vonarskarð in Iceland (photo by S.K. Pétursdóttir, 2008); (b) A microbial mat in a sulfide-rich hot spring in Seltun in Krisuvik in Iceland. The color of the mat is white underneath the gray clay particles (photo by G. O. Hreggvidsson, 2008)

Fig. 9. *Left*, a water-rich alkaline hot spring in Fludir in Iceland (photo by H. Jóhannesson); *right*, thick microbial mats of cyanobacteria, chloroflexi and "sulfur bacteria," well visible in an alkaline hot spring effluent in Iceland (photo by M. Schmid)

hot springs are often affected seasonally and can become very low during dry periods.

Colorful microbial mats are common in circumneutral and alkaline hot springs, where different microbes dominate, and are often associated with characteristic colors: green (cyanobacteria), red-orange (Chloroflexi), pink (*Thermocrinis ruber*, USA), gray, bluish (*Thermocrinis alba*, Iceland), white, yellowish, and black (*Sulfurihydrogenibium*). Sharp boundaries can be observed where one dominant organism replaces another. The color change usually reflects differences in growth temperature ranges of the species (Fig. 9, right). It may also reflect radical changes in chemistries of the hot spring system. Certain chemicals exert strong selective pressures on mat communities as energy sources or by their toxicities. These chemicals include hydrogen sulfide and arsenic compounds (see below).

Fig. 10. *Left*, steam vents in Jarðbaðsholar in Myvatnssveit in Iceland. *Right*, a closer look at a steam vents opening (photos by S.K. Pétursdóttir)

Ward and coworkers studied photosynthetic microbial mats common in alkaline hot springs. Using culture-independent molecular methods they showed that the common *Synechococcus* morphologies in the temperature range of 65–73°C mask a considerable genetic diversity. Also, that diversity decreased with increasing temperature and that genetically distinct clades lived in alkaline hot springs in different geographical regions of the world. They maintain that this distribution pattern cannot be explained by different chemical conditions, suggesting that geographical isolation is involved in diversification of hot spring cyanobacteria (Papke et al. 2003).

Steam vents are yet another type of geothermal surface manifestations in geothermal areas. These are relatively rare, but sometimes found in low-temperature lava fields. Steam emerges from the hot groundwater below up above the water table and through the porous lava, blowing slowly out from the openings (Fig. 10). The temperature is in the range of 55–85°C and the pH between 7 and 8 in the soil around the openings. The soil is relatively untransformed and the steam usually does not contain geothermal gases.

3 Ecology of thermophiles

Enormous diversity of chemical and physical properties influences and determines what kind of life can exist in a geothermal environment. It is an extreme habitat characterized by high temperature, high or low pH, and a relatively high ionic strength. The species diversity in hot springs is generally low as estimated by culture-independent methods. Extreme conditions of temperature and pH create environmental pressures resulting in fewer species capable of coping with the relatively harsh environment. Regular temperature fluctuations appear to be common in hot springs, and up to 10°C differences with approximately

6 h periodicity have been observed between the lowest and the highest temperature within a 24–60 h period in many hot springs in Iceland. Periodic disturbances of this kind may help to maintain a stable community structure, consisting of functionally near equivalent groups. Other diverse environmental factors in hot springs are water content, the content of gases, salts and minerals, dissolved oxygen, chemical compounds, and light. Environmental gradients are common, especially in the effluents of the alkaline type of hot springs where water is abundant. These add still other variables for life to cope with or exploit in this environment (Kristjansson et al. 2000; Hreggvidsson and Kristjansson 2003).

3.1 Temperature

The lower temperature limit of a hot spring has been set at 50°C. However, the upper limit is above 100°C in deep sea hydrothermal vents where pressure comes into play. The temperature of mud pools, fumaroles, and solfataras is generally higher than that of alkaline hot springs and steam vents.

The temperature of the surrounding environment is considerably lower than the temperature of a hot spring. Therefore, gradients form in all directions from the main heat source. These temperature gradients are usually very steep, of the order of 1–10°C per mm and 10–50°C per m in a flowing hot water. This means that only a few millimeters away from the site where the temperature is too high microbes thrive and build up massive mats (Kristjansson and Hreggvidsson 1995). Temperature fluctuations lead to periods of fast growth and buildup of a profuse biomass at permissive temperatures, then, periods of too high temperatures lead to death and decomposition of the same biomass providing an abundant source of nutrients for more thermophilic heterotrophs.

3.2 pH

The pH levels of hot springs are determined by the origin and amount of water available as well as the concentration of the volcanic gases such as H_2S. In the high-temperature areas the H_2S is oxidized to H_2SO_4. As water is usually scarce in these areas the H_2SO_4 builds up in the pools or in the solfataric humid soil stabilizes at pH values from 2 to 2.5. In sulfide-rich hot springs at temperatures around 70°C with abundant water the acid is washed away in the effluent so the pH stabilizes around 6. In alkaline hot springs the H_2S content of the source water is much lower than in the acidic geothermal areas. The chemical compounds directing the pH of the alkaline hot springs are HCO_3^-/CO_3^{2-} and SiO_2. The pH becomes rather high or in the range of pH $\geq$ 7–10. Thermoacidophily and thermoalkalophily are concepts reflecting the adaptation of microbes to extreme pH at high temperatures

(Kristjansson and Hreggvidsson 1995; Kristjansson et al. 2000; Hreggvidsson and Kristjansson 2003).

3.3 Energy sources as a selective pressure

The classification of organisms within an ecosystem into primary producers and consumers is well known. The primary production generally occurs by means of photosynthesis creating organic matter for the consumers of the ecosystem. However, as photosynthesis does not occur above 74°C the primary production above this limit is provided by chemolithoautotrophs, which can use inorganic chemicals as energy sources. These organisms often form conspicuous microbial filaments, streamers, or microbial mats, similar to photosynthetic mats at lower temperatures.

When different thermal habitats are analyzed in relation to available energy sources and metabolic types of the organisms present we find a characteristic pattern of community structures. Clearly the bacterial diversity in geothermal sites is established along the environmental gradients of temperature, pH, and the available electron donors and acceptors with similar communities found in similar conditions in far apart regions.

Terrestrial lithoautotrophic species of Aquificales often form conspicuous thick microbial mats at high temperatures. The particular species composition depends on the physicochemical conditions in the hot spring, e.g., temperature, pH, and available chemical energy sources (Skirnisdottir et al. 2000; Hjorleifsdottir et al. 2001; Reysenbach et al. 2005). The most important electron donors appear to be the geothermal gases H_2S or H_2, and their absolute concentrations and ratios may have marked effects on the microbial mat diversity. For example, microbial mats in alkaline (~pH 8) low-sulfide hot springs ~80–92°C are predominately populated by hydrogen-oxidizing *Thermocrinis* species, whereas circumneutral high-sulfide springs around 70°C are characterized by gray or yellowish microbial mats dominated by *Sulfurihydrogenibium* species (Skirnisdottir et al. 2000; Hjorleifsdottir et al. 2001; Reysenbach et al. 2005). Sulfide as an energy source, selects for *Sulfurihydrogenibium* in the latter case, but its toxicity may also select against *Thermocrinis* species (Skirnisdottir et al. 2000; Hjorleifsdottir et al. 2001). Genetically distinct but related Aquificales groups appear to occupy similar niches in far apart regions. *T. ruber* is the dominant species of the pink filament mats in Yellowstone National Park (Huber et al. 1998), and a related species, *T. alba*, dominates the commonly observed blue threads or gray streamers in Icelandic hot springs above 80°C (Skirnisdottir et al. 2000; Eder and Huber 2002).

In Tables 1–4, some examples are given of organisms that are important representatives of certain metabolic groups.

Table 1. Community structure in freshwater, alkaline hot springs

Energy source[a]	Representative organisms	T_{max}	T_{opt}	pH_{opt}
Primary producers				
H_2O/light	Synechococcus lividus	73	65	8.0
H_2S/light	Chloroflexus aurantiacus	70	56	8.0
H_2/O_2	Hydrogenobacter thermophilus	77	72	6.8
H_2S/O_2	Thermocrinis ruber	89	80	7–8.5
H_2/CO_2	Methanobacterium thermoautotrophicum	75	65	7.4
H_2/SO_4^{2-}	Desulfovibrio thermophilus	85	65	7.5
Consumers				
Org.m[b]/O_2	Thermus sp.	80	60–72	7.2
Org.m[b]/NO_3	Thermus sp.	80	60–72	7.2
Org.m[b]/O_2	Bacillus sp.	85	60–80	7.0
Org.m[b]/Org.m[b]	Clostridium sp.	80	58–68	5.7
Org.m[b]/Org.m[b]	Thermoanaerobacter sp.	78	60–69	7.2
Org.m[b]/SO_4^{2-}	Desulfotomaculum sp.	65	55–60	7.0

[a]External sources of organic matter. Sources of organic matter are from the surrounding vegetation of mesophilic and thermophilic algae and cyanobacteria
[b]*Org.m* organic materials or organic compounds other than C1 compounds, without further specification in each case

Table 2. Community structure in acidic solfatara fields

Energy source[a]	Representative organisms	T_{max}	T_{opt}	pH_{opt}
Primary producers (autotrophs)				
S/O_2	Sulfolobus acidocaldarius	90	75	2.5
S/O_2	Acidianus infernus	96	90	2.0
H_2/S	Acidianus infernus	96	90	2.0
Consumers (heterotrophs)				
Org.m[a]/O_2	Thermoplasma volcanium	67	60	2.0
Org.m[a]/O_2	Sulfolobus acidocaldarius	90	75	2.5

[a]See Table 1

Table 3. Community structure in anaerobic geothermal mud and soil

Energy source[a]	Representative organisms	T_{max}	T_{opt}	pH_{opt}
Primary producers (autotrophs)				
H_2/CO_2	Methanothermus fervidus	97	83	6.5
H_2/CO_2	Methanococcus igneus	91	88	5.7
H_2/S	Thermoproteus tenax	96	88	5.5
CO/S	Thermoproteus tenax	96	88	5.5
H_2/S	Pyrodictium occultum	110	105	6.5
H_2/SO_4^{2-}	Archaeoglobus fulgidus	92	83	7.0
Consumers (heterotrophs)				
Org.m[a]/org.m[a]	Thermotoga maritima	90	80	6.5
Org.m[a]/S	Thermotoga neapolitana	90	80	7.0
Org.m[a]/S	Pyrobaculum islandicum	102	100	6.0
Org.m[a]/SO_4^{2-}	Archaeoglobus fulgidus	92	83	7.0
Org.m[a]/S	Pyrobaculum islandicum	102	100	6.0
Org.m[a]/org.m[a]	Pyrococcus furiosus	103	100	7.0
Org.m/S	Thermoproteus tenax	96	88	5.5
Org.m[a]/SO_4^{2-}	Archaeoglobus profundus	90	82	6.0

[a]See Table 1

Table 4. Community structure in sulfide-rich hot springs

Energy source[a]	Representative organisms	T_{max}	T_{opt}	pH_{opt}
Primary producers (autotrophs)				
H_2S/O_2	Sulfurihydrogenibium spp.	78	70	6.5
Consumers (heterotrophs)				
Org.m[a]/O_2	Thermus scotoductus and Thermus oshimai (Iceland), and Thermus aquaticus (USA)	65	79	6,5

[a]See Table 1

4 Biogeography of thermophiles

Geothermal areas have yielded a large number of highly diverse thermophilic and hyperthermophilic genera of bacteria and archaea. Not surprisingly many of these genera, such as *Thermus, Thermoplasma, Rhodothermus, Bacillus, Sulfolobus,* and *Hydrogenobacter,* to name just a few, have a worldwide distribution. On the species level the situation is more complicated, owing to the phenotypic similarity of thermophilic species and also because the genetic structures are in most cases unknown. However, the methods of molecular systematics have started to reveal clear endemic patterns in the distribution of some thermophiles at and below the level of species, and there are cases of conspicuous absence of a particular species in certain geothermal areas. It can be expected that culture-independent studies will continue to be of major importance in examining the global distribution of thermophiles (Hreggvidsson and Kristjansson 2003).

Geographical isolation might be expected to be a significant factor in causing and enhancing the divergence of microbes. The geographical structure underlying the thermal biosphere promotes the separation of populations, disrupts the cohesive force and subsequently accelerates genetic divergence by various mechanisms. Founder effects and genetic drifts may occur and genetically separate lineages and different geothermal regions may have different selective pressures leading to local environmental adaptations by periodic selection. Geographical isolation also creates opportunities for isolated evolutionary events e.g. local lateral gene transfer occurrences opening new niches for separate populations within the species (Cohan 2001, 2002; Hreggvidsson and Kristjansson 2003).

The global geothermal ecosystem might be expected to harbor unique species compositions in far apart habitats. A particularly striking pattern of geographically influenced distribution is found for species of *Thermus* (Fig. 11). Distinct differences in species compositions seem to exist between widely separated locations, with mixtures of unique endemic lineages and more cosmopolitan species (Hreggvidsson et al. 2006). Another example of a discontinuous global distribution at the species level is found among the thermophilic cyanobacteria. Different clades of *Synechococcus* bacteria are found in far apart geothermal locations around the world (Papke et al. 2003) and some of them appear to represent local environmental adaptations. Thus, high-temperature *Synechococcus* species are abundant and easily visible in North American hot springs, where they form greenish mats together with *Chloroflexus* at temperatures up to 73°C. These species have not been detected in alkaline and neutral hot springs in Iceland between 65 and 73°C. This temperature interval is dominated by the anoxygenic photoautotroph *Chloroflexus* that forms instead colorful pink and orange mats. Therefore, there appears to be an unoccupied niche for high-temperature oxygenic photosynthetic bacteria in Iceland. A similar, but perhaps less striking, distribution pattern is seen for the *Aquificales*. Different *Thermocrinis* species appear to occupy comparable niches of low-sulfide, high-temperature hot springs in Iceland and in the Yellowstone National Park. Also, microbial mats in high-sulfide hot springs in Japan, America, and Iceland are dominated by distinct *Sulfurihydrogenibium*

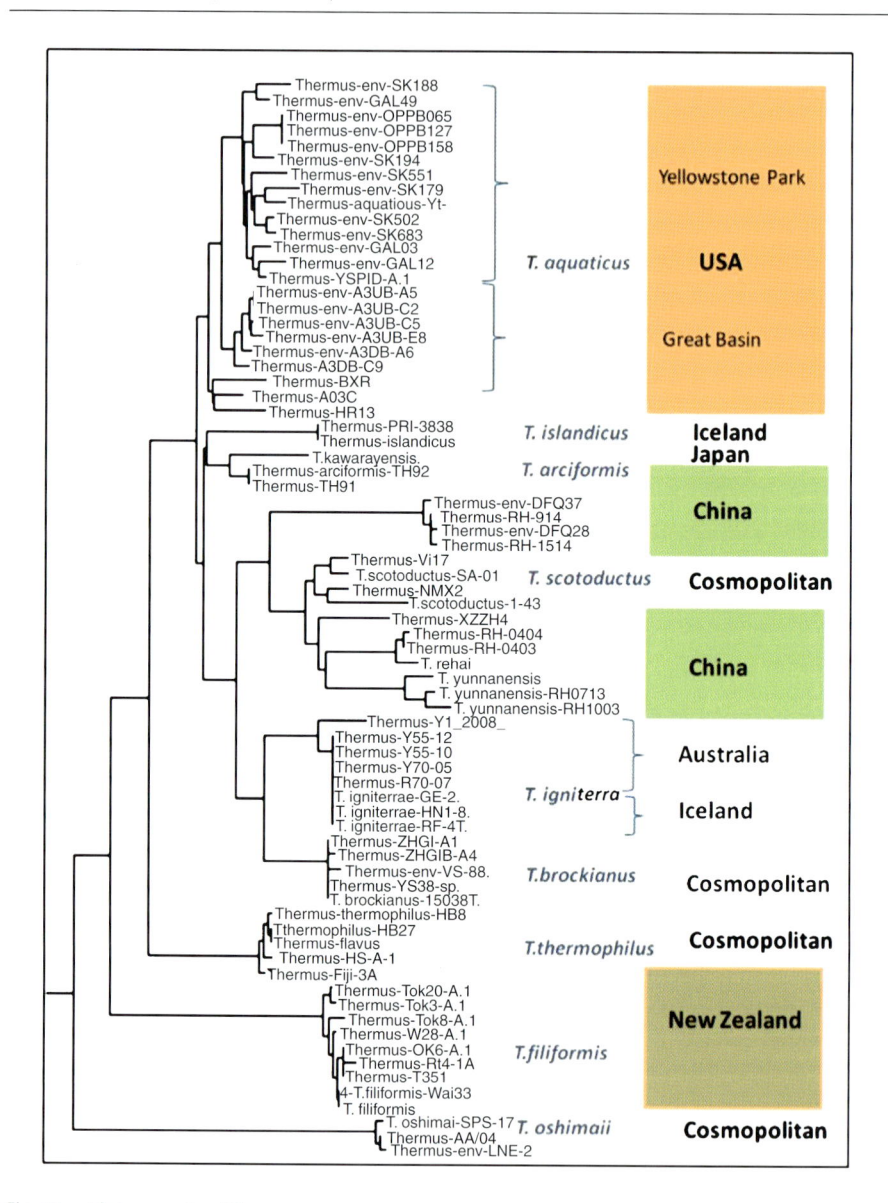

Fig. 11. Phylogeography of *Thermus*

genospecies differing by at least 3% in 16S rRNA distance (Skirnisdottir et al. 2000; Hreggvidsson and Kristjansson 2003; Reysenbach et al. 2005).

The genetic structure of a species is shaped by geography even on a local scale. Multilocus studies of *Rhodothermus marinus* and *Thermus thermophilus* in coastal hot springs in Iceland clearly revealed that the populations at different sites were evolving independently of each other (Petursdottir et al. 2000; Hreggvidsson et al. 2006). This is despite the fact that both are marine species and microbial transfer between sites should not be impeded by a completely different physical medium of passage as would be the case with microorganisms in terrestrial hot springs passing through air.

The terrestrial sulfur-oxidizing species *Sulfolobus islandicus* is adapted to and thrives in acidic geothermal springs at around 80°C. The very low pH ($\sim$2) of these sulfuric hot springs is another factor restricting the distribution of the species. A multilocus study by Whitaker and coworkers clearly showed genetic differences between populations in different regions in the Northern hemisphere, in Kamchatka (Russia), Yellowstone National Park, Alaska, and Iceland (separated by 250 to $\sim$6000 km), but also between populations within areas separated by as little as 15 km. It could be concluded that *S. islandicus* strains clustered by a geographical locale rather than by hot spring character, temperature or pH and that the genetic distances between populations increased proportionally with the geographical distance (Whitaker et al. 2003).

Genome sequencing gives evidence that lateral gene flow across species, phyla, and even domain boundaries has occurred to a considerable extent between thermophilic microbial lineages. The concept of a dispensable fraction of a genome in a particular species implies that genes have been lost and gained since the separation from a common ancestor. The genome of the thermophile *Sulfolobus solfataricus* contains a larger number of transposable elements than any other sequenced prokaryotic genome and, as a consequence, may have been particularly susceptible to recombination and rearrangements (Redder and Garrett 2006).

Reno et al. (2009) examined pan-genomic structure of *S. islandicus* isolates from far apart locations in order to test whether barriers to dispersal or ecological selection were primarily responsible for shaping its population structure. Most of the peripheral genes came from viruses and plasmids and about one-third was specific to a geographical location. The viruses and plasmids that had lent their genes to *Sulfolobus* in one site were different from those found in another. Also, much of the variation was found in genes devoted to the microbe's defence system against foreign genetic elements, indicating that *S. islandicus* is evolving largely in response to the assault of local pathogens such as viruses.

4.1 Dispersal of thermophiles

Various factors may cause and maintain a discontinuous distribution pattern of a thermophilic species. Obviously, a large distance between geothermal sites reduces the number of migration events from one area to another. However, the nature of the surroundings if it is hostile or harmful to the organisms may also hinder distribution. For example, marine migration routes may be barred for some terrestrial thermophiles and oxygen may be harmful to obligate thermophilic anaerobes.

Alkaline hot springs in low-temperature regions may be interconnected below the surface by streams or they may be fed by a single subterranean reservoir. The subsurface temperature of the water is higher than in openings at the surface where the water rapidly cools down. The temperature of the underground water stream may in this case act as a physical dispersal threshold preventing dissemination. Depending on the temperature, distribution of some species, but not all, might be obstructed. In this context it bears mentioning that topological features may play a role as migration follows waterways from watersheds, both above and below ground. Places at higher altitudes may therefore be different and perhaps less diverse than comparable sites in lowlands where migration events are recurrent.

Geothermal regions may be connected across long distances by hot underwater streams. The presence of apparently purely terrestrial thermophiles in the submarine hydrothermal vent in Eyjafjördur, a fjord at the north coast of Iceland, confirmed the freshwater origin of the water. Surprisingly, no terrestrial *Thermus* strains were isolated from the samples indicating the presence of a high-temperature or hydrological barriers along the way. The water could be traced to high inland mountains located about 100 km south of the cone on the basis of stable oxygen and hydrogen isotopic ratios (Marteinsson et al. 2001).

4.1.1 Dispersal capabilities of thermophiles

Dispersal mechanisms or capabilities may differ between microorganisms. Those that endure desiccation or form metabolically inactive resting bodies resistant to harsh environmental conditions, may be scattered by winds all over the world. Thus, spore-forming *Geobacillus* species might be expected to have wider distribution than non-spore-forming species, such as *Thermus*. The cyanobacterial species *Mastigocladus laminosus* lives in alkaline hot spring below 57°C. It is found in geothermal areas all over the world in contrast to the high-temperature *Synechococcus* species. The species consists of at least six genetically different groups. Some of them appear to be endemic such as the population in the Waitangi hot springs in New Zealand, whereas others are widespread. *M. laminosus* is very tolerant to desiccation and freezing, which may facilitate airborne dispersal and explain its phylogeographical structure (Miller et al. 2002). Increased ecological specialization may also be an agent of geographical isolation. High-temperature *Synechococcus* isolates have been reported to be sensitive to freezing and desiccation, factors likely to be important in dispersal (Miller and Castenholz 2000) and possibly a trade-off for the adaptation to higher temperatures. Such sensitivity and a surrounding sea may be sufficient to explain the absence of these strains in Iceland. Similarly, temperature shifts from the high temperature of the habitat to the lower temperature of the surroundings may induce cell cycle arrest and chromosomal DNA degradation as observed in *Sulfolobus* cultures (Hjort and Bernander 1999), thus limiting their dispersal range.

4.1.2 Chemical and biological barriers

Distinctive geochemical or physical characteristics of a particular region may also exert their influence through biological barriers. A particular hot spring chemistry may be characteristic of a certain geological region affecting the colonization success of migrating species. Local biota or ecotypes of a particular species adapted to the existing physicochemical conditions may out-compete incoming strains and species. For example, high arsenic levels in geothermal springs may influence the species composition and overall diversity both directly and indirectly. Arsenic toxicity is believed to be mediated by arsenite reacting with thiol functional groups in enzymes and possibly inhibiting their activities and by arsenate through substitution of phosphate (Stolz and Oremland 1999).

The arsenic-rich Champagne Pool is a large hot spring in the Waiotapu region of the North Island in New Zealand formed by a hydrothermal eruption 900 years ago. The area has the sulfur chemistry features of a high-temperature region, hydrogen sulfide gas discharge, sulfur precipitation and buildup of sulfuric acid that are manifested in solfatara fields, mud pits, and acidic hot springs. The Champagne Pool has a pH of 5.5 which is relatively constant due to buffering by the high flux of CO_2. Gases other than carbon dioxide (73%) are nitrogen (16.2%), methane (6.4%), hydrogen (2.3%), and hydrogen sulfide (1.7%). The pool is fed by chloride water directly from a deep hydrothermal reservoir at 230°C, but the temperature in the pool is around 75°C. It has high concentrations of silica and metalloid ion–sulfide complexes. Precipitates formed within the pool are enriched in arsenic, antimony, thallium, and mercury and the pool water contains high concentration of arsenic ions and compounds (5.6 mg/l; Hedenquist 1991; Giggenbach et al. 1994). The Champagne Pool water is apparently a rather toxic cocktail for living organisms. Microbial density and diversity were studied by Hetzer et al. (2007), using a combination of culture and culture-independent methods. The cell density appeared to be low compared to other geothermal springs within New Zealand and the overall diversity likewise. Gene clone library analyses of environmental DNA indicated the presence of few but apparently novel microbes related to *Sulfurihydrogenibium*, *Sulfolobus*, and *Thermofilum* species. The phylogenetic relationships primarily indicated hydrogen-oxidizing and sulfur-dependent metabolism and little else. The authors proposed that a unique chemical character, the presence of arsenic and other metallic compounds, is the limiting factor for the microbial diversity and biomass and that only metal ion-tolerant or metal ion-resistant microorganisms survive these conditions. A new species has been described from this hot spring, *Venenivibrio stagnispumantis*. The species grew in the presence of arsenite, arsenate and antimonite at considerably higher concentrations than found in the Champagne Pool spring water (Hetzer et al. 2008).

It is interesting to compare the Champagne Pool with arsenic-rich hot springs in a completely different geological setting of the Alvord desert in North America. They are located in the Great Basin where the highest temperatures below ground reach levels which are not much higher than 200°C. The geothermal water is not heated up by magmatic heat as in volcanic areas, but rather the geothermal activity reflects loci of shear transfer between fault systems in the geological strata. This accounts for relatively low levels of hydrogen sulfide in the water and the absence of sulfuric acid-buffered springs in the region. Recently the microbiology and geochemistry of a typical arsenic-rich hot spring system in this area were studied by Connon et al. (2008). The total arsenic concentration in the spring was relatively stable at 4.5 mg/l while the ratio of As(III) to As(V) decreased down the efflux channel. The pH was also relatively narrow, in the range of 6.77–6.81, and increased along the flow. Temperature was close to 80°C at the source. The thermophilic biota was manifested in chemolithoautotrophically driven microbial communities of a quite different visual character along the temperature gradient. The biomass was abundant compared to the Champagne Pool in New Zealand. The clone libraries obtained were small; therefore, the rarefaction curves from the biomass samples did not reach plateaus. Still, they were informative about the dominant species in these communities. Close to the source, at a temperature range of 74–78°C, the microbial diversity was low and was displayed in a thin colorless biofilm on rock pebbles. It was dominated by *Sulfurihydrogenibium* spp. as reflected by 73.5% of clones while 23.5% and 3% of the clones belonged to *Thermus* and *Thermocrinis*, respectively. A profuse orange microbial mat at the lower sampling temperature was more diverse. The clone library included not only sequences belonging to the same clades of *Sulfurihydrogenibium* (26%), *Thermocrinis* (26%), and *Thermus* (17%), but also clades belonging to OP10 (12%), *Bacteroidetes* (14%) and a clade possibly related to *Acidobacteria* (5%). Metabolic activities construed from the phylogenetic analysis indicated hydrogen and sulfide as the primary sources of energy.

In the Yellowstone Park the geology is completely different and the chemistry is dominated by sulfur. Arsenic hot springs are widespread. Similar organisms are present in such hot springs and their distribution is apparently determined by the same environmental variables. Thus, *Sulfurihydrogenibium* bacteria are detected in hot springs with high-sulfide concentrations and similarly, *Thermocrinis* at low-sulfide levels (Hamamura et al. 2009).

Taken together these studies demonstrate very clearly differences in bacterial diversity along geochemical gradients of temperature, of pH, of the energy sources sulfide and hydrogen, and of the toxic compounds arsenite and antimonite. Substantial presence of *Thermocrinis* spp., *Thermus*, and other heterotrophs and the absence of *Sulfolobus*, as in the Alvord system, indicate little relevance of sulfide for a particular ecosystem.

Arsenite does not serve as an energy source in the thermophiles studied so far. Rather, arsenite oxidase in these organisms serves a role in a detoxification process. The enzyme has been identified in some *Thermus* spp. and is also present in some *Sulfurihydrogenibium* species. The presence of arsenite oxidase appears related to the source of the isolate, i.e., if it came from an arsenic-rich environment or not. Thus, none have been detected in Icelandic *Sulfurihydrogenibium* strains. Even more interesting is the fact that arsenite oxidases are found in some but not all strains within a particular *Sulfurihydrogenibium* species, implying that specialized ecotypes are found within the species. One such strain appeared to have acquired the gene by lateral gene transfer from *Thermus* (Hanamura et al. 2010).

The relatively large abundance of *Thermus* in the Alvord system is of particular interest and indicates an important role for this organism in arsenite detoxification in this ecosystem. *Thermus* spp. have been shown to be able to oxidize arsenite and an arsenite-oxidizing *Thermus* strain (A03C) was isolated from the Alvord hot springs. (Gihring and Banfield 2001; Gihring et al. 2001). The *Thermus* bacteria in the Alvord hot spring system form a phylogenetic cluster of their own, separated from their relatives in the Yellowstone Park, while belonging to the *Thermus aquaticus* clade (Fig. 11). The question arises whether only distance keeps these groups genetically apart, or if the populations adapted to their respective environments and play an "active" part in restricting migration between the areas.

4.1.3 Historical barriers – time

Those factors where time plays a role in influencing distributions may be termed historical. Successful colonization from an incoming species depends on both the abundance and frequency of migration events, factors that depend on geographical distance and time. Measured on a geological time scale "everything may be everywhere" in the sense that a particular species or strain migration may have occurred at some point in time but then not "intensively enough" to overcome biological or physicochemical barriers. Also, the geological period when a successful migration event takes place may be sufficiently long to establish a distinct genetic difference between the colony and the source populations.

Major geological events such as rare major crustal upheavals or other momentous geological events can also be termed historic on both local and global scales. They may result in the creation of new faults in the geological strata as conduits of heated water and routes of dissemination. Also, new geothermal habitats may be created that need time to develop to their full extent. Conversely, volcanic eruptions may also spread microbes from existing habitats or from the subsurface over long distances in a sufficient magnitude to enable colonization of distant regions. The different examples given below stress in some way the historical dimension of dispersal.

A volcanic eruption and its aftermath may have a significant effect on the dispersal and geographical distribution patterns of hyperthermophiles in the sea by releasing site bound subsurface dwellers. The presence of hyperthermophilic archaea in low-temperature hydrothermal fluids from the Juan de Fuca Ridge, that were not detected in the ambient seawater, was reported by Holden et al. (1998). This suggested that they had grown below the seafloor at permissive temperatures. Similarly, hyperthermophiles have been enriched from cold plume waters shortly after an eruption (Huber et al. 1990; Delaney et al. 1998). A historic event of this kind may explain the distribution of a particular microbial lineage and theoretically at least the place of origin and approximate time since the "scattering" may be inferred from gene histories and phylogeographical distributions.

Miller et al. (2002) have speculated on various historical aspects of the distribution of *M. laminosus*. On the basis of 16S rRNA and metabolic gene divergence analysis they attempted to reconstruct the evolutionary history of the species and the timing of diversification events, and tried to identify the site of origin of a relatively recent expansion of a particular subgroup of the species. They discuss how this history may relate to geological events.

Historical reasons have also been suggested to explain the distribution of microbes belonging to Aquificales in the Yellowstone National Park. Molecular phylogenetic approaches and dispersal – vicariance analyses combined with environmental data were used to examine the distribution of the members of *Sulfurihydrogenibium* in thermal springs in the region. A clear pattern of geographically isolated microbial populations was found. The distribution correlated with the boundary of Yellowstone's calderas (or volcanic craters) and suggested that volcanic eruptions of the past 2 million years explained more of the DNA sequence divergence than contemporary factors, such as habitat preferences or geographical distance (Takacs-Vesbach et al. 2008).

4.2 Biogeography of *Thermus*

Presently, 10 *Thermus* species have been validly described. These are *T. aquaticus* (Brock and Freeze 1969), *T. thermophilus* (Oshima and Imahori 1974), *T. filiformis* (Hudson et al. 1987), *T. scotoductus* (Kristjansson et al. 1994), *T. brockianus* (Williams et al. 1995), *T. oshimai* (Williams et al. 1996), *T. igniterra* and

T. antranikianii (Chung et al. 2000), the recently described *Thermus islandicus* (Björnsdottir et al. 2009), and *T. arciformis* (Zhang et al. 2010). Additionally, three species that have not been validly described are "*T. kawarayensis*", "*T. rehai*," and "*T. yunnanensis*" from Japan, Tibet, and China, respectively. Another putative species, "*T. eggertsonii*", also seems to be abundant in Iceland (unpublished).

Geographical isolation appears to be an important factor in species divergence within the genus *Thermus* (Fig. 11). The *T. aquaticus* lineage has only been found in the North American continent. Interestingly, as noted above, the *T. aquaticus* linage apparently shows phylogeographical clustering within the continent as well as with genetically different populations in the Great Basin and the Yellowstone Park. Only one other *Thermus* species, the cosmopolitan *T. brockianus*, has been detected in Yellowstone Park despite intensive studies over decades.

T. filiformis has only been isolated or detected in New Zealand and is so far the only known terrestrial *Thermus* species. Recently, distinct new lineages, apparently confined to Tibet and China, have been described. One species, *T. thermophilus* seems to be adapted to marine environments and is found in submarine and coastal hot springs worldwide (Kristjansson et al. 1986, 2000; Williams and Sharp 1995). The discordant note is the distribution of *T. igniterra* so far found only in Iceland and then in Australia. Globally, certain regions appear to be more diverse than others. It appears that Iceland harbors more *Thermus* species than any other region.

Species boundaries should ideally circumscribe ecologically distinct populations. This is not clear for any of the genotypically different *Thermus* species. A particular *Thermus* genospecies may, therefore, comprise different ecotypes or local adaptations having distribution patterns depending on the geographical area and presence or absence of other species or strains. *Thermus* isolates from widely distant places grouped on the basis of phenotypic traits apparently reflected the geographical origin rather than affinities to a specific genospecies (Hudson et al. 1989; Santos et al. 1989). Also, some but not all populations within the *T. aquaticus* clade in North America appear to be capable of oxidizing arsenite and relates to whether they come from an arsenite-rich environment or not. In this context it is noteworthy that the phylogenetic depth of the *T. aquaticus* lineage in North America is relatively large (Fig. 11), indicating that the lineage may have a guild structure consisting of different ecotypes. This could be an example of adaptive radiation of specific lineages into unoccupied niches in this geothermal area, whereas similar niches in Iceland are populated by different *Thermus* species.

4.3 Ecological adaptations of *Thermus*

Perhaps *T. thermophilus* comes closest as an example for describing an ecologically distinct population as a whole, since it has been found almost exclusively in marine and coastal hot springs. All other *Thermus* species are found in terrestrial hot springs and their niches, or the preferences of habitats have not been determined.

Today we have a rather fragmented picture of the ecology of *Thermus* in terrestrial hot springs. Under laboratory conditions the different species are similar in phenotypic properties, such as pH and temperature optima as well as utilization patterns of carbon sources. Frequently we isolated different *Thermus* species from the same hot spring. With reference to the competitive exclusion principle this raises the question of how phenotypically and physiologically very similar *Thermus* species coexist in a particular hot spring. However, we cannot really distinguish between representatives of stable populations and transient species, using traditional isolation techniques. The observed species diversity may be partly maintained by migration events from microhabitats within the hot spring, sediments or microbial mats, from different patches of environmental gradients, by periodic disturbances or by environmental fluctuations.

Adaptive traits important for the habitat preferences of many thermophilic species may involve small physiological differences or marginally different responses to physicochemical adverse conditions and may therefore be difficult to evaluate. For example, a particular species could have a slight competitive edge for limiting resources under oligotrophic conditions, or there could be a small but significant difference in growth rates between species at the limits of their temperature or pH growth ranges. Subtle combinations of such traits might determine the niche position of a *Thermus* species and consequently influence the diversity and species abundance in a particular hot spring. It follows that such differences between species would be difficult to detect and evaluate in the laboratory.

Two important niche parameters that may explain growth in a particular habitat are pH and temperature. In a recent study based on cultivated isolates, we observed that the specific temperatures of the isolation sites did not relate to any of the *Thermus* species lineages examined (Hreggvidsson et al. 2006). However, the pH of the isolation sites indicated a relationship to lineage formation. *T. igniterrae* and *T. brockianus* were generally isolated from hot springs with pH > 8.0, while *T. scotoductus* was more commonly isolated from hot springs of neutral and lower pH. Furthermore *T. igniterrae* appeared to be very sensitive to ionic strength. However, no clear phenotype could be attributed to the genetic lineages based on carbon utilization abilities. A number of isolates from each *Thermus* species were also genotyped and the species were shown to be genotypically tight, except for *T. scotoductus* that formed several deep subclusters, which perhaps indicated different ecotypes (Hreggvidsson et al. 2006). In other studies we established that at least some of the *Thermus* species may harness energy by oxidizing sulfur compounds (Skirnisdottir et al. 2001; Björnsdottir et al. 2009) and there are reports of *T. scotoductus* strains using nitrate, $Fe(III)$,

Mn(IV), or S° as terminal electron acceptors in the respiratory chain (Balkwill et al. 2004). Similarly, an isolate *Thermus* HR13 has been reported (Gihring and Banfield 2001) from arsenite-rich hot spring in California that is able to use arsenate as a terminal electron donor. Different abilities to exploit these compounds in their electron transport chain may explain differential distribution of some *Thermus* species. Water activity in the habitats may also be important in explaining the distribution, e.g., a requirement, ability or inability to tolerate an ionic strength above a certain level. Differences in the tolerance to salt concentration have been observed between the different genospecies. The highest salt tolerance was found among strains of the *T. thermophilus* lineage, which is explained by its habitat of marine and coastal hot springs. The opposite was found for the *T. igniterrae* lineage that appeared to be very sensitive to ionic strength. *T. igniterrae* shows a discontinuous and very unexpected distribution. So far it has only been found in Iceland and north-eastern Australia. In Australia it was detected and isolated as the dominant *Thermus* species at the higher temperatures in a nonvolcanic hot runoff from a borehole in the Great Artesian Basin aquifer in Central Queensland (Spanevello and Patel 2004). The physicochemical characteristic of this habitat of *T. igniterrae* in Australia agrees well with our findings, which indicate that the species prefers highly alkaline and high-temperature hot springs low in mineral salt content, and is sensitive to NaCl, bivalent cations and sulfur compounds. In this case apparently "the environment selects".

The conclusion is that the genospecies of *Thermus* may be ecologically distinct populations distinguished by only few adaptive traits to physicochemical conditions and that their distribution ranges are dictated by both geographical distances and environmental variables. It remains to be seen if adaptive traits observed in one region for many genospecies are universal. The niche space may have been defined for some of them at least in one or few dimensions, e.g., tolerance and intolerance to salts for *T. thermophilus* and *T. igniterrae*, respectively, or adaptation to high alkalinity for *T. brockianus* and *T. igniterrae*. An ecological approach in conjunction with genomic analysis and multilocus population studies will definitely reveal more about the different adaptations of *Thermus*, the evolutionary history of the genus and possibly enable the positioning and tracing of ecological radiations of different species in time and space.

References

Amann R, Ludwig W, Schleifer KH (1995) Phylogenetic identification and in situ detection of individual microbial cells without cultivation. Microbiol Rev 59:143–169

Arnórsson S (2003) Arsenic in surface- and up to 90°C ground waters in a basalt area, N-Iceland: processes controlling its mobility. Appl Geochem 18:1297–1312

Balkwill DL, Kieft TL, Tsukuda T, Kostandarithes HM, Onstott TC, Macnaughton S, Bownas J, Fredrickson JK (2004) Identification of iron-reducing *Thermus* strains as *Thermus scotoductus*. Extremophiles 8:37–44

Björnsdottir SH, Petursdottir SK, Hreggvidsson GO, Skirnisdottir S, Hjorleifsdottir S, Arnfinnsson J, Kristjansson JK (2009) *Thermus islandicus* sp. nov., a mixotrophic sulfur-oxidizing member of the genus *Thermus*. Int J Syst Evol Microbiol 59:2962–2966

Bödvarsson G (1961) Physical characteristics of natural heat resources in Iceland. Jökull 11:29–38

Brock TD, Freeze H (1969) *Thermus aquaticus* gen. nov. and sp. nov., a non-sporulating extreme thermophile. J Bacteriol 98:289–297

Chung AP, Rainey FA, Valente M, Nobre MF, da Costa MS (2000) *Thermus igniterrae* sp. nov. and *Thermus antranikianii* sp. nov., two new species from Iceland. Int J Syst Evol Microbiol 50:209–217

Cohan FM (2001) Bacterial species and speciation. Syst Biol 50:513–524

Cohan FM (2002) What are bacterial species? Annu Rev Microbiol 56:457–487

Cohan FM (2006) Towards a conceptual and operational union of bacterial systematics, ecology, and evolution. Philos Trans R Soc Lond B Biol Sci 361:1985–1996

Connon SA, Koski AK, Neal AL, Wood SA, Magnuson TS (2008) Ecophysiology and geochemistry of microbial arsenic oxidation within a high arsenic, circumneutral hot spring system of the Alvord Desert. FEMS Microbiol Ecol 64:117–128

Delaney JR, Kelley DS, Lilley MD, Butterfield DA, Baross JA, Wilcock WSD, Embley RW, Summit M (1998) The quantum event of oceanic crustal accretion: impacts of diking at mid-ocean ridges. Science 281:222–230

Eder W, Huber R (2002) New isolates and physiological properties of the Aquificales and description of *Thermocrinis albus* sp. nov. Extremophiles 6:309–318

Giggenbach WF, Sheppard DS, Robinson BW, Stewart MK, Lyon GL (1994) Geochemical structure and position of the Waiotapu geothermal field, New Zealand. Geothermics 23:599–644

Gihring TM, Banfield JF (2001) Arsenite oxidation and arsenate respiration by a new *Thermus* isolate. FEMS Microbiol Lett 204:335–340

Gihring TM, Druschel GK, McCleskey RB, Hamers RJ, Banfield JF (2001) Rapid arsenite oxidation by *Thermus aquaticus* and *Thermus thermophilus*: field and laboratory investigations. Environ Sci Technol 35:3857–3862

Gogarten JP, Doolittle WF, Lawrence JG (2002) Prokaryotic evolution in light of gene transfer. Mol Biol Evol 19:2226–2238

Hamamura N, Macur RE, Korf S, Ackerman G, Taylor WP, Kozubal M, Reysenbach AL, Inskeep WP (2009) Linking microbial oxidation of arsenic with detection and phylogenetic analysis of arsenite oxidase genes in diverse geothermal environments. Environ Microbiol 11:421–431

Hamamura N, Macur RE, Liu Y, Inskeep WP, Reysenbach AL (2010) Distribution of aerobic arsenite oxidase genes within the Aquificales. In: Hamamura N, Suzuki S, Mendo S, Barroso CM, Iwata H, Tanabe S (eds) Interdisciplinary studies on environmental chemistry – biological responses to contaminants. TERRAPUB, Tokyo, Japan, pp 47–55

Hedenquist JW (1991) Boiling and dilution in the shallow portion of the Waiotapu geothermal system, New Zealand. Geochim Cosmochim Acta 55:2753–2765

Hetzer A, Morgan HW, McDonald IR, Daughney CJ (2007) Microbial life in Champagne Pool, a geothermal spring in Waiotapu, New Zealand. Extremophiles 11:605–614

Hetzer A, McDonald, IR, Morgan HW (2008) *Venenivibrio stagnispumantis* gen. nov., sp. nov., a thermophilic hydrogen-oxidizing bacterium isolated from Champagne Pool, Waiotapu, New Zealand. Int J Syst Evol Microbiol 58:398–403

Hjorleifsdottir S, Skirnisdottir S, Hreggvidsson GO, Holst O, Kristjansson JK (2001) Species composition of cultivated and non-cultivated bacteria from short filaments in an Icelandic hot spring at 88°C. Microb Ecol 42:117–125

Hjort K, Bernander R (1999) Changes in cell size and DNA content in *Sulfolobus* cultures during dilution and temperature shift experiments. J Bacteriol 181:5669–5675

Holden JF, Summit M, Baross JA (1998) Thermophilic and hyperthermophilic microorganisms in 3–30°C hydrothermal fluids following a deep-sea volcanic eruption. FEMS Microbiol Ecol 25:33–41

Hreggvidsson GO, Kristjansson JK (2003) Thermophily. In: Gerday C, Glansdorff N (eds) Extremophiles – Encyclopedia of Life Support Systems (EOLSS) Developed under the Auspices of the UNESCO. Eolss Publishers, Oxford, UK. http://www.eolss.net

Hreggvidsson GO, Skirnisdottir S, Smit B, Hjorleifsdottir S, Marteinsson, VT, Petursdottir SK, Kristjansson JK (2006) Polyphasic analysis of *Thermus* isolates from geothermal areas in Iceland. Extremophiles 10:563–575

Huber R, Stoffers P, Cheminee JL, Richnow HH, Stetter KO (1990) Hyperthermophilic archaebacteria within the crater and open-sea plume of erupting Macdonald Seamount. Nature 345:179–181

Huber R, Eder W, Heldwein S, Wanner G, Huber H, Rachel R, Stetter KO (1998) *Thermocrinis ruber* gen. nov., sp. nov., a pink-filament-forming hyperthermophilic bacterium isolated from Yellowstone National Park. Appl Environ Microbiol 64:3576–3583

Hudson JA, Morgan HW, Daniel, RM (1987) *Thermus filiformis* sp. nov., a filamentous caldoactive bacterium. Int J Syst Bacteriol 37:431–436

Hudson JA, Morgan HW, Daniel RM (1989) Numerical classification of *Thermus* isolates from globally distributed hot springs. Syst Appl Microbiol 11:250–256

Kristjansson JK, Hreggvidsson GO (1995) Ecology and habitats of extremophiles. World J Microbiol Biotechnol 11:17–25

Kristjansson JK, Hreggvidsson GO, Alfredsson GA (1986) Isolation of halotolerant *Thermus* spp. from submarine hot springs in Iceland. Appl Environ Microbiol 52:1313–1316.

Kristjansson JK, Hjorleifsdottir S, Marteinsson VT, Alfredsson GA (1994) *Thermus scotoductus*, sp. nov., a pigment-producing thermophilic bacterium from hot tap water in Iceland and including *Thermus* sp. X-1. Syst Appl Microbiol 17:44–50

Kristjansson JK, Hreggvidsson GO, Grant WD (2000) Taxonomy of extremophiles. In: Priest FG, Goodfellow M (eds) Applied microbial systematics. Kluwer Academic Publishers, pp 231–292

Langner HW, Jackson CR, McDermott TR, Inskeep WP (2001) Rapid oxidation of arsenite in a hot spring ecosystem, Yellowstone National Park. Environ Sci Technol 35:3302–3309

Lawrence JG (2002) Gene transfer in bacteria: speciation without species? Theor Popul Biol 61:449–460

Levin R (1981) Periodic selection, infectious gene exchange and the genetic structure of *E. coli* populations. Genetics 99:1–23

Marteinsson VT, Kristjánsson JK, Kristmannsdóttir H, Dahlkvist M, Smundsson K, Hannington M, Petursdottir SK, Geptner A, Stoffers P (2001) Discovery and description of giant submarine smectite cones on the seafloor in Eyjafjordur, Northern Iceland, and a novel thermal microbial habitat. Appl Environ Microbiol 67:827–833

Miller SR, Castenholz RW (2000) Evolution of thermotolerance in hot spring cyanobacteria of the genus *Synechococcus*. Appl Environ Microbiol 66:4222–4229

Miller SR, Castenholz RW, Pedersen D (2002) Phylogeography of the thermophilic cyanobacterium *Mastigocladus laminosus*. Appl Environ Microbiol 73:4751–4759

Oshima T, Imahori K (1974) Description of *Thermus thermophilus* (Yoshida and Oshima) comb. nov., a nonsporulating thermophilic bacterium from a Japanese thermal spa. Int J Syst Bacteriol 24:102–112

Palmason G (2005) Jardhiti – Edli og nyting audlindar. Reykjavik, Hid islenska bokmenntafelag (Geothermal energy – its nature and applications). Reykjavik, Icelandic Literary Society, 298 pp

Papke RT, Ramsing NB, Bateson MM, Ward, DM (2003) Geographical isolation in hot spring cyanobacteria. Environ Microbiol 5:650–659

Petursdottir SK, Hreggvidsson GO, da Costa MS, Kristjansson JK (2000) Genetic diversity analysis of *Rhodothermus* reflects geographical origin of the isolates. Extremophiles 4:267–274

Redder P, Garrett RA (2006) Mutations and rearrangements in the genome of *Sulfolobus solfataricus* P2. J Bacteriol 188:4198–4206

Reno ML, Held NL, Fields CJ, Burke PV, Whitaker RJ (2009) Biogeography of the *Sulfolobus islandicus* pan-genome. Proc Natl Acad Sci USA 106:8605–8610

Reysenbach AL, Banta A, Civello S, Daly J, Mitchel K, Lalonde S, Konhauser KO, Rodman A, Rusterholtz K, Takacs-Vesbach C (2005) The Aquificales of Yellowstone National Park, In: Inskeep WP, McDermott TR (eds) Geothermal biology and geochemistry in Yellowstone National Park. Thermal Biology Institute, Montana State University, Bozeman, pp 129–142

Romero L, Alonso H, Campano P, Fanfani L, Cidu R, Dadea C, Keegan T, Thornton I, Farago M (2003) Arsenic enrichment in waters and sediments of the Rio Loa (Second Region, Chile). Appl Geochem 18:1399–1416

Santos MA, Williams RAD, da Costa MS (1989) Numerical taxonomic study of *Thermus* isolates from hot springs in Portugal. Syst Appl Microbiol 12:10–15

Skirnisdottir, S, Hreggvidsson GO, Hjorleifsdottir S, Marteinsson VT, Petursdottir SK, Holst O, Kristjansson JK (2000) Influence of sulfide and temperature on species composition and community structure of hot spring microbial mats. Appl Environ Microbiol 66:2835–2841

Skirnisdottir S, Hreggvidsson GO, Holst O, Kristjansson JK (2001) Isolation and characterization of a mixotrophic sulfur oxidizing *Thermus scotoductus*. Extremophiles 5:45–51

Spanevello MD, Patel BKC (2004) The phylogenetic diversity of *Thermus* and *Meiothermus* from microbial mats of an Australian subsurface aquifer runoff channel. FEMS Microbiol Ecol 50:63–73

Stolz JF, Oremland RS (1999) Bacterial respiration of arsenic and selenium. FEMS Microbiol Rev 23:615–627

Takacs-Vesbach C, Mitchell K, Jackson-Weaver O, Reysenbach AL (2008) Volcanic calderas delineate biogeographic provinces among Yellowstone thermophiles, Environ Microbiol 10:1681–1689

Ward DM (1998) A natural species concept for prokaryotes. Curr Opin Microbiol 1:271–277

Ward DM, Cohan FM (2005) Microbial diversity in hot spring cyanobacterial mats: pattern and prediction. In: Inskeep WP, McDermott TR (eds) Geothermal biology and geochemistry in Yellowstone National Park. Thermal Biology Institute, Montana State University, Bozeman, pp 185–201

Ward DM, Bateson MM, Ferris MJ, Nold SC (1998) A natural view of microbial biodiversity within hot spring cyanobacterial mat communities. Microbiol Mol Biol Rev 62:1353–1370

Ward DM, Cohan FM, Bhaya D, Heidelberg JF, Kühl, M, Grossman A (2008) Genomics, environmental genomics and the issue of microbial species. Heredity 100:207–219

Whitaker RJ, Grogan DW, Taylor JW (2003) Geographic barriers isolate endemic populations of hyperthermophilic archaea. Science 301:976–978

Williams R, Sharp R (1995) The taxonomy and identification of *Thermus*. In: Sharp R, Williams R (eds) *Thermus* species. Plenum, New York, pp 1–42

Williams RAD, Smith KE, Welch SG, Micallef J, Sharp RJ (1995) DNA relatedness of *Thermus* strains, description of *Thermus brockianus* sp. nov., and proposal to reestablish *Thermus thermophilus* (Oshima and Imahori). Int J Syst Bacteriol 45:795–499

Williams RAD, Smith KE, Welch SG, Micallef J (1996) *Thermus oshimai* sp. nov., isolated from hot springs in Portugal, Iceland, and the Azores and comment on the concept of a limited geographical distribution of *Thermus* species. Int J Syst Bacteriol 46:403–408

Zhang XQ, Ying Y, Ye Y, Xu XW, Zhu XF, Wu M (2010) *Thermus arciformis* sp. nov., a thermophilic species from a geothermal area. Int J Syst Evol Microbiol 60:834–839

Bacterial adaptation to hot and dry deserts

Thierry Heulin, Gilles De Luca, Mohamed Barakat, Arjan de Groot, Laurence Blanchard, Philippe Ortet and Wafa Achouak

Laboratory of Microbial Ecology of the Rhizosphere and Extreme Environment (LEMIRE), UMR 6191 CNRS-CEA-Aix-Marseille Univ., Institute of Environmental Biology and Biotechnology (iBEB), CEA/Cadarache, St-Paul-lez-Durance, France

The Moula-moula bird

The Moula-moula bird is a wheatear of the Desert. It is the only living soul that might enliven the desolation of volcanic areas with its two, black and white, clean-cut colours. I often used to follow it with my eyes, as it would tinkle its faint trill above the bald heights of the hillocks. It cannot stay still as its feet would get burned by the rock. The sun, which would elsewhere be life, here is devastating furnace.

Only, there is the Moula-moula bird to pour a note of freshness on the inferno. Its faint trill furtively gives the illusion of greenness; it falls upon the ardent rocks as drops of dew. One only has to close one's eyes and listen to the crystal shrill then one forgets this bare world reduced to its vertebral column and also forgets the surveying mineral world.

Tahar Djaout (1954–1993)
L'invention du desert (Éditions du Seuil, 1987)

Kindly translated from French by Gaëlle Catois

1 Introduction

Prokaryotic microorganisms are known to be highly adaptable to diverse environmental conditions and to thrive in harsh environments. Halophilic microorganisms (Bacteria and Archaea) tolerate and grow in the presence of salt concentrations 10 times higher than seawater, whereas acidophiles withstand a pH of 1, and hyperthermophiles face temperatures above 85°C. Bacteria are able to sense changing environmental parameters such as temperature, pressure, pH, ionic strength, solute concentrations and water availability, and to adapt by protecting biological molecules and adjusting biochemical reactions in response to extreme conditions.

These extreme conditions may be transient or permanent and will greatly influence the various adaptation mechanisms. Actually, four strategies might be used to overcome environmental stresses: compensation, conservation, protection and damage repair.

Facing transient extreme conditions, compensatory responses seek to restore equilibrium and to maintain normal functions, such as up-regulation of efflux pumps to extrude toxic metals when the concentration of heavy metals increases. Compensation can also be engaged for long-term adaptation, as exemplified by *Helicobacter pylori*, the extremely acid resistant and unique microorganism able to thrive within the human stomach, by producing copious amounts of urease (Mobley et al. 1995). Another response of bacteria to stress is to assume 'non-growth' states. When spore-forming bacteria face a strong or prolonged stress, they achieve conservation responses by entering a non-dividing state and shifting to a dormancy state under spore form. It is a reversible state of reduced basal metabolic rate in a unit that maintains viability (Barer 2003). Protective responses are used to maintain the physical integrity of living organisms. Stresses such as desiccation may cause water loss resulting in cell-volume collapse and in serious damages of cellular macromolecules such as proteins and nucleic acids. To protect themselves against desiccation, many bacterial cells accumulate solutes, including carbohydrates, amino acids, quaternary amines and tetrahydropyrimidines (Takagi 2008). Lastly, members of *Deinococcaceae* show an exceptional ability to withstand the lethal effects of DNA-damaging agents, and to repair efficiently hundreds of DNA double- and single-strand breaks as well as other types of DNA damages following desiccation or gamma radiation.

Bacterial adaptive responses to permanent harsh environments may include survival at the surface of hot arid desert soils that are exposed to heat (up to 58°C in summer), desiccation and intense ultraviolet (UV) radiation. To thrive in such extreme conditions, specific and/or unusual adaptive mechanisms may be involved.

2 Characteristic of hot and dry deserts with emphasis on Sahara

Hot and dry deserts can be considered as a paradigm of extreme environment for life because, in these conditions, the main limiting factor is water combined with drastic and highly contrasted temperatures. In extreme desert environments such as the Sahara, the mean annual rainfall is less than 50 mm, and years without any rainfall event are not an exception (e.g. Le Houérou 1997). During the day water is almost missing at the surface of desert, due to high temperature, and, except for some rare rainfall events, water can be present only at the end of some nights when the difference of temperatures between soil and air is suitable for dew. In these conditions, there is a direct link between light intensity and water availability.

The main characteristics of hot and dry deserts are the following (Le Houérou 1986, 1997):

- scarce and irregular rainfalls with an annual average lower than 100 mm;
- dew at the end of some nights/early in the morning;
- great amplitude of temperatures between night and day;
- wind responsible for 'soil' erosion;
- very low diversity of plants and animals;
- very low organic content of 'soil' mainly constituted of sand.

Besides the hot and dry deserts, the polar regions are also considered cold deserts with similar limiting factors (water/ice). Hot and dry deserts are the largest desert regions, mainly located at the tropics of Cancer and Capricorn, representing about 50 millions km^2 (one-third of the continental surfaces). In Africa, the largest ones are the Namib and Kalahari deserts in the South (Capricorn tropic) and the Sahara in the North (Cancer tropic). Sahara is the largest hot and dry deserts extended over 9 millions km^2 from Mauritania to Egypt.

The climate of the Sahara greatly varied along with the geological time. The first recorded desertification phase is dated to 7 million years BP (Miocene-Pliocene; Schuster et al. 2006). By the end of the Pliocene (ca. 2.5 My), the Sahara was submitted to recurrent arid–humid episodes. Between 200,000 and 70,000 years BP, the Sahara was wetter than today and covered by savannas, but desert conditions occurred between 70,000 and 40,000 years BP with the increase of dunes of sand. Between 40,000 and 30,000 years BP, Sahara was characterized again by a humid period followed by an arid phase (from 30,000 to 12,000 years BP). During this period, very large dunes invaded a region including present Senegal and Chad. During the Holocene, the last humid periods of the 'Green Sahara' occurred between 11,000–9000 and 7000–4500 years BP, and they were characterized by the progression of the Sahelian zone to the North. The rivers Hoggar and Aïr were again active as tributaries of the Niger River. The landscape was transformed into savannas populated by elephants, giraffes and antelopes, as evidenced by the rock engravings (Ain Tassili n-Ajjer, Koudiat Abd El Hak, Fezzan). After several climatic crises, the desert came back gradually with the present highly arid environmental conditions from 2700 years BP (Kröpelin et al. 2008).

3 Mechanisms for desiccation tolerance

The production of bacterial spores (endospores of *Firmicutes*, exospores of *Actinobacteria*) and akinetes (*Cyanobacteria*) is the most extensively documented mechanism explaining bacterial desiccation tolerance. Another bacterial cell differentiation allowing a tolerance of dryness has been described for a very long time: cysts of *Azotobacter* (*Gamma-Proteobacteria*) and myxospores of *Myxococcus* (*Delta-Proteobacteria*). A completely different mechanism was more recently described: no cell differentiation is involved in *Deinococcus* but a highly efficient and rapid ability to repair DNA damage (Mattimore and Battista 1996). This mechanism for tolerance of desiccation which is correlated to tolerance of gamma radiation is discussed in Sect. 5.

Using a range of complementary microbiological approaches (culture-dependent and -independent methods), we and others showed the presence of an extensive diversity of bacterial species in nutrient-poor environments such as deserts. These data confirmed that *Firmicutes* (*Bacillus, Paenibacillus,* etc.) and *Actinobacteria* (*Arthrobacter,* etc.) represented the dominant bacterial communities in deserts such as Sahara (Chanal et al. 2006; Benzerara et al. 2006; Gommeaux et al. 2010) and Namib (Prestel et al. 2008). These studies also revealed the abundance and a huge diversity of *Proteobacteria* belonging to the four subgroups (*Alpha, Beta, Gamma* and *Delta*) and the absence of *Myxococcus* and *Azotobacter* isolates (or 16S-rDNA sequences).

Several publications reported the presence of rhizobia in deserts; especially strains able to nodulate the legume tree *Acacia* (Zerhari et al. 2000; Khbaya et al. 1998). The mechanism mentioned to explain the adaptation of rhizobia to dryness conditions is the synthesis of osmoprotectants (glycine-betaine, sucrose and ectoine). More recently, it was demonstrated that mannosucrose and trehalose are also involved in the desiccation tolerance of *Rhizobium* (Essendoubi et al. 2007). Exploring the diversity of extracellular polymer-producing bacteria in desert soils of South Algeria, Kaci et al. (2005) showed that *Rhizobium* strains producing heteroglycan constitute the most abundant population able to colonize plant roots. The synthesis of such extracellular polymers, potentially able to limit water loss, is probably an important mechanism explaining the desiccation tolerance of rhizobia, as shown in terrestrial vegetative cells of cyanobacteria (Potts 1994; Billi and Potts 2002).

The cyanobacteria are classical inhabitants of desert surfaces (Garcia-Pichel and Belnap 1996; Karnieli et al. 1999). For instance, Garcia-Pichel and Pringault (2001) showed that the filamentous cyanobacterium *Oscillatoria* could migrate to the soil surface in response to wetting events or retreating below in response to drying event. This nomadic behaviour seems to be the main explanation of *Oscillatoria* adaptation to desert life. An interesting work on cyanobacterial diversity of crust formation in Utah desert (Colorado Plateau, USA) showed the complementarity of culture-dependent and -independent techniques in revealing a new cluster named 'Xeronema' grouping cyanobacteria closely related to *Phormidium* (Garcia-Pichel et al. 2001).

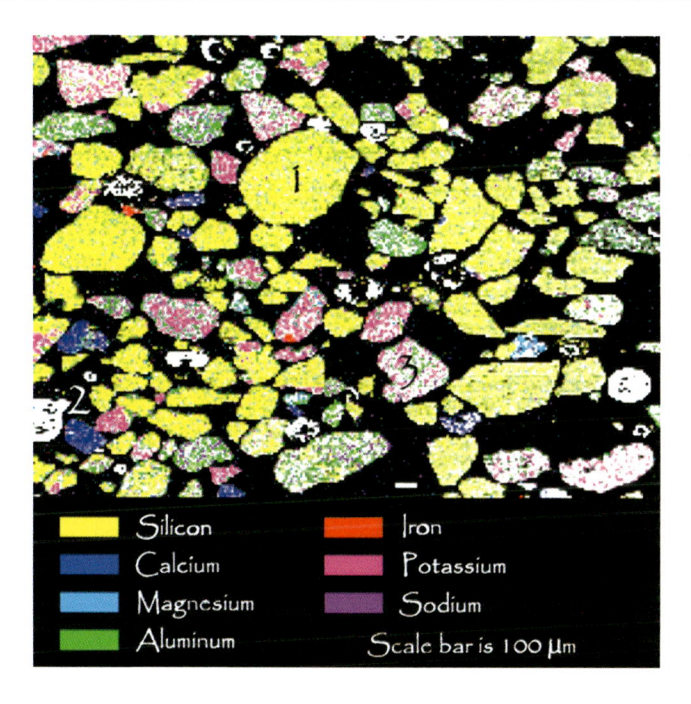

Fig. 1. Scanning electron microscopic (SEM) observation of a thin section of Merzouga sand. The superimposed colours correspond to energy-dispersive X-ray spectroscopy mapping of selected elements. At least three types of grains were distinguished (quartz grains as 1, carbonate as 2, silicate as 3), according to the bulk chemistry of the grains (from Gommeaux et al. 2010, with permission)

4 Counting and describing bacterial populations in desert environments

Here we present two recent activities carried out in the frame of a French research project (Treasures of Sahara[1]). The first one was dedicated to the mineral and microbial analysis of sand sampled in the Merzouga dunes (Morocco) (Gommeaux et al. 2010). The second one was performed in a semi-arid region of South Tunisia (Tataouine) with a special focus on gamma radiation/desiccation tolerant bacteria (Benzerara et al. 2006; Chanal et al. 2006).

Considering the Merzouga dunes, the mineral analysis revealed mostly quartz grains, pure or plated with iron oxides, some carbonates and other minor silicates (Fig. 1; see Gommeaux et al. 2010). The only source of carbon in these conditions

---

[1]French laboratories from Cadarache, Marseille, Orsay, Rennes in collaboration with laboratories in Rabat (Morocco) and in El-Fjé-Medenine (Tunisia). This work was supported by the GEOMEX program grant from the Centre National de la Recherche Scientifique (CNRS).

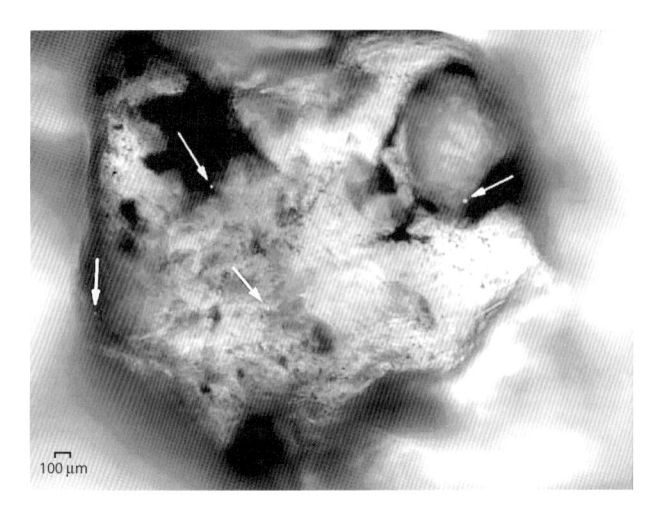

Fig. 2. Observation of Syto9 labelled bacteria on the grain surface. Composite image of Syto9 fluorescence spots superimposed on the white-light image using the software Adobe Photoshop. The dots were interpreted as individual cells whereas the spots were interpreted as groups of a few cells, or microcolonies (*arrows*; from Gommeaux et al. 2010, with permission)

was the occasional presence of cyanobacterial crusts. From a series of 160 sand grains randomly chosen from sand treated with Syto9 (a green fluorescent nucleic acid stain), a number of 10.4 ± 1.0 fluorescent spots per grain were obtained (Fig. 2; Gommeaux et al. 2010). Considering the number of 2.1×10^4 grains per gram of sand, a minimum of $2.2 \pm 0.2 \times 10^5$ bacteria per gram were present on the surface of sand grains. An original method based on a 'grain-by-grain' cultivation (direct plating of grains on solid $0.1 \times$ TSA medium) revealed about 14% of the numbers obtained by Syto9 direct counting, constituting a greatly more efficient method to reveal 'culturable' bacteria compared to the classical method based on suspension–dilution and plating (1.6% of the numbers obtained by Syto9 direct counting; see Gommeaux et al. 2010). Both culture- and molecular-based (cloning-sequencing of 16 S rDNA) analyses of bacterial diversity revealed that *Firmicutes*, *Actinobacteria*, *Proteobacteria* and '*Cytophaga–Flexibacter–Bacteroides*' (CFB) were the most frequent groups. Green-non-sulfur bacteria, *Acidobacteria* and *Planctomycetes* were present, but less frequently. Dividing the fluorescence cell-count by the number of OTUs (Operational Taxonomic Units) yielded an estimated 1560 ± 140 cells per OTU per gram of sand, in the same range as bacterial diversity of arable and pasture soils (Torsvik et al. 2002). Thus, the very low biomass in the Merzouga sand is not correlated with a reduction in bacterial diversity, but with a very low number of cells per taxon (Gommeaux et al. 2010).

Table 1. List of new bacterial species isolated from desert validly published since 2003

Title	Authors	References
Actinoalloteichus spitiensis sp. nov., a novel actinobacterium isolated from a cold desert of the Indian Himalayas	Singla AK, Mayilraj S, Kudo T, Krishnamurthi S, Prasad GS, Vohra RM	Int J Syst Evol Microbiol 2005, 55:2561–2564
Actinomadura namibiensis sp. nov.	Wink J, Kroppenstedt RM, Seibert G, Stackebrandt E	Int J Syst Evol Microbiol 2003, 53:721–724
Agrococcus lahaulensis sp. nov., isolated from a cold desert of the Indian Himalayas	Mayilraj S, Suresh K, Schumann P, Kroppenstedt RM, Saini HS	Int J Syst Evol Microbiol 2006, 56:1807–1180
Amycolatopsis australiensis sp. nov., an actinomycete isolated from arid soils	Tan GY, Robinson S, Lacey E, Goodfellow M	Int J Syst Evol Microbiol 2006, 56:2297–2301
Caldanaerovirga acetigignens gen. nov., sp. nov., an anaerobic xylanolytic, alkalithermophilic bacterium isolated from Trego Hot Spring, Nevada, USA	Wagner ID, Ahmed S, Zhao W, Zhang CL, Romanek CS, Rohde M, Wiegel J	Int J Syst Evol Microbiol 2009, 59:2685–2691. Epub 2009 Jul 22
Citricoccus alkalitolerans sp. nov., a novel actinobacterium isolated from a desert soil in Egypt	Li WJ, Chen HH, Zhang YQ, Kim CJ, Park DJ, Lee JC, Xu LH, Jiang CL	Int J Syst Evol Microbiol 2005, 55:87–90
Deinococcus gobiensis sp. nov., an extremely radiation-resistant bacterium	Yuan M, Zhang W, Dai S, Wu J, Wang Y, Tao T, Chen M, Lin M	Int J Syst Evol Microbiol 2009, 59:1513–1517
Deinococcus xinjiangensis sp. nov., isolated from desert soil	Peng F, Zhang L, Luo X, Dai J, An H, Tang Y, Fang C	Int J Syst Evol Microbiol 2009, 59:709–713
Deinococcus peraridilitoris sp. nov., isolated from a coastal desert	Rainey FA, Ferreira M, Nobre MF, Ray K, Bagaley D, Earl AM, Battista JR, Gómez-Silva B, McKay CP, da Costa MS	Int J Syst Evol Microbiol 2007, 57:1408–1412
Deinococcus deserti sp. nov., a gamma-radiation-tolerant bacterium isolated from the Sahara Desert	de Groot A, Chapon V, Servant P, Christen R, Saux MF, Sommer S, Heulin T	Int J Syst Evol Microbiol 2005, 55:2441–2446
Extensive diversity of ionizing-radiation-resistant bacteria recovered from Sonoran Desert soil and description of nine new species of the genus *Deinococcus* obtained from a single soil sample	Rainey FA, Ray K, Ferreira M, Gatz BZ, Nobre MF, Bagaley D, Rash BA, Park MJ, Earl AM, Shank NC, Small AM, Henk MC, Battista JR, Kämpfer P, da Costa MS	Appl Environ Microbiol 2005, 71:5225–5235. Erratum in: Appl Environ Microbiol 2005, 71:7630
Description of *Dietzia lutea* sp. nov., isolated from a desert soil in Egypt	Li J, Chen C, Zhao GZ, Klenk HP, Pukall R, Zhang YQ, Tang SK, Li WJ	Syst Appl Microbiol 2009, 32:118–123. Epub 2009 Jan 20
Dyadobacter alkalitolerans sp. nov., isolated from desert sand	Tang Y, Dai J, Zhang L, Mo Z, Wang Y, Li Y, Ji S, Fang C, Zheng C	Int J Syst Evol Microbiol 2009, 59:60–64
Hymenobacter deserti sp. nov., isolated from the desert of Xinjiang, China	Zhang L, Dai J, Tang Y, Luo X, Wang Y, An H, Fang C, Zhang C	Int J Syst Evol Microbiol 2009, 59:77–82
Hymenobacter xinjiangensis sp. nov., a radiation-resistant bacterium isolated from the desert of Xinjiang, China	Zhang Q, Liu C, Tang Y, Zhou G, Shen P, Fang C, Yokota A	Int J Syst Evol Microbiol 2007, 57:1752–1756

(*continued*)

Table 1 (*continued*)

Title	Authors	References
Jiangella gansuensis gen. nov., sp. nov., a novel actinomycete from a desert soil in north-west China	Song L, Li WJ, Wang QL, Chen GZ, Zhang YS, Xu LH	Int J Syst Evol Microbiol 2005, 55:881–884
Kineococcus xinjiangensis sp. nov., isolated from desert sand	Liu M, Peng F, Wang Y, Zhang K, Chen G, Fang C	Int J Syst Evol Microbiol 2009, 59:1090–1093
Kocuria aegyptia sp. nov., a novel actinobacterium isolated from a saline, alkaline desert soil in Egypt	Li WJ, Zhang YQ, Schumann P, Chen HH, Hozzein WN, Tian XP, Xu LH, Jiang CL	Int J Syst Evol Microbiol 2006, 56:733–737
Lechevalieria atacamensis sp. nov., *Lechevalieria deserti* sp. nov. and *Lechevalieria roselyniae* sp. nov., isolated from hyperarid soils	Okoro CK, Bull AT, Mutreja A, Rong X, Huang Y, Goodfellow M	Int J Syst Evol Microbiol 2010, 60:296–300
Mesorhizobium gobiense sp. nov. and *Mesorhizobium tarimense* sp. nov., isolated from wild legumes growing in desert soils of Xinjiang, China	Han TX, Han LL, Wu LJ, Chen WF, Sui XH, Gu JG, Wang ET, Chen WX	Int J Syst Evol Microbiol 2008, 58:2610–2618
Nocardiopsis alkaliphila sp. nov., a novel alkaliphilic actinomycete isolated from desert soil in Egypt	Hozzein WN, Li WJ, Ali MI, Hammouda O, Mousa AS, Xu LH, Jiang CL	Int J Syst Evol Microbiol 2004, 54:247–252
Paenibacillus harenae sp. nov., isolated from desert sand in China	Jeon CO, Lim JM, Lee SS, Chung BS, Park DJ, Xu LH, Jiang CL, Kim CJ	Int J Syst Evol Microbiol 2009, 59:13–17
Paenibacillus gansuensis sp. nov., isolated from desert soil of Gansu Province in China	Lim JM, Jeon CO, Lee JC, Xu LH, Jiang CL, Kim CJ	Int J Syst Evol Microbiol 2006, 56:2131–2134
Paenibacillus tarimensis sp. nov., isolated from sand in Xinjiang, China	Wang M, Yang M, Zhou G, Luo X, Zhang L, Tang Y, Fang C	Int J Syst Evol Microbiol 2008, 58:2081–2085
Planobacterium taklimakanense gen. nov., sp. nov., a member of the family Flavobacteriaceae that exhibits swimming motility, isolated from desert soil	Peng F, Liu M, Zhang L, Dai J, Luo X, An H, Fang C	Int J Syst Evol Microbiol 2009, 59:1672–1678
Planococcus stackebrandtii sp. nov., isolated from a cold desert of the Himalayas, India	Mayilraj S, Prasad GS, Suresh K, Saini HS, Shivaji S, Chakrabarti T	Int J Syst Evol Microbiol 2005, 55:91–94
Pontibacter korlensis sp. nov., isolated from the desert of Xinjiang, China	Zhang L, Zhang Q, Luo X, Tang Y, Dai J, Li Y, Wang Y, Chen G, Fang C	Int J Syst Evol Microbiol 2008, 58:1210–1214
Pontibacter akesuensis sp. nov., isolated from a desert soil in China	Zhou Y, Wang X, Liu H, Zhang KY, Zhang YQ, Lai R, Li WJ	Int J Syst Evol Microbiol 2007, 57:321–325
Pseudomonas xinjiangensis sp. nov., a moderately thermotolerant bacterium isolated from desert sand	Liu M, Luo X, Zhang L, Dai J, Wang Y, Tang Y, Li J, Sun T, Fang C	Int J Syst Evol Microbiol 2009, 59:1286–1289

(*continued*)

Table 1 (continued)

Title	Authors	References
Pseudomonas duriflava sp. nov., isolated from a desert soil	Liu R, Liu H, Feng H, Wang X, Zhang CX, Zhang KY, Lai R	Int J Syst Evol Microbiol 2008, 58:1404–1408
Ramlibacter tataouinensis gen. nov., sp. nov., and *Ramlibacter henchirensis* sp. nov., cyst-producing bacteria isolated from subdesert soil in Tunisia	Heulin T, Barakat M, Christen R, Lesourd M, Sutra L, De Luca G, Achouak W	Int J Syst Evol Microbiol 2003, 53:589–594
Rhodococcus kroppenstedtii sp. nov., a novel actinobacterium isolated from a cold desert of the Himalayas, India	Mayilraj S, Krishnamurthi S, Saha P, Saini HS	Int J Syst Evol Microbiol 2006, 56:979–982
Saccharibacillus kuerlensis sp. nov., isolated from a desert soil	Yang SY, Liu H, Liu R, Zhang KY, Lai R	Int J Syst Evol Microbiol 2009, 59:953–957
Salinicoccus luteus sp. nov., isolated from a desert soil	Zhang YQ, Yu LY, Liu HY, Zhang YQ, Xu LH, Li WJ	Int J Syst Evol Microbiol 2007, 57:1901–1905
Skermanella xinjiangensis sp. nov., isolated from the desert of Xinjiang, China	An H, Zhang L, Tang Y, Luo X, Sun T, Li Y, Wang Y, Dai J, Fang C	Int J Syst Evol Microbiol 2009, 59:1531–1534

The genera *Chelatococcus* and *Saccharothrix* are strongly attached to sand grains, considering their exclusive isolation by the 'grain-by-grain' method (Gommeaux et al. 2010). Another important conclusion of this work deals with the correlation between mineral and bacterial diversity. For instance, *Arthrobacter* appears to be well adapted to this harsh environment with a preference for grains other than the dominant mineral quartz.

The analysis of microbial diversity was also performed on Tataouine sand grains (Benzerara et al. 2006; Chanal et al. 2006). The main difference to the Merzouga dunes was the number of cultivated bacteria in $0.1 \times$ TSA medium, which was 10-fold higher in the Tataouine sample. This result can be explained by steppic vegetation in Tataouine (semi-arid region) and a correlated higher carbon content of the sandy soil. The classical culture-dependent technique revealed that the most frequent isolates belonged to *Firmicutes*, *Actinobacteria*, *Proteobacteria* and CFB, in agreement with the Merzouga data. The culture-independent technique (cloning-sequencing of 16S rDNA) revealed clone sequences corresponding to *Proteobacteria*, *Actinobacteria* and *Acidobacteria*. It was also shown that strains tolerant to gamma radiation, which were present in the Tataouine sample, belong to *Firmicutes* (*Bacillus*), *Deinococcus* and *Chelatococcus* (*Alpha-Proteobacteria*) genera. *Chelatococcus* were isolated as desiccation tolerant strains from the surface soil of a shrub-steppe (Fredrickson et al. 2008) and Merzouga sand grains (Gommeaux et al. 2010).

A surprising result was the absence of known thermotolerant bacteria (for example, in both studies no isolate or clone sequences related to *Thermus* was obtained). The evidence of clone sequences of non-thermophile *Crenarchaeota* in Tataouine (Chanal et al. 2006) reinforced the notion that the most important adaptive microbial trait to face desert conditions is the tolerance of desiccation, more than that of temperature: water, which is allowing bacterial activity, is present only when the temperature of the sand is low (below 20°C).

Considering the cyanobacterial diversity in Tataouine samples, sequences of *Oscillatoria, Anabaena, Nostoc* and *Symploca* were identified after cloning-sequencing of 16S rDNA (Benzerara et al. 2006). Such large diversity of cyanobacteria in deserts was previously described by Garcia-Pichel et al. (2001).

A more complete list of newly described bacterial species isolated from desert regions is presented in Table 1.

5 *Ramlibacter* and its life cycle

The *Ramlibacter* story started in 1931 with the fall of a meteorite near Tataouine (Tunisia) (Lacroix 1931). One important point of this story is the collection of the largest fragments the day after the fall, which were sent to the Muséum National d'Histoire Naturelle (Paris, France) and used as negative control of meteorite alteration studies (Barrat et al. 1998). Several weathered fragments of the meteorite were collected in 1994 in the superficial sandy soil. Scanning electron microscopic observations revealed alteration zones at the surface of the meteorite crystals (pyroxene and chromite) and also secondary calcite crystals resulting from terrestrial weathering (Barrat et al. 1998, 1999). We investigated the isolation of the bacterial agent responsible for this mineral weathering. We isolated and characterized a bacterial strain (TTB310) from a meteorite fragment embedded in sandy soil (Gillet et al. 2000) that we assigned to a new genus and a new species, *Ramlibacter tataouinensis*, belonging to the *Beta-Proteobacteria* (Heulin et al. 2003). In the same paper we described a second species, *Ramlibacter henchirensis*, with the same morphological characteristics. The presence of *Ramlibacter* in the sandy soil of Tataouine was confirmed using a culture-independent method (cloning-sequencing of 16S-rDNA) (Chanal et al. 2006).

Since the description of this new genus in 2003, the presence of *Ramlibacter* was demonstrated in various environments using culture-independent methods (PCR amplification of 16S rDNA followed by cloning or fingerprinting). For instance, *Ramlibacter* was found associated to spores of *Gigaspora* (Long et al. 2008), in a semi-arid soil sample (Rutz and Kieft 2004), in soil crusts of the Colorado Plateau (Gundlapally and Garcia-Pichel 2006), in soil from agricultural systems (Jangid et al. 2008), in heavy-metal-contaminated acidic waters from zinc mine residues (Almeida et al. 2009), as a potential *m*-xylene degrader (Xie et al. 2010) and during a propane-stimulated bioremediation process in trichloroethene-contaminated groundwater (Connon et al. 2005). Using a classical method, *Ramlibacter* strains were also isolated from soil (Shrestha et al. 2007). All these data (about seventy 16S-rDNA sequences and a few isolates) indicate that *Ramlibacter* is adapted to various environments including hot and cold deserts.

R. tataouinensis presents an original life cycle characterized by the coexistence of spherical and rod-shaped cells (Heulin et al. 2003). We demonstrated that two types of cell division occurred during the life cycle of this bacterium. The first one consists of a spherical cell division generating spherical daughter cells leading to the formation of bacterial colonies (Heulin et al. 2003). The second one, at the border of these colonies, consists of a spherical cell division into rod-shaped daughter cells corresponding to the dissemination form (Gommeaux et al. 2005). The reversion of rod-shaped cells into spherical cells occurred at a distance from mother colony (Benzerara et al. 2004a; Gommeaux et al. 2005). Colonization of orthopyroxene, which is the major meteorite-forming mineral, by *R. tataouinensis* and its subsequent alteration was described (Benzerara et al. 2004a) as well as biomineralization of calcium phosphate by this bacterium (Benzerara et al. 2004b).

In this way, *R. tataouinensis* is an example of a novel mechanism for desiccation tolerance, illustrated by its ability to divide as a desiccation tolerant form (spherical cells). This mechanism is novel when compared to the long-term storage and desiccation-resistant forms (spores, akinetes and cysts) or to the highly efficient DNA repair mechanisms of vegetative cells of *Deinococcus*, resulting in their desiccation tolerance. This desiccation-resistant form resembles mostly the vegetative cells of terrestrial cyanobacteria, which withstand constraints resulting from multiple cycles of drying and wetting and/or prolonged desiccation (Potts 1994; Billi and Potts 2002). On the one hand, the spherical cells of *R. tataouinensis* present some traits of *Azotobacter* cysts, notably the spherical morphotype, absence of motility, cells embedded in a thick capsular material, presence of spherical polyhydroalkanoate lipid granules in the cytoplasm and long-term resistance to desiccation (Heulin et al. 2003). On the other hand, spherical cells of *R. tataouinensis* are not resting cells such as the cysts of *Azotobacter* or *Rhodospirillum* (Berleman and Bauer 2004), but are vegetative cells able to divide. Spherical cells present a larger diameter ($0.85\,\mu m$) and a higher volume ($0.34\,\mu m^3$) than the motile rod-shaped cells ($0.24\,\mu m \times 2.9\,\mu m$, $0.13\,\mu m^3$) with the same cell surface area ($2.20\,\mu m^2$) (Gommeaux et al. 2005). Therefore, the spherical cell division into motile rod-shaped daughter cells, which takes about 3 h, appears to be very different from the mechanism of cyst germination of *Azotobacter* (Heulin et al. 2003; Gommeaux et al. 2005; Sadoff 1975).

Associated with the potential trehalose metabolism revealed by genomic analysis (unpublished data), the original life cycle of *R. tataouinensis* displaying an ability to divide as a desiccation tolerant form (cyst), is probably responsible for its adaptation to an extreme environment (desert).

6 *Deinococcus* and DNA repair

Deinococcus radiodurans has been isolated from a can of corned beef that was supposed to have been sterilized by irradiation. This species and other members of the genus *Deinococcus* appear to be tolerant to high doses of gamma and UV radiation. This extreme tolerance results from their ability to repair massive DNA damages, including hundreds of DNA double-strand breaks, generated during irradiation. However, natural sources of ionizing radiation on earth exist at levels much lower than those used to generate large numbers of double-strand breaks, suggesting that radiation tolerance is a consequence of the bacterial response to natural non-radioactive DNA-damaging conditions such as desiccation, to which deinococci are also tolerant (Mattimore and Battista 1996; Cox and Battista 2005; Blasius et al. 2008). Indeed, several radiation tolerant *Deinococcus* strains have been isolated from arid desert soil (de Groot et al. 2005; Rainey et al. 2005; Chanal et al. 2006), and desiccation of *Deinococcus* induces DNA double-strand breaks, which are rapidly repaired upon rehydration (Mattimore and Battista 1996; de Groot et al. 2009). Moreover, various radiation sensitive *D. radiodurans* mutants were also found to be sensitive to desiccation (Mattimore and Battista 1996), and desiccation and gamma irradiation of *D. radiodurans* induced the expression of a common set of genes (Tanaka et al. 2004).

For *D. radiodurans*, it was shown that efficient reconstitution of an intact genome is dependent on extended synthesis-dependent strand annealing (ESDSA) and homologous recombination (Zahradka et al. 2006). This DNA repair mechanism requires homologues of proteins that are found in radiation sensitive bacteria, such as RecA and RecFOR, but several *Deinococcus*-specific proteins are probably also involved (Tanaka et al. 2004; Bentchikou et al. 2010). Deinococci possess a highly condensed nucleoid, which may contribute to efficient DNA repair by limiting diffusion of free DNA ends (Levin-Zaidman et al. 2003; Zimmerman and Battista 2005).

Not only DNA, but also other components of the cell will be damaged by irradiation or desiccation. For survival, at least some components should be protected. Deinococci and other radiation-tolerant bacteria possess a high intracellular manganese concentration, linked to protection of proteins (but not DNA) from oxidative damage by decreasing the concentration of reactive oxygen species formed during irradiation or desiccation (Daly et al. 2007). Moreover, deinococci contain genes coding for homologues of plant proteins associated with desiccation tolerance. Inactivation of two of these proteins sensitizes *D. radiodurans* to desiccation, but not to ionizing radiation (Battista et al. 2001).

The genome sequence of *D. radiodurans* was published in 1999 (White et al. 1999) and that of the slightly thermophilic *Deinococcus geothermalis*, isolated from a hot spring, in 2007 (Makarova et al. 2007). We recently reported the genome sequence and proteome analysis of *Deinococcus deserti* VCD115, isolated from Sahara surface sand (de Groot et al. 2009). Comparative analysis revealed a set of about 200 proteins specifically conserved in *Deinococcus*. Some of these proteins are or may be involved in desiccation and radiation tolerance. After growth of *D. deserti* under standard laboratory conditions, 1,348 proteins were detected by proteome analysis (40% of the entire theoretical proteome). Amongst these are three homologues of plant desiccation tolerance-associated proteins, suggesting their importance under conditions of rapid dehydration as occurring in the desert. Compared to the other sequenced deinococci, *D. deserti* possesses several supplementary genes involved in manganese and nutrient import, as well as additional DNA repair genes, including two extra *recA* and three translesion DNA polymerase genes. The three *recA* genes code for two different RecA proteins, both of which contribute to recombinational repair of massive DNA damage (Dulermo et al. 2009). However, only one RecA allows induction of two translesion polymerases that are involved in mutagenic bypass of UV-induced DNA lesions (Dulermo et al. 2009). The supplementary nutrient import and DNA repair genes are probably important for survival and adaptation of *D. deserti* to its hostile environment.

7 Conclusions

A surprising result was the absence of known thermotolerant bacteria such as *Thermus* in both studies. The evidence of clone sequences of non-thermophilic *Crenarchaeota* in Tataouine (Chanal et al. 2006) reinforced the idea that the most important adaptive microbial trait to face desert conditions is the tolerance of desiccation rather than of temperature: water allowing bacterial activity is present only when the temperature of the sand is low (below $20°C$).

As demonstrated for other environments, the culture-independent analysis of bacterial diversity in desert revealed that 70–80% of 16S-rDNA sequences do not match with described bacterial species, suggesting the presence of a majority of 'new' species and genera to be described. Considering the culture-dependent approach, the grain-by-grain cultivation represents an original and efficient technique allowing the isolation of bacteria strongly attached to sand grain surface. In Sahara samples, most of the bacterial isolates or clone sequences belong to phyla and genera for which the mechanisms of adaptation to desiccation are known: *Firmicutes* and *Actinobacteria* (sporulation), *Deinococcus* and *Rubrobacter* (DNA repair). On the other hand, the mechanisms underlying the adaptation to desiccation of most *Proteobacteria* remain to be described. Surprisingly, the very low density of bacteria in desert sand compared to non-arid soils is not due to a much lower diversity but to a very low number of bacteria per taxon. In these hot and dry environments, the number of bacteria is limited by water availability (only few hours per year), and by the very low content in carbon and nutrients (oligotrophic environment). The main conclusion is that numerous bacterial species were able to adapt in response to extreme conditions prevailing in hot and dry deserts. Many new mechanisms of desiccation tolerance remain to be unravelled, especially considering proteobacterial species. The ability of all these bacteria to tolerate desiccation, UV radiation, heat, oligotrophy and so many other adverse constraints constitute a real treasure that can be exploited for biotechnology applications.

Acknowledgements

We thank Frédéric MEDAIL (Institut Méditerranéen d'Ecologie et de Paléoécologie, Université Paul Cézanne, Aix-Marseille III) and Mylène SALINI for comments and critical reading of the manuscript.

References

Almeida WI, Vieria RP, Machado Cardoso A, Silveira CB, Costa RG, Gonzalez AM, Paranhos R, Medeiros JA, Freitas FA, Albano RM, Martins OB (2009) Archaeal and bacterial communities of heavy metal contaminated acidic waters from zinc mine residues in Sepetiba Bay. Extremophiles 13:263–271

Barer MR (2003) Physiological and molecular aspects of growth, non-growth, culturability and viability in bacteria. In: Anthony R.M. Coates (ed) Dormancy and low-growth states in microbial disease. Cambridge University Press, Cambridge, pp 1–35

Barrat JA, Gillet P, Lécuyer C, Sheppard SM, Lesourd M (1998) Formation of carbonates in the Tatahouine meteorite. Science 280:412–414

Barrat JA, Gillet P, Lesourd M, Blichert-Toft J, Poupeau GR (1999) The Tatahouine diogenite: mineralogical and chemical effects of sixty-three years of terrestrial residence. Meteor Planet Sci 34:91–97

Battista JR, Park MJ, McLemore AE (2001) Inactivation of two homologues of proteins presumed to be involved in the desiccation tolerance of plants sensitizes *Deinococcus radiodurans* R1 to desiccation. Cryobiology 43:133–139

Bentchikou E, Servant P, Coste G, Sommer S (2010) A major role of the RecFOR pathway in DNA double-strand-break repair through ESDSA in *Deinococcus radiodurans*. PLoS Genet 6:e1000774

Benzerara K, Barakat M, Menguy N, Guyot F, De Luca G, Audrain C, Heulin T (2004a) Experimental colonization and alteration of orthopyroxene by the pleomorphic bacteria *Ramlibacter tataouinensis*. Geomicrobiol J 21:341–349

Benzerara K, Menguy N, Guyot F, Skouri F, De Luca G, Barakat M, Heulin T (2004b) Biologically controlled precipitation of calcium phosphate by *Ramlibacter tataouinensis*. Earth Planet Sci Lett 228:439–449

Benzerara K, Chapon V, Moreira D, LopeZ-Garcia P, Guyot F, Heulin T (2006) Microbial diversity on the Tatahouine meteorite. Meteor Planet Sci 41:1249–1265

Berleman JE, Bauer CE (2004) Characterization of cyst cell formation in the purple photosynthetic bacterium *Rhodospririllum centenum*. Microbiology-SGM 150:383–390

Billi D, Potts M (2002) Life and death of dried prokaryotes. Res Microbiol 153:7–12

Blasius M, Sommer S, Hübscher U (2008) *Deinococcus radiodurans*: what belongs to the survival kit? Crit Rev Biochem Mol Biol 43:221–238

Chanal A, Chapon V, Benzerara K, Barakat M, Christen R, Achouak W, Barras F, Heulin T (2006) The desert of Tataouine: an extreme environment that hosts a wide diversity of microorganisms and radiotolerant bacteria. Environ Microbiol 8:514–525

Connon SA, Tovanabootr A, Dolan M, Vergin K, Giovannoni SJ, Semprini L (2005) Bacterial community composition determined by culture-independent and -dependent methods during propane-stimulated bioremediation in trichlorethene-contaminated groundwater. Environ Microbiol 7:165–178

Cox MM, Battista JR (2005) *Deinococcus radiodurans* – the consummate survivor. Nat Rev Microbiol 3:882–892

Daly MJ, Gaidamakova EK, Matrosova VY, Vasilenko A, Zhai M, Leapman RD, Lai B, Ravel B, Li SM, Kemner KM, Fredrickson JK (2007) Protein oxidation implicated as the primary determinant of bacterial radioresistance. PLoS Biol 5:e92

Dulermo R, Fochesato S, Blanchard L, de Groot A (2009) Mutagenic lesion bypass and two functionally different RecA proteins in *Deinococcus deserti*. Mol Microbiol 74:194–208

Essendoubi M, Brhada F, Eljamali JE, Filali-Maltouf A, Bonnassie S, Georgeault S, Blanco C, Jebbar M (2007) Osmoadaptative responses in the rhizobia nodulating *Acacia* isolated from south-eastern Moroccan Sahara. Environ Microbiol 9:603–611

Fredrickson JK, Li SM, Gaidamakova EK, Matrosova VY, Zhai M, Sulloway HM, Scholten JC, Brown MG, Balkwill DL, Daly MJ (2008) Protein oxidation: key to bacterial desiccation resistance? ISME J 2:393–403

Garcia-Pichel F, Belnap J (1996) Microenvironments and microscale productivity of cyanobacterial desert crusts. J Phycol 32:774–782

Garcia-Pichel F, Pringault O (2001) Cyanobacteria track water in desert soils. Nature 143:380–381

Garcia-Pichel F, López-Cortés A, Nübel U (2001) Phylogenetic and morphological diversity of cyanobacteria in soil desert crusts from the Colorado plateau. Appl Environ Microbiol 67:1902–1910

Gillet P, Barrat JA, Heulin T, Achouak W, Lesourd M, Guyot F, Benzerara K (2000) Bacteria in the Tatahouine meteorite: nanometric-scale life in rocks. Earth Planet Sci Lett 175:161–167

Gommeaux M, Barakat M, Lesourd M, Thiéry J, Heulin T (2005) A morphological transition in the pleiomorphic bacterium *Ramlibacter tataouinensis* TTB310. Res Microbiol 156:1026–1030

Gommeaux M, Barakat M, Montagnac G, Christen R, Guyot F, Heulin T (2010) Mineral and bacterial diversities of desert sand grains from South-East Morocco. Geomicrobiol J 27:76–92

de Groot A, Chapon V, Servant P, Christen R, Fisher-Le Saux M, Sommer S, Heulin T (2005) *Deinococcus deserti* sp. nov., a gamma-radiation tolerant bacterium isolated from the Sahara desert. Int J Syst Evol Microbiol 55:2441–2446

de Groot A, Dulermo R, Ortet P, Blanchard L, Guérin P, Fernandez B, Vacherie B, Dossat C, Jolivet E, Siguier P, Chandler M, Barakat M, Dedieu A, Barbe V, Heulin T, Sommer S, Achouak W,

Armengaud J (2009) Alliance of proteomics and genomics to unravel the specificities of Sahara bacterium *Deinococcus deserti*. PLoS Genet 5:e1000434.

Gundlapally SR, Garcia-Pichel F (2006) The community and phylogenetic diversity of biological soil crusts in the Colorado Plateau studied by molecular fingerprinting and intensive cultivation. Microb Ecol 52:345–357

Heulin T, Barakat M, Christen R, Lesourd M, Sutra L, De Luca G, Achouak W (2003) *Ramlibacter tataouinensis* gen. nov., and *Ramlibacter henchirensis* sp. nov., cyst-producing bacteria isolated from sub-desert soil in Tunisia. Int J Syst Evol Microbiol 53:589–594

Jangid K, Williams MA, Franzluebbers AJ, Sanderlin JS, Reeves JH, Jenkins MB, Endale DM, Coleman DC, Whitman WB (2008) Relative impacts of land-use, management intensity and fertilization upon soil microbial community structure in agricultural systems. Soil Biol Biochem 40:2843–2853

Kaci Y, Heyraud A, Barakat M, Heulin T (2005) Isolation and identification of an EPS-producing *Rhizobium* strain from arid soil (Algeria): characterization of its EPS and the effect of inoculation on wheat rhizosphere soil structure. Res Microbiol 156:522–531

Karnieli A, Kidron GJ, Glaesser C, Eyal Ben-Dor E (1999) Spectral characteristics of cyanobacteria soil crust in semiarid environments. Remote Sens Environ 69:67–75

Khbaya B, Neyra M, Normand P, Zerhari K, Filali-Maltouf A (1998) Genetic diversity and phylogeny of rhizobia that nodulate acacia spp. in morocco assessed by analysis of rRNA genes. Appl Environ Microbiol 64:4912–4917

Kröpelin S, Verschuren D, Lézine AM, Eggermont H, Cocquyt C, Francus O, Cazet JP, Fagot M, Ramus B, Russell JM, Conley DJ, Schuster M, von Suchodoletz H, Engstrom DR (2008) Climate-driven ecosystem succession in the Sahara: the past 6000 years. Science 320:765–768

Lacroix A (1931) Sur la chute récente (27 juin 1931) d'une météorite asidérite dans l'extrême Sud Tunisien. C R Acad Sci Paris 193:305–309

Le Houérou HN (1986) The desert and arid zones of Northern Africa. In: Evenari M, Noy-Meir E, Goodall DW (eds) Hot deserts and arid shrublands. Ecosystems of the world, vol 12B. Elsevier, Amsterdam, pp 101–147

Le Houérou HN (1997) Climate, flora and fauna changes in the Sahara over the past 500 million years. J Arid Environ 37:619–647

Levin-Zaidman S, Englander J, Shimoni E, Sharma AK, Minton KW, Minsky A (2003) Ringlike structure of the *Deinococcus radiodurans* genome: a key to radioresistance? Science 299:254–256

Long LK, Zhu HH, Yao Q, Ai YC (2008) Analysis of bacterial communities associated with spores of *Gigaspora margarita* and *Gigaspora rosea*. Plant Soil 310:1–9

Makarova KS, Omelchenko MV, Gaidamakova EK, Matrosova VY, Vasilenko A, Zhai M, Lapidus A, Copeland A, Kim E, Land M, Mavrommatis K, Pitluck S, Richardson PM, Detter C, Brettin T, Saunders E, Lai B, Ravel B, Kemner KM, Wolf YI, Sorokin A, Gerasimova AV, Gelfand MS, Fredrickson JK, Koonin EV, Daly MJ (2007) *Deinococcus geothermalis*: the pool of extreme radiation genes shrinks. PLoS One 2:e995

Mattimore V, Battista JR (1996) Radioresistance of *Deinococcus radiodurans*: functions necessary to survive ionizing radiation are also necessary to survive prolonged desiccation. J Bacteriol 178:633–637

Mobley HL, Island MD, Hausinger RP (1995) Molecular biology of microbial ureases. Microbiol Rev 59:451–480

Potts M (1994) Desiccation tolerance of prokaryotes. Microbiol Lett 58:755–805

Prestel E, Salamitou S, DuBow MS (2008) An examination of the bacteriophages and bacteria of the Namib desert. J Microbiol 46:364–372

Rainey FA, Rau K, Ferreira M, Gatz BZ, Nobre MF, Bagaley D, Rash BA, Park MJ, Earl AM, Shank NC, Small AM, Henk MC, Battista JR, Kämpfer P, da Costa MS (2005) Extensive diversity of ionizing-radiation-resistant bacteria recovered from Sonoran desert soil and description of nine new species of the genus *Deinococcus* obtained from a single soil sample. Appl Environ Microbiol 71:5225–5235

Rutz BA, Kieft TL (2004) Phylogenetic characterization of dwarf archaea and bacteria from a semiarid soil. Soil Biol Biochem 36:825–833

Sadoff HL (1975) Encystment and germination in *Azotobacter vinelandii*. Bacteriol Rev 39:516–539

Schuster M, Duringer P, Ghienne JF, Vignaud P, Taisso-Mackaye H, Likius A, Brunet M (2006) The age of the Sahara desert. Science 311:821

Shrestha PM, Noll M, Liesack W (2007) Phylogenetic identity, growth-response time and rRNA operon copy number of soil bacteria indicate different stages of community succession. Environ Microbiol 9:2464–2474

Takagi H (2008) Proline as a stress protectant in yeast: physiological functions, metabolic regulations, and biotechnological applications. Appl Microbiol Biotechnol 8:211–223

Tanaka M, Earl AM, Howell HA, Park MJ, Eisen JA, Peterson SN, Battista JR (2004) Analysis of *Deinococcus radiodurans*'s transcriptional response to ionizing radiation and desiccation reveals novel proteins that contribute to radioresistance. Genetics 168:21–33

Torsvik T, Øvreas L, Thingstad TF (2002) Prokaryotic diversity – magnitude, dynamics, and controlling factors. Science 296:1064–1066

White O, Eisen JA, Heidelberg JF, Hickey EK, Peterson JD, Dodson RJ, Haft DH, Gwinn ML, Nelson WC, Richardson DL, Moffat KS, Qin H, Jiang L, Pamphile W, Crosby M, Shen M, Vamathevan JJ, Lam P, McDonald L, Utterback T, Zalewski C, Makarova KS, Aravind L, Daly MJ, Minton KW, Fleischmann RD, Ketchum KA, Nelson KE, Salzberg S, Smith HO, Venter JC, Fraser CM (1999) Genome sequence of the radioresistant bacterium *Deinococcus radiodurans* R1. Science 286: 1571–1577

Xie S, Sun W, Luo C, Cupples AM (2010) Stable isotope probing identifies novel *m*-xylene degraders in soil microcosm from contaminated and uncontaminated sites. Water Air Soil Pollut. DOI: 10.1007/s11270-010-0326-z

Zahradka K, Slade D, Bailone A, Sommer S, Averbeck D, Petranovic M, Lindner AB, Radman M (2006) Reassembly of shattered chromosomes in *Deinococcus radiodurans*. Nature 443:569–573

Zerhari K, Aurag J, Khbaya B, Kharchaf D, Filali-Maltouf A (2000) Phenotypic characteristics of rhizobia isolates nodulating acacia species in the arid and Saharan regions of Morocco. Lett Appl Microbiol 30:351–357

Zimmerman JM, Battista JR (2005) A ring-like nucleoid is not necessary for radioresistance in the *Deinococcaceae*. BMC Microbiol 5:17

Extremophiles in Antarctica: life at low temperatures

David A. Pearce

British Antarctic Survey, Natural Environment Research Council, Cambridge, UK

Abstract

In this chapter, we will explore the different adaptations of extremophiles to life in the extreme cold. We generally forget that the Earth is mostly cold and that most ecosystems are exposed to temperatures that are permanently below 5°C. Such low mean temperatures mainly arise from the fact that ∼70% of the Earth's surface is covered by oceans that have a constant temperature of 4–5°C (below a depth of 1000 m), irrespective of the latitude. The polar regions account for another 15% of the surface, to which the glacier and alpine regions must also be added. Here, we will take an illustrated look in particular at the Antarctic environment, as it is by far the coldest environment on Earth – the lowest temperature on the surface of the Earth (−89.2°C) was recorded at the Russian Vostok Station, at the centre of the East Antarctic ice sheet. Antarctica is a place where organisms are often subjected to combined stresses including desiccation, limited nutrients, high salinity, adverse solar radiation and low biochemical activity. The incredibly harsh environment of the Antarctic continent precludes life in most of its forms, and the microorganisms are therefore dominant.

Generally, conditions include air temperatures that average well below freezing all year round, strong winds that increase the effects of the cold, light which varies from months of total darkness to total sunlight and little free available water, with all but 2% of the continent covered with ice. Given this combination of extremes, it is surprising that anything lives on the continent at all, let alone thrives there. For the organisms that do manage to adapt, however, the benefits from a lack of competition in the extreme cold are enormous, and this is often seen in very large population sizes. So how do some organisms survive extreme cold, and others exploit and take advantage of it? As organisms have been subjected to these stresses over extremely long periods of time, a range of adaptations have evolved; some species have adapted to live at the limits, some produce specific compounds such as antifreeze, some remain viable but frozen in a state of suspended animation whilst higher organisms can adapt their life cycles in such a way that when conditions are harsh they die, leaving the next generation to recover as conditions improve. By looking at a range of different strategies in a wide variety of organisms, it is hoped to bring together general mechanisms of adaptation to life in the extreme cold.

1 Introduction

Extremophiles are microorganisms that survive or thrive in extreme environments. There remains much debate as to what actually constitutes an extreme environment, as evidenced by the recent coordination action for research activities on life in extreme environments initiative (CAREX, www.carex-eu.org), and definitions can vary according to the perspective, i.e. what is extreme to a human is not necessarily extreme for a microorganism, and also the scale, so for example, many people view the whole of the Antarctic continent as an extreme environment, when many different types of condition exist. However, extremophiles can thrive in ice, boiling water, acid, the water core of nuclear reactors, salt crystals and toxic waste and in a range of other extreme habitats that were previously thought to be inhospitable for life (Cavicchioli 2002). For the purposes of this discussion, I shall therefore define all microorganisms living in the Antarctic as living in an extreme environment, but will highlight those that are specifically classified as extremophiles. A range of different factors make the Antarctic environment extreme including the general physical characteristics: extremes of temperature, desiccation and osmotic stress, low-nutrient concentrations, high levels of UVB radiation (under the Antarctic ozone hole) and a highly variable photoperiod (from no light at all to continous light during a 24-h period).

The continent of Antarctica is large, approximately the size of Europe, and so there are many additional localized extremes, and these include areas for which specific extremophiles may have adapted – such as regions of volcanic activity, hypersaline lakes, subglacial lakes and even within the ice itself. As a result, there are numerous examples of microorganisms that have special adaptations to cope with these often unique combinations of selection pressures, and these lead to a wide range of novel biodiversity, much of which is yet to be described. Another key feature of the Antarctic ecosystem is the extreme variation within the physical conditions that exist, for example, from freshwater lakes (some of the most oligotrophic environments on Earth) to hypersaline lakes (Laybourn-Parry 2009). Microorganisms found under these extreme environmental conditions in the Antarctic are therefore ideal candidates to study the ecophysiological and biochemical adaptations of living at the limits for life.

Despite the relative isolation of the Antarctic continent, microorganisms still arrive constantly from around the globe and particularly from the rest of the atmosphere (Pearce et al. 2009). Whether they remain as permanent members of the Antarctic microbiota or not depends on their rate of arrival, ability to compete with existing populations, potential rate of evolution and rate of removal. Often microorganisms arriving on the Antarctic continent are pre-adapted to survival in this type of extreme atmosphere, as they have been pre-filtered through the atmosphere by similar extreme conditions to those found in the Antarctic. Indeed, it has been observed that the harsher the Antarctic environment, the weedier the species found (compare studies by Hughes et al. 2004; Pearce et al. 2010). Human impact can also influence this delicate balance. Ah Tow and Cowan (2005) found that the predominant commensal microorganism occurring on human skin, *Staphylococcus epidermidis*, could be detected by PCR in Dry Valley mineral (DVM) soils collected from heavily impacted areas, but could not be detected in DVM soils collected from low impact and pristine areas. Cell viability of this non-enteric human commensal appeared to be rapidly lost in DVM soil. However, *S. epidermidis* can persist for long periods in this soil as non-viable cells and/or naked DNA.

The Antarctic continent is also one of the most remote places on Earth. It is isolated by distance (the tip of the Antarctic Peninsula is >1000 km from the Southern tip of South America), air movement (the prevailing movement is towards the coast), the Antarctic circumpolar current which moves in an anti-clockwise direction around the pole, the polar front where a significant temperature difference prevents free mixing and a relative lack of human activity to introduce new colonists. For this reason, the microorganisms present there are also relatively isolated from the rest of the biosphere. The observed balance between mesophiles growing sub-optimally and true psychrophiles may reflect in some way the balance between indigenous microorganisms which have evolved *in situ* and those which have subsequently colonized the area; however, as yet there is no data to support this theory. There is also the issue of psychrotolerant (organisms which are tolerant of low growth temperatures from $-15°C$ to $+10°C$) versus psychrophilic (organisms with a requirement for low growth temperatures) isolates. For example, Clocksin et al. (2007) isolated eight strains of chemoorganotrophic bacteria from the water column of Lake Hoare, in the McMurdo Dry Valleys of Antarctica, using cold enrichment temperatures. All isolates grew at $0°C$, and all but one grew at subzero temperatures characteristic of the water column of Lake Hoare. However, the growth temperature optima varied amongst isolates, with the majority showing optima near $15°C$, indicative of cold-active phenotypes. However, only one isolate was truly psychrophilic, growing optimally around $10°C$ and not above $20°C$.

In contrast to the terrestrial realm, the seas around Antarctica are extremely productive and diverse. One of the reasons the seas around Antarctica support so many living things, despite the extreme cold, is the abundance of nutrients in the water. Within the Antarctic environment, it is the marine ecosystem which is the most stable, and this is reflected in the biodiversity of the organisms found there. The Antarctic marine ecosystem bucks the trend of decreasing biodiversity with increasing latitude. It can also be said to harbour truly psychrophilic microorganisms as the temperature is consistently low. However, even within the marine system, niches exist for extremophiles, such as hydrothermal vent systems, mud volcanoes and cold seeps, and in the relatively unexplored deep-sea environment in general, there is also very high pressure. Understanding the biodiversity of marine bacterioplankton is crucial for predicting changes in other organisms at the ends of marine food chains. Most bacteria live in microenvironments below the scale of millimetres such as across physical discontinuities of the sea-ice or seawater column or on plant, animal or detritus surfaces. Heterogeneity exists at microscale levels, and changes in microorganisms will produce a cascade of changes through the marine food chain.

Some controversy still exists over the 'everything is everywhere but the environment selects' debate; however, data are increasingly becoming available, particularly for the cyanobacteria, which suggest that endemic Antarctic forms occur which have not been found elsewhere, and there may indeed be an endemic Antarctic microflora. In contrast to the rather limited diversity of plants and animals to be found in the Antarctic, the microbial diversity of this continent has been shown to be surprisingly diverse (Tindall 2004). Exploring the biodiversity of Antarctic extreme environments is often targeted, particularly the unusual and less explored more extreme environments.

2 Environmental extremes associated with the Antarctic

Many of the environmental extremes in Antarctica have the same characteristics and features as extreme environments that are found elsewhere, but which in addition tend to be much colder. Key characteristics are:

Strong winds – these dry the Antarctic environments, and they also increase the effects of the cold.

Temperature – Antarctica contains the coldest environments on Earth. Environmental temperature has a profound impact on species distributions, abundance and survival, and the importance of temperature in limiting species distributions can be ascribed in part to its powerful effects on subcellular and molecular systems (Hochachka and Somero 2002). Indeed, temperature is a critical environmental factor controlling the evolution and biodiversity of life on Earth. The majority of the Earth's biosphere is permanently below 5°C, dominated by ocean depths, glaciers, alpine and polar regions (Feller and Gerday 2003). Surprisingly, however, although Antarctic temperatures are predominantly cold, in the austral summer, on a sunny day temperatures can reach in excess of 30°C on dark surfaces and these alternate with extremes of cold in the winter.

Desiccation – although the Antarctic continent contains a large percentage of the Earths' freshwater, the vast majority of this water is in solid form and therefore unavailable to organisms. All but around 2% of the continent is covered with ice and so the Antarctic is very much a polar desert. However, despite these challenging low temperatures, as long as liquid water is present, life is possible. Even the coldest environments on Earth have enough liquid water to sustain life (Laybourn-Parry 2009).

The physical properties of water – they lead to a mechanical impact of external ice (water expands in volume as estimates vary considerably on freezing); this affects biochemical processes, can directly damage cells and membranes and disrupt water relations (controlled by water status and availability).

Limited nutrients – chemical analysis has shown that Antarctic soils are relatively low in nutrient content: 1.95–0.33% carbon, 0.20–0.03% nitrogen, 8.00–0.06% phosphorus and 0.22–0.20% potassium (Newsham et al. 2010), with an organic matter content of $\sim$2.56% (Lawley et al. 2004).

Patchiness of high nutrients – this can lead to localized extremes, for example, exposure to high acidity can occur due to immersion in penguin guano (Fig. 1).

Significant chemical gradients – such patchiness of nutrients and stressors in general can lead to strong gradients in physicochemical parameters at a wide range of spatial scales – of the order of metres (Chong et al. 2009), kilometres (Chong et al. 2010) or hundreds of kilometres (Yergeau et al. 2008).

Fig. 1. Ornithogenic soils on South Thule Island, South Sandwich Islands (photo by Dr. D. Pearce, British Antarctic Survey)

High salinity – many of the saline lakes of coastal oases such as the Vestfold Hills are marine derived, formed from pockets of seawater trapped in closed basins when the land rose during isostatic rebound flowing the last glaciation. This suite of lakes ranges from slightly brackish (approximately 4–5‰) to hypersaline (240‰) (Laybourn-Parry and Pearce 2007). There is also the potential for alternate immersion in both salt and freshwater due to weather and tides at the coast.

Adverse solar radiation – intense ultraviolet light radiation can occur as a result of the hole in the ozone layer. The photoperiod also varies from months of total darkness to total sunlight.

Low biochemical activity – due to low temperatures and the Q10 effect.

Instability and variability – in terms of severity, seasonality and the changes in seasonality with latitude. These physical and chemical conditions can change annually, and these can be exacerbated by climate change (Pearce 2008). In Antarctic freshwater lakes, bacterial community structures have been shown to be influenced by temperature, at temperatures lower than that already experienced as a result of climate change (Pearce 2005). This means that Antarctic environments can be variable or unpredictable, and this in itself imposes a further stress.

Biotic interactions – such as competition and predation can produce stress. Duarte et al. (2005) showed that bacteria – chlorophyll and bacterial production – primary production relationships in the Southern Ocean differed from the typical relationships applicable to aquatic ecosystems elsewhere. Bacteria responded to phytoplankton blooms, but they responded so weakly that the bacterial production represented a small percentage of primary production (1–10%). Although other mechanisms might also contribute to the weak bacterial response to phytoplankton blooms, they demonstrated that the reason was likely to be the tight control of bacterial populations by their predators.

Combined stresses – many Antarctic environments such as terrestrial, freshwater, aerial and ice combine more than one of the above stresses. For example, Antarctic soils can have high salinity, low nutrients, low temperature and adverse solar radiation, and these may result in a unique suite of selection pressures.

3 Extreme environments that also occur in the Antarctic

So what type of extreme environment also occurs in the Antarctic? Within the Antarctic continent there are specific locations in which adapation to a variety of potential extremes is possible. For this reason, many have been specifically targeted for study in detail.

Fig. 2. Aerial transfer of microorganisms into the Antarctic (photo by Dr. Pearce, British Antarctic Survey)

Aerial – the aerial environment is very similar to the Antarctic environment in terms of selection pressures, with temperatures around 0°C in the stratosphere, where life has been shown to exist (Sattler et al. 2001). It is also low nutrient, desiccating and with high UVB radiation. As a result, resistant spore-forming bacteria are often found here (Fig. 2).

Volcanic – a number of regions of volcanic activity are found both in and around the Antarctic environment, these include Mount Erebus, Deception Island, Montague Island and South Thule (Fig. 3).

Hydrothermal – the discovery of the Indian Ocean hydrothermal vent communities raised many questions about the biogeography of these ecosystems, for example, how has the distribution of the vent fauna evolved over geological time and how has this been influenced by continental drift and past climate change. One of the obvious routes for the dispersal of hydrothermal vent organisms between the Atlantic, Pacific and Indian Ocean is via the Southern Ocean. To address this problem a Natural Environment Research Council (NERC) funded Chemosynthetic Communities of the Southern Ocean (CHESSO) Consortium has been exploring areas of the Southern Ocean where chemical signals in the ocean and the shape of the seabed indicate that hydrothermal vents might be present.

Sea ice – Antarctic sea ice is highly variable and can double the size of the Antarctic continent in winter (Fig. 4). The sea ice itself is extremely productive with a whole trophic cascade based around the algae that grow on or under the surface of the ice.

Fig. 3. Deception island, a region of volcanic activity to the North of the Antarctic Peninsula (62°58′S 60°39′W) which last experienced an eruption in 1987 (photo by Dr. Pearce, British Antarctic Survey)

Fig. 4. Sea ice showing colouration due to the presence of associated microorganisms (photo by Dr. Pearce, British Antarctic Survey)

Beneath the snow lies a unique habitat for a group of bacteria and microscopic plants and animals that are encased in an ice matrix at low temperatures and light levels, with the only liquid being pockets of concentrated brines. Survival in these conditions requires a complex suite of physiological and metabolic adaptations, but sea-ice organisms thrive in the ice (Thomas and Dieckmann 2002). In the Sea

Ice Microbial Communities (SIMCO) study, an extensive array of bacterial isolates were obtained from bottom ice assemblages as well as from algae-free 'clean' ice and under-ice seawater to determine the influence diatom blooms have on the bacterial inhabitants. Previous results had indicated that biodiversity increased three- to sevenfold between ice free of algae and under-ice seawater to ice samples rich in algae (>100 mg chlorophyll m^3). From this, 35–95% of the bacterial biomass from diatom assemblages was estimated to be psychrophilic. By comparison ice lacking algae and under-ice seawater samples are nearly devoid of psychrophiles which were almost always isolates of *Psychrobacter glacincola* (Bowman et al. 1997).

Fast ice – *Psychrobacter salsus* sp. nov. and *Psychrobacter adeliensis* sp. nov. have also been isolated from fast ice from Adelie Land, Antarctica (Shivaji et al. 2004).

Radiation – on the snow surface of glaciers and polar caps, psychrophiles are exposed to strong ultraviolet radiation (Fig. 5). Antarctic lakes are also unproductive, with typical levels of photosynthesis in the region of 0.5–30 $\mu g\, l^{-1}\, day^{-1}$. This results from low annual levels of photosynthetically active radiation and ice-covers that attenuate light to the water column, continuous low temperatures and the lack of any significant input of inorganic nutrients and organic carbon (Laybourn-Parry 2002).

Microbial mats – *Planococcus antarcticus* and *Planococcus psychrophilus* sp. nov. were isolated from cyanobacterial mat samples collected from ponds in Antarctica (Reddy et al. 2002) along with psychrophilic *Planococcus maitriensis* sp. nov. and an orange pigmented bacterium, which was isolated from a cyanobacterial mat sample collected in the vicinity of Schirmacher Oasis, Maitri, the Indian station, in

Fig. 5. Microorganisms growing on the surface of Antarctic snow (photo by Dr. P. Bucktrout, British Antarctic Survey)

Antarctica (Alam et al. 2003). Van Trappen et al. (2002) isolated 746 heterotrophic bacteria from within microbial mats from 10 Antarctic lakes. The frequency of such isolations suggests that this is a habitat conducive to the growth of microorganisms.

Permanently cold marine sediments – a 16S ribosomal DNA (rDNA) clone library from permanently cold Antarctic marine sediments was established (Ravenschlag et al. 1999). Screening 353 clones by dot blot hybridization with group-specific oligonucleotide probes suggested a predominance of sequences related to bacteria of the sulphur cycle (43.4% potential sulphate reducers). Within this fraction, the major cluster (19.0%) was affiliated to *Desulfotalea* sp. and other closely related psychrophilic sulphate reducers isolated from the same habitat. The cloned sequences showed between 93% and 100% similarity to these bacteria. Two additional groups were frequently encountered: 13% of the clones were related to *Desulfuromonas palmitatis*, and a second group was affiliated to *Myxobacteria* spp. and *Bdellovibrio* spp. Many clones (18.1%) belonged to the gamma subclass of the class *Proteobacteria* and were closest to symbiotic or free-living sulphur oxidizers. Rarefaction analysis suggested that the total diversity assessed by 16S rDNA analysis was very high in these permanently cold sediments and was only partially revealed by screening of 353 clones.

Lakes – in Antarctic lakes, organisms are confronted by continuous low temperatures as well as a poor light climate and nutrient limitation. Such extreme environments support truncated food webs with no fish, few metazoans and a dominance of microbial plankton (Laybourn-Parry 2002). The key to success lies in entering the short Antarctic summer with actively growing populations. In many cases, the most successsful organisms continue to function throughout the year (Laybourn-Parry and Pearce 2007).

Ornithogenic soils – the amount of penguin guano deposited by a breeding population on the shores of Admiralty Bay is about 6.4 tons (dry weight) per day (Rakusa-Suszczewski 1980), causing a significant impact on the local budgets of carbon, nitrogen, phosphorus and other minerals (Tatur 2002) such instances are seen wherever significant penguin colonies exist.

Endolithic, sublithic and chasmolithic – the endolithic microbial communities that are found in rocks of the Antarctic dry deserts, which comprise lichens, yeasts, cyanobacteria and heterotrophic bacteria, survive low water and nutrient availability (Fig. 6). Pointing et al. (2009) demonstrated that life has adapted to form highly specialized communities in distinct lithic niches. Endoliths and chasmoliths in sandstone displayed greatest diversity, whereas soil was relatively depauperate and lacked a significant photoautotrophic component, apart from isolated islands of

Fig. 6. An endolithic cyanobacterial community (photo by Dr. C. Gilbert, British Antarctic Survey)

hypolithic cyanobacterial colonization on quartz rocks in soil contact (Pointing et al. 2009). The findings showed that biodiversity near the cold-arid limit for life is more complex than previously appreciated, but communities lack variability probably due to the high selective pressures of this extreme environment.

Cryoconite holes – cryoconite holes are a water filled cylindrical melt-holes on glacial ice surface (Christner et al. 2003).

Lake sediments – a bacterial community has been described from along a historic lake sediment core of Ardley Island, West Antarctica (Li et al. 2006). There are clearly many such examples from around the continent, so the Antarctic harbours a diverse range of highly specialized niches for extremophile growth.

4 Extreme environments particular to the Antarctic

So what are the extreme environments particular to the Antarctic?

Antarctic subglacial lakes are one of the few remaining unexplored environments on Earth, and from a microbiological perspective, perhaps one of the most interesting. They have the potential to be one of the most extreme environments on Earth, and are certainly one of the least accessible, with combined stresses of high pressure, low but probably stable temperature, permanent darkness, predominantly low-nutrient availability (potentially one of the most oligotrophic environments on Earth, particularly in the absence of geothermal activity) and highly variable oxygen concentrations (derived from the ice that provided the original meltwater). Indeed, the predominant mode of nutrition is likely to be chemoautotrophic (Siegert et al.

2003). Lake Vostok, the largest subglacial lake found has shown encouraging signs of containing life, as all of the Vostok accreted ice samples between 3541 and 3611 m have been found to contain both prokaryotic and eukaryotic microorganisms (Poglazova et al. 2001). The identification of significant subglacial bacterial activity (Sharp et al. 1999) as well as the work on permafrost communities (Gilichinsky et al. 1995) suggests that life can survive and potentially thrive at low temperatures. Microbes have been detected in the two Antarctic subglacial lakes sampled to date; accreted ice from subglacial Lake Vostok in East Antarctica (e.g. Karl et al. 1999; Priscu et al. 1999; Christner et al. 2006; Bulat et al. 2009) and saturated till from beneath the Kamb Ice Stream, West Antarctica (Priscu et al. 2005; Lanoil et al. 2009). Data from the accretion ice support the working hypothesis that a sustained microbial ecosystem is present in this subglacial lake environment, despite high pressure, constant cold, low-nutrient input, potentially high oxygen concentrations and an absence of sunlight (Christner et al. 2006). The small numbers of microbes found so far within the accreted ice have DNA profiles similar to those of contemporary surface microbes.

The arid soils of the Antarctic Dry Valleys – these constitute some of the oldest, coldest, dryest and most oligotrophic soils on Earth. Recent applications of molecular methods have revealed a very wide diversity of microbial taxa, many of which are uncultured and taxonomically unique, and a community that seems to be structured solely by abiotic processes (Cary et al. 2010).

Marine hypersaline deeps – habitats on the sea floor include sediments of varying geology, mineral nodule fields, carbonate mounds, cold seeps, hydrocarbon seeps, saturated brines and hydrothermal vents. Below sea subfloor and deep biosphere, marine sediments merge into the below sea subfloor and deep biosphere.

Ice cave environments – such as that on Mount Erebus and elsewhere.

Permafrost (though this has mainly been studied in the Northern hemisphere) – in a study of permafrost, Vishnivetskaya et al. (2007) found that half of the isolates were spore-forming bacteria unable to grow or metabolize at subzero temperatures. Other Gram-positive isolates metabolized, but never exhibited any growth at $-10°$C. One Gram-negative isolate metabolized and grew at $-10°$C, with a measured doubling time of 39 days. Metabolic studies of several isolates suggested that as temperature decreased below $+4°$C, the partitioning of energy changed with much more energy being used for cell maintenance as the temperature decreased. In addition, cells grown at $-10°$C exhibited major morphological changes at the ultrastructural level. Bacteria of the genus *Exiguobacterium* have been repeatedly isolated from ancient permafrost sediments of the Kolyma lowland of Northeast Eurasia. Vishnivetskaya et al. (2007) reported that the Siberian permafrost isolates *Exiguobacterium sibiricum*

255-15, *E. sibiricum* 7-3, *E. undae* 190-11 and *Exiguobacterium* sp. 5138, as well as *Exiguobacterium antarcticum* DSM 14480, isolated from a microbial mat sample of Lake Fryxell (McMurdo Dry Valleys, Antarctica), were able to grow at temperatures ranging from −6°C to 40°C. In comparison to cells grown at 24°C, the cold-grown cells of these strains tended to be longer and wider. There are also habitats within the permafrost such as cryopegs.

Given their relative isolation and unique conditions, extreme environments in Antarctica offer unique opportunities for bioprospecting. Although relatively unexplored, the potential exists for new biosubstances such as novel psychrophilic enzymes. However, much of the work to date has focussed on those microorganisms capable of degrading hydrocarbons associated with low-temperature hydrocarbon spills.

5 Key Antarctic extremophiles

Psychrophiles are organisms capable of growth and reproduction at temperatures from −15°C to +10°C. The ability of psychrophiles to survive and proliferate at low temperatures implies that they have overcome key barriers inherent to permanently cold environments. These challenges include: reduced enzyme activity; decreased membrane fluidity; altered transport of nutrients and waste products; decreased rates of transcription, translation and cell division; protein cold-denaturation; inappropriate protein folding and intracellular ice formation. Cold-adapted organisms have successfully evolved features, genotypic and/or phenotypic, to surmount the negative effects of low temperatures and to enable growth in these extreme environments. Psychrophiles are true extremophiles as they are adapted not only to low temperatures, but frequently also to further environmental constraints. They occur throughout the Antarctic continent, but particularly in the Southern Ocean.

Thermophiles are organisms capable of growth and reproduction at temperatures between 45°C and 80°C. They occur in regions of geothermal activity in Antarctica.

Piezophiles are organisms capable of growth and reproduction at high pressures. Microbial life has been shown to function at gigapascal pressures (Nogi et al. 1998; Sharma et al. 2002) and bacteria recovered from the deep ocean at around 4000 m have been shown to retain both structural integrity and metabolic activity (Pearce 2009). They occur in Antarctica in the deep sea and deep within the ice.

Acidophiles are organisms capable of growth and reproduction under highly acidic conditions where pH less than or equal to 2.0.

Alkaliphiles are organisms capable of growth and reproduction under alkaline conditions of pH > 9.0; both pH extremes occur in Antarctica for example lake ecosystems such as Blood Falls (Mikucki and Priscu 2007).

Halophiles are organisms capable of growth and reproduction at high salinities, for example, *Halomonas alkaliantarctica* sp. nov., isolated from saline lake in Cape Russell in Antarctica, an alkalophilic moderately halophilic, exopolysaccharide-producing bacterium (Poli et al. 2006). Euryhaline halophilic bacteria have been described from Suribati Ike, a meromictic lake in East Antarctica (Naganuma et al. 2005). Further, the microbial communities that are found in sea ice, are exposed to salt concentrations of several molar in brine veins, and are therefore halo-psychrophiles.

6 Biodiversity

Microbial communities in the Antarctic are characterized by low numbers of dominant species but have been found to have a high diversity of subdominant species. This observation has implications for the total biodiversity of these environments, which could be much higher than was originally thought. A large percentage of microorganisms detected in Antarctic environments are new to science. Up to 80% of the dominant members of the microbial community can be detected but have yet to be studied as they have not been cultured. A high proportion of uncultivated microbes are common, generally, but particularly so for Antarctic extreme environments. A bacterial phylogenetic survey of three environmentally distinct Antarctic Dry Valley soil biotopes showed a high proportion of so-called 'uncultured' phylotypes, with a relatively low diversity of identifiable phylotypes (Smith et al. 2006). One thing is certain, however that there is generally a much higher biodiversity than was once thought, and this is coming through in increasing numbers of studies. For example, high 16S rDNA bacterial diversity was described in glacial meltwater lake sediment from Bratina Island, Antarctica (Sjöling and Cowan 2003).

Particularly abundant groups of microorganisms, identified in microbiological studies in the Antarctic, include the cyanobacteria (photosynthetic bacteria), lichens (symbiotic associations between a cyanobacterium or an algae and a fungus – particularly well adapted for life in the most extreme environments as the symbiosis makes them widely trophically independent of substrate type), psychrophilic bacteria, psychrotolerant bacteria and psychrotrophs. Obligate psychrophilic bacteria are unable to grow above $20°C$, facultative psychrophiles are able to grow at low temperature and above $20°C$. Amongst cold-adapted microorganisms, psychrotrophs tend to dominate in environments that undergo thermal fluctuations. True psychrophiles have a more restricted growth range and are found in permanently cold habitats. However, even in permanently cold environments $>50\%$ of the bacteria are not psychrophilic. In global terms, psychrotrophs are much more widely distributed than psychrophiles. From permanently cold habitats in the polar regions, the psychrophilic alga *Raphidonema nivale* and the fungus *Sclerotinia borealis* have been isolated. Amongst terrestrial psychrophiles are members of the genera *Pseudomonas, Cytophaga* and *Flavobacterium* (each commonly encountered in the freshwater lakes of Signy Island). However, the potential exists for all forms of microbial life to exist in Antarctic ecosystems. Microorganism groups are likely to include:

Viruses – it is now well established that viruses are ubiquitous in aquatic ecosystems worldwide (Wilson et al. 2000; Pearce and Wilson 2003), and bacteriophage are common in polar inland waters (Säwström et al. 2007).

Cyanobacteria are photosynthetic bacteria which fix nitrogen and carbon in Antarctic soils.

Fungi – basidiomycetous yeasts have been isolated from glacial and subglacial waters of northwest Patagonia (Brizzio et al. 2007) and *Penicillium* in Arctic subglacial ice (Sonjak et al. 2006).

Archaea – Archaea have been isolated from the basal ice of the Greenland Ice Core (GISP2 core).

Eukaryotes – Willerslev et al. (1999), using PCR amplification of fragments of the 18S rRNA gene, identified a diversity of fungi, plants, algae and protists extracted from 2000- to 4000-year-old ice-core samples from North Greenland. Colouring of snow can also be caused by a variety of photosynthetic eukaryotes such as *Chlamydomonas nivalis, Chloromonas (Scotiella), Ankistrodesmus, Raphionema, Mycanthococcus* and certain dinoflagellates (Prescott 1978) and represents a well-known illustration of cold adaptation. *Heteromita* is a very common soil microflagellate with a worldwide distribution with an optimum temperature for growth around 23°C. Under Antarctic conditions it demonstrates adaptations which permit survival of freeze – thaw cycles (by rapid and temperature-sensitive encystment and excystment) and by optimal utilization of resources during short periods, which allow this temperate species to grow actively at mean temperatures close to zero (Hughes and Smith 1989). The ciliate *Holosticha* sp. was reported to be unable to divide above $-2°C$ (Lee and Fenchel 1972).

The common view that 'everything is everywhere', first formulated at the beginning of the last century by the biologist Beijerinck (Brock 1961) and reintroduced by Fenchel and Finlay in a series of publications (Finlay et al. 1999; Finlay and Clarke 1999; Finlay 2002; Fenchel and Finlay 2004) focussed on the marine microbial community, support for the ubiquitous distribution of bacterioplankton came from group- or species-specific analyses. Because of their small size, great abundance and easy dispersal, it is often assumed that marine planktonic microorganisms have a ubiquitous distribution that prevents any structured assembly into local marine bacterioplankton communities. Pommier et al. (2007) examined coastal communities at nine locations distributed worldwide through the use of comprehensive clone libraries of 16S ribosomal RNA genes and showed that there were marked differences in the composition and richness of operational taxonomic units (OTUs) between locations. Remarkably, the global marine bacterioplankton community showed a high degree of endemism, and conversely included few cosmopolitan OTUs. In addition a number of new Antarctic species are published on an annual basis (Bowman and McCuaig 2003; Busse et al. 2003; Reddy et al. 2003; Sheridan et al. 2003; Spring et al. 2003; Van Trappen et al. 2003, 2004a, b, c, 2005; Donachie et al. 2004; Hirsch et al. 2004a, b; Jung et al. 2004; Montes et al. 2004; Pocock et al. 2004; Yi et al. 2005a, b; Smith et al. 2006; Yu et al. 2008; Labrenz et al. 2009) along with an increasing number of studies from new extremophilic Antarctic environments.

Biodiversity studies are also of increasing relevance due to the effects of climate change. Significant changes in microbial community structure as a result of environmental change can be detected at levels below that already experienced as a result of climate change. The climate around the Antarctic Peninsula is warming at a rate $3\times$faster than elsewhere (Turner et al. 2002). These communities might therefore act as indicators of potential effects of climate change.

7 Methodology

Extremophiles in Antarctica have been investigated using standard techniques, which can be classified into a number of generic categories:

Culture and physiology – the growth and activity of microorganisms.

Microscopy – visualization using both fluorescent and electron microscopy (sometimes used in combination with specific gene probes).

Biochemistry (and biogeochemical cycling)–many gene probes are available to assay samples for the presence of specific biogeochemical activities.

Molecular biology – genomic DNA (using gene probes coupled with FISH – fluorescent in situ hybridization) and the construction and analysis of metagenomic libraries are used to screen for novel physiologies.

Physical – for example, infrared Raman spectrophotometry (used to detect biomolecules).

Recently, epifluorescence microscopy and flow cytometry were used to show that an ice sample contained over 6×10^7 cells ml^{-1} from a Greenland ice core that had remained at $-9°C$ for over 100,000 years. Anaerobic enrichment cultures inoculated with melted ice were also grown and maintained at $-2°C$. Genomic DNA extracted from these enrichments was used for the PCR amplification of 16S rRNA genes with both bacterial and archaeal primers and also used in the preparation of clone libraries, and this illustrates a standard polyphasic approach to the study of such environments. In addition, contemporary biological studies are being fuelled by the increasing availability of genome sequences and associated functional studies of extremophiles. This is leading to the identification of new biomarkers, an accurate assessment of cellular evolution, insight into the ability of microorganisms to survive in meteorites and during periods of global extinction, and knowledge of how to process and examine environmental samples to detect viable life forms (Cavicchioli 2002). Most of these aspects are illustrated in Sect. 9 on *Pseudoalteromoans haloplanktis* and the methanogenic Archaea.

A critical assessment of detection strategies is important in any community analysis and the description of Antarctic extremophiles can be hampered by low cell numbers, the methodological difficulty involved in culturing and the detection limits of the assays used. Advances in molecular technology have vastly improved life detection limits, such that microscopy and PCR are now capable of detecting individual cells per ml, or the DNA itself at $0.1-0.2$ ng l^{-1}. Adopting a culture-based approach from Antarctic ice cores, 0, 2 and 10 cfu ml^{-1} have been isolated from Dyer Plateau, Siple Station and Taylor Dome, respectively (Christner et al. 2000), and $1-16$ cfu ml^{-1} from a Dronning Maud Land ice core (Pearce, unpublished data). Radiolabelled substrates can yield uptake rates at the level of several hundred cells (Karl et al. 1999). However, no approach is likely to provide a complete unbiased picture of the microorganisms residing in a sample or their relative numbers, and the design of specific, clean sampling strategies is extremely important. Stingl et al. (2008) used a dilution-to-extinction culturing technique for psychrotolerant planktonic bacteria from permanently ice-covered lakes in the McMurdo Dry Valleys, Antarctica. PCR theoretically can detect a single cell; realistically PCR detection limit is $2-8$ cells ml^{-1} (Bulat et al. 2002) such that a minimum DNA concentration for PCR is about 17 ng/2.5 ng or $0.8-1.6$ ng DNA in the starting sample, for whole genome amplification. ($1-10$ ng input DNA genomiPhi from GE Healthcare produces $4-7$ µg in 1.5 h.) Using a *Flavobacterium* estimate of 8.4 fg DNA per cell (bacteria $6-25$ fg DNA per cell); one would need 10^{5-6} cells for genomiPhi to work; if 10^3 cells ml^{-1} are present in the sample, then 1 l would be necessary.

8 Adaptations

A number of adaptations have been described for coping with the extreme cold, for a range of different types of organisms and these include anatomical, behavioural, physiological and genetic, which can apply from bacteria to humans and from algae to grasses. The main benefit from adapting to life in the cold is a lack of competition. Most cellular adaptations to low temperatures and the underlying molecular mechanisms are not fully understood and are still being investigated. In multicellular organisms strategies include: modification of metabolism, elimination of nucleators and the accumulation of cryoprotectants to allow supercooling (freezing resistance).

Specific adaptations include:

The heat shock response – the response to heat stress in six yeast species isolated from Antarctica was examined (Deegenaars and Watson 1998). The yeasts were classified into two groups: one psychrophilic, with a maximum growth temperature of 20°C, and the other psychrotrophic, capable of growth at temperatures above 20°C. Elsewhere, molecular cloning and expression analysis of a cytosolic Hsp70 gene have been conducted in the Antarctic ice algae *Chlamydomonas* sp. ICE-L (Liu et al. 2010).

Cold-active secondary metabolites – production of a cold-active killer toxin by *Mrakia frigida* 2E00797 isolated from sea sediment in Antarctica has been reported (Hua et al. 2010).

Cold-active enzymes – numerous studies have shown that psychrophilic enzymes are generally characterized by high turnover rates and catalytic efficiency at low temperatures along with reduced stability at moderate and high temperatures. Reduction in the activation energy is also achieved, resulting from an increased structural flexibility of either the selected residues located at the active site or in the overall protein structure. The reduced heat stability has been correlated with subtle changes to their sequences compared with mesophilic enzymes, such as decreased levels of Pro and Arg residues, increased numbers or clustering of Gly residues, weakening of intramolecular interactions, increased solvent interactions, decreased number of charged residue interactions and of disulphide bonds (Mavromatis et al. 2003). A cold-active DnaK of an Antarctic psychrotroph *Shewanella* sp. Ac10 has been isolated, supporting the growth of dnaK-null mutant of *Escherichia coli* at cold temperatures (Yoshimune et al. 2005). Siddiqui and Cavicchioli (2006) reported a systematic comparative analysis of 21 psychrophilic enzymes belonging to different structural families from prokaryotic and eukaryotic organisms. The sequences of these enzymes were multiply aligned to 427 homologous proteins from mesophiles and thermophiles. The net flux of amino acid exchanges from meso/thermophilic to psychrophilic enzymes was measured. Other

Fig. 7. Dark pigments produced by Antarctic lichens for protection from UV radiation (photo by Dr. C. Gilbert, British Antarctic Survey)

specific studies of psychrophilic enzymes are also reported in the literature, for example: Feller et al. (1997) and de Pascale et al. (2008). To compensate for reduced metabolic activity at low temperatures, it has been suggested that psychrophiles also synthesize elevated levels of enzymes (Herbert 1989). The high production of a key enzyme to counterbalance its poor catalytic efficiency at low temperature could constitute a novel type of adaptive mechanism to cold environments, as for example, RUBISCO (Devos et al. 1998). Much of the adaptation is achieved through structural adaptations at the active site.

Pigment production (Fig. 7) is a common adaptation to high UVB radiation in Antarctic microorganisms (Dieser et al. 2010).

Freeze – thaw tolerance – as mentioned previously, bacteria of the genus *Exiguobacterium*, have been repeatedly isolated from ancient permafrost sediments of the Kolyma lowland of Northeast Eurasia, as well as *E. antarcticum* DSM 14480, isolated from a microbial mat sample of Lake Fryxell (McMurdo Dry Valleys, Antarctica). All are able to grow at temperatures ranging from $-6°C$ to $40°C$. In comparison with cells grown at $24°C$, the cold-grown cells of these strains tended to be longer and wider. Bacteria grown in broth at $4°C$ showed markedly greater survival following freeze – thaw treatments (20 repeated cycles) than bacteria grown in broth at $24°C$. Surprisingly, significant protection to repeated freeze – thaw was also observed when bacteria were grown on agar at either $4°C$ or $24°C$ (Vishnivetskaya et al. 2007).

Photosynthesis – the stress of low temperatures on life is exacerbated in organisms that rely on photoautotrophic production of organic carbon and energy sources. Phototrophic organisms must coordinate temperature-independent reactions of light absorption and photochemistry with temperature-dependent processes of electron transport and utilization of energy sources through growth and metabolism. Photoautotrophs rely on low-temperature acclimative and adaptive strategies that have been described for other low-temperature-adapted heterotrophic organisms, such as cold-active proteins and maintenance of membrane fluidity. Psychrophilic (organisms tolerant of growth temperatures of $-15°C$ to $+10°C$) versus psychrophilic (optimal growth and reproduction at $-15°C$ to $+10°C$) photoautotrophs rely on low-temperature acclimative and adaptive strategies that have been described for other low-temperature-adapted heterotrophic organisms, such as cold-active proteins and maintenance of membrane fluidity (Morgan-Kiss et al. 2006).

Symbiosis – lichens have been found only 400 km from the South Pole. They recover very slowly from freezing after the winter with photosynthesis and respiration levels not reaching high levels until late spring. Lichens are able to function with less light and water than other plants and have a high concentration of pigments and acids. Last year a New Zealand research team measured a lichen photosynthesizing at $-20°C$, the lowest so far recorded (http://www.anta.canterbury.ac.nz/resources/adapt.html).

Regulation of membrane fluidity – cold adaptation in *Methanococcoides burtonii* was shown to involve growth-temperature-regulated membrane-lipid unsaturation. The studies on *M. burtonii* and *H. lacusprofundi* indicate that a general feature of cold adaptation in psychrophilic Archaea might be to increase the abundance of unsaturated lipids at low temperature to ensure that membrane fluidity, and thereby membrane function, is maintained (see Sect. 9).

Cold-shock response – the cold-shock response is a specific pattern of gene expression in response to abrupt changes to lower temperatures. One effect of reducing temperature is to block initiation of protein synthesis. Cold-shock proteins can stabilize mRNA and re-initiate protein production. Others are also linked to maintaining the fluidity of the membrane such as inducible desaturases.

Cold-shock proteins and cold-acclimation proteins (CAPs) – the structure and function of cold-shock proteins in Archaea have been investigated (Giaquinto et al. 2007).

Cold-inducible proteins – proteomic studies of an Antarctic cold-adapted bacterium, *Shewanella livingstonensis* Ac10, have been conducted for global identification of cold-inducible proteins (Kawamoto et al. 2007). Temperature downshift also induces an antioxidant response in fungi isolated from Antarctica (Gocheva et al. 2009).

Antifreeze proteins – these seem to have less of a role in bacteria. Sea-ice bacteria have been shown to be active at temperatures down to $-20°C$ and motile down to $-10°C$. It is thought that cryoprotectants such as antifreeze may be produced to inhibit ice crystal formation. The antifreeze proteins produced by bacteria are possibly part of the large pool of extracellular polymeric substances (EPSs) located on the cell surface.

Ice-active substances (IASs) – are produced by sea-ice diatoms; they may be glycoproteins that bind preferentially to ice crystals causing pitting.

Membrane fluidity – changes in membrane composition can lead to increased flexibility at low temperatures, for example, increase in fatty acid unsaturation, decrease in fatty acid chain length or increase in methyl branching of fatty acids increase in the ratio of anteiso branching relative to iso branching. Overall, the changes in temperature tend to alter the balance between protein flexibility and stability; higher temperatures can make a protein overly flexible, reducing substrate affinity by disrupting the active site and ultimately leading to denaturation. Colder temperatures can make an enzyme overly stable, reducing catalytic rates below the range needed to maintain metabolic homeostasis in the cell (Fields 2001). Temperature can influence the response of microorganisms either directly, by its effects on growth rate, enzyme activity, cell composition and nutritional requirements, or indirectly by its effects on the solubility of solute molecules, ion transport and diffusion, osmotic effects on membranes, surface tension and density (Herbert 1986).

We have learned a great deal about the mechanisms of adapation to cold environments through the studies of the genome sequences and proteomics of three specific organisms and their comparisons. The most significant finding is that both microbes have flexible proteins, which allow their cells to survive cold temperatures and carry out basic cell functions under extreme conditions. These proteins are more rigid and stable in bacteria that live at higher temperatures. In *Methanogenium frigidum*, researchers also identified cold-shock proteins that are not found in heat-loving Archaea. Cold-shock proteins are known to help other organisms, such as Bacteria, adapt to cold environments.

8.1 *Pseudoalteromonas haloplanktis*

P. haloplanktis is a versatile marine bacterium that grows in the Antarctic and was isolated near the French Dumont d'Urville research station on the Antarctic continent. A collaboration between Genoscope in Evry, the University of Hong

Kong, and researchers from the Universities of Liège, Naples and Stockholm, analysed the genome of this interesting bacterium and was able to reconstruct the adaptations and metabolic mechanisms that allow it to flourish in such an extreme environment. They found out how the bacterium was able to adapt to the presence of atmospheric oxygen, which is very soluble in cold water and then becomes a highly toxic reactive element. They discovered how it was able to keep its membrane fluid and prevent it from freezing at low temperatures, through modification of its lipid composition. Specifically, they found concerted changes in the amino acid composition of its proteins. Organisms living at moderate temperatures relatively infrequently use a delicate amino acid, asparagine, in spite of its advantageous properties, because it deteriorates chemically over time and is thus one of the most significant factors in ageing. The *P. haloplanktis* bacterium, protected from ageing by the cold, was found to use asparagine to a larger degree in its proteins. They also found kinetic and structural optimization of catalysis at low temperatures in a psychrophilic cellulase; no structural alteration related to cold-activity could be found in the catalytic cleft, whereas several structural factors in the overall structure could explain the weak thermal stability, suggesting that the loss of stability provides the required active-site mobility at low temperatures (Médigue et al. 2005).

8.2 *M. burtonii*

M. burtonii, a cold-adapted archaeon, has been investigated using proteomics (Campanaro et al. 2010). Cellular processes that are important for cold adaptation were derived from studies of cells grown at 4°C. The genome sequence has also been studied (Allen et al. 2009). Key aspects of cold adaptation were found to be related to transcription, protein folding and metabolism, including specific roles for RNA polymerase subunit E, a response regulator and peptidyl prolyl *cis/trans* isomerase. A heat shock protein (DnaK) was found to be expressed during growth at optimal temperatures, indicating that growth at such temperatures was stressful for this cold-adapted organism. Thermal regulation in *M. burtonii* was found to be achieved through complex gene expression events involving gene clusters and operons, through to protein modifications.

8.3 *M. frigidum*

M. frigidum sp. nov. was isolated from the perennially cold, anoxic hypolimnion of Ace Lake in the Vestfold Hills of Antarctica. This was the first report of a psychrophilic methanogen growing by CO_2 reduction (Franzmann et al. 1997). The cells are psychrophilic, exhibiting most rapid growth at 15°C and no growth at temperatures above 18–20°C. Comparative genomics revealed trends in amino acid and tRNA composition, and structural features of proteins. Proteins from this cold-adapted *Archaeon* were characterized by a higher content of noncharged polar amino acids, particularly Gln and Thr and a lower content of hydrophobic amino acids, particularly Leu.

A cold-shock domain (CSD) protein (CspA homologue) was also identified in *M. frigidum*, two hypothetical proteins with CSD-folds in *M. burtonii*, and a unique winged helix DNA-binding domain protein in *M. burtonii*. This suggests that these types of nucleic-acid-binding proteins have a critical role in cold-adapted Archaea. Structural analysis of tRNA sequences from the Archaea indicated that GC content is the major factor influencing tRNA stability in hyperthermophiles, but not in the psychrophiles, mesophiles or moderate thermophiles. Below an optimal growth temperature of 60°C, the GC content in tRNA was largely unchanged, indicating that any requirement for flexibility of tRNA in psychrophiles is mediated by other means. This was the first time that comparisons have been performed with genome data from Archaea spanning the growth temperature extremes from psychrophiles to hyperthermophiles (Saunders et al. 2003).

9 Discussion and future perspectives

We can learn a lot from the study of microorganisms that have adapted to life in the cold. Many of the stress response mechanisms are universal. There appears to be no single response to cold, but a combination of responses at different levels and these differ by organism. Extremophiles in Antarctica include representatives of all three domains (Bacteria, Archaea and Eukarya); however, the majority are microorganisms, and a high proportion of these are Archaea. Recently, however, new groups of abundant, uncultivated Archaea have been found to be widespread in more pedestrian biotopes, including marine plankton, terrestrial soils, lakes, marine and freshwater sediments, and in association with metazoa. Research efforts are presently focussed on characterizing the physiology, biochemistry and genetics of these abundant and cosmopolitan but poorly understood Archaea (DeLong 1998). Knowledge of extremophile habitats is expanding the number and types of extraterrestrial locations that may be targeted for exploration (Cavicchioli 2002), and this is certainly true for Antarctica. Extremophiles do not seem to possess any outrageous measures of adaptation to the cold in Antarctica – it appears to be a combination of more stable proteins, modified enzyme activity, more or less fluid membranes, etc. So is the Antarctic really extreme? Probably not for the organisms that have adapted to thrive there.

As the latest techniques are applied to the study of Antarctic extremophiles, for example, the new study of cold adaptation in the marine bacterium, *Sphingopyxis alaskensis* using quantitative proteomics (Ting et al. 2010), there is an increasing realization of the potential for new and radically innovative biotechnology, including a good potential source of novel biochemicals. Despite the remarkable opportunities that these uncommon organisms present for biotechnological applications, only few instances can be reported for actual exploitation. This lack of progress from the research findings at a laboratory-scale to the actual development of pilot and large-scale production is correlated with the difficulties encountered in the cultivation of extremophiles. Schiraldi and De Rosa (2002) report recent achievements in the production of biomass and related enzymes and biomolecules from extremophile sources, especially focussing on the application of novel fermentation strategies. Extremophiles are a potential source for novel enzymes, a particular interest is for low-temperature enzyme activity, cold-shock induction and ice-active substances (Ferrer et al. 2007). Novel physiological adaptations could also suggest evolutionary separation – biofilm formation and synergy may be two physiological strategies for nutrient acquisition in these systems. The low concentration of nutrient has led to nitrogen fixation levels of 1 g m^{-2} year^{-1} in cyanobacterial mats, so nitrogen availability is a key nutritional factor controlling microbial production in Antarctic freshwater habitats (Olson et al. 1998). Low temperatures might also induce the viable but non-culturable (VBNC) state in Antarctic lake microorganisms and the VBNC state of some bacteria, collected from Antarctic lakes, has been reported (Chattopadhyay 2000).

We may also learn a great deal from analogy to higher organisms. However, care is required in developing generalizations because one element can serve in many different ways. For example, cryoprotectants can have multiple functions: they appear to favour supercooling or disable nucleators, resist or protect against desiccation, protect frozen membranes or other cell constituents directly, modify the freezing process, mitigate cellular damage from ice crystallization or recrystallization whilst tissues are frozen or whilst they are thawing, and repair damage after thawing. In the same way, developmental delays can serve multiple purposes: they may conserve energy, protect against adversity, synchronize the feeding stage with food resources, optimize the timing of reproduction, synchronize individuals with one another, prevent a risky generation late in the year, or assist in further life-cycle programming by allowing the environment to be monitored for a longer period (Danks 2002).

The fact that microflora have been found in the basal strata of an Antarctic ice core above the Vostok lake is extremely encouraging in the search for relict forms of life on the Earth and also as a model for solving a number of problems of exobiology, for instance for development of methods to penetrate into the under-ice sea at Europa – Jupiter's satellite (Abyzov et al. 2001). Mars has extensive regions rich in clay minerals. Even at low near-surface temperatures, a thin layer of unfrozen water almost certainly coats mineral grains (Möhlmann 2003), accounting for the survival of water detected within a metre of the Mars surface despite the high vapour pressure of solid water (Price 2006). Antarctic subglacial lakes offer an excellent opportunity to develop technologies for exploring these new worlds, because of their hydraulic isolation, proximity to Antarctic infrastructure, and because their location can provide an analogue to a Martian polar cap (Price et al. 2002). Such potential habitats include Mars and icy satellites in outer space. Titan, Europa and Callisto have all shown evidence of previously unknown bodies of water that might be home to unique life forms. Indeed, cold environments are common throughout the Galaxy (Reid et al. 2006).

Antarctic extremophiles could therefore provide a rich source of unexpected and unique adaptive mechanisms, particularly the double or triple extremophiles. They provide molecular insights into psychrophilic lifestyles and an understanding of specific adaptive mechanisms. The genomic era has pushed back the frontiers of discovery of Antarctic extremophiles over the past decade and the study of extremophiles in the field using techniques of molecular biology will undoubtedly generate a great deal of excitement over the coming years.

Acknowledgements

This contribution was funded by the Terrestrial and Coastal Work Package of the Ecosystems Programme, Polar Science for Planet Earth, British Antarctic Survey (BAS), Natural Environment Research Council.

References

Abyzov SS, Mitskevich IN, Poglazova MN, Barkov IN, Lipenkov VY, Bobin NE, Koudryashov BB, Pashkevich VM, Ivanov MV (2001) Microflora in the basal strata at Antarctic ice core above the Vostok lake. Adv Space Res 28:701–706

Ah Tow L, Cowan DA (2005) Dissemination and survival of non-indigenous bacterial genomes in pristine Antarctic environments. Extremophiles 9:385–389

Alam SI, Singh L, Dube S, Reddy GS, Shivaji S (2003) Psychrophilic *Planococcus maitriensis* sp.nov. from Antarctica. Syst Appl Microbiol 26:505–510

Allen MA, Lauro FM, Williams TJ, Burg D, Siddiqui KS, De Francisci D, Chong KWY, Pilak O, Chew HH, De Maere MZ, Ting L, Katrib M, Ng C, Sowers KR, Galperin MY, Anderson IJ, Ivanova N, Dalin, E, Martinez M, Lapidus A, Hauser L, Land M, Thomas T, Cavicchioli R (2009) The genome sequence of the psychrophilic archaeon, *Methanococcoides burtonii*: the role of genome evolution in cold-adaptation. ISME J 3:1012–1035

Bowman JP, McCammon SA, Brown MV, Nichols DS, McMeekin TA (1997) Diversity and association of psychrophilic bacteria in Antarctic sea ice. Appl Environ Microbiol 63:3068–3078

Bowman JP, McCuaig RD (2003) Diversity and biogeography of prokaryotes dwelling in Antarctic continental shelf sediment. Appl Environ Microbiol 69:2463–2484

Brizzio S, Turchetti B, de García V, Libkind D, Buzzini P, van Broock M (2007) Extracellular enzymatic activities of basidiomycetous yeasts isolated from glacial and subglacial waters of northwest Patagonia (Argentina). Can J Microbiol 53:519–525

Brock TD (1961) Milestones in microbiology. Prentice-Hall, Englewood Cliffs, NJ

Bulat SA, Alekhina IA, Krylenkov VA, Lukin VV (2002) Molecular biological studies of microbiota in subglacial lake Vostok (the Antarctic). Adv Curr Biol 122:211–221 [in Russian]

Bulat SA, Alekhina IA, Lipenkov VYa, Lukin VV, Marie D, Petit JR (2009) Cellular concentrations of microorganisms in glacial and lake ice of the Vostok ice core, East Antarctica. Microbiology 78:808–810

Busse H-J, Denner EBM, Buczolits S, Salkinoja-Salonen M, Bennasar A, Kämpfer P (2003) *Sphingomonas aurantiaca* sp. nov., *Sphingomonas aerolata* sp. nov. and *Sphingomonas faeni* sp. nov., air- and dustborne and Antarctic, orange-pigmented, psychrotolerant bacteria, and emended description of the genus *Sphingomonas*. Int J Syst Evol Microbiol 53:1253–1260

Campanaro S, Williams TJ, De Francisci D, Treu L, Lauro FM, Cavicchioli R (2010) Temperature-dependent global gene expression in the Antarctic archaeon *Methanococcoides burtonii*. Environ Microbiol. DOI: 10.1111/j.1462-2920.2010.02367.x [online November 8]

Cary SC, McDonald IR, Barrett JE, Cowan DA (2010) On the rocks: the microbiology of Antarctic Dry Valley soils. Nat Rev Microbiol 8:129–138

Cavicchioli R (2002) Extremophiles and the search for extraterrestrial life. Astrobiology 2:281–292

Chattopadhyay MK (2000) Cold adaptation of Antarctic microorganisms – possible involvement of viable but non culturable state. Polar Biol 23:223–224

Chong CW, Dunn MJ, Convey P, Annie Tan GY, Wong, RCS, Tan IKP (2009) Environmental influences on bacterial diversity of soils on Signy Island, maritime Antarctic. Polar Biol 32:1571–1582

Chong CW, Pearce DA, Convey P, Tan GYA, Wong RCS, Tan IKP (2010) High levels of spatial heterogeneity in the biodiversity of soil prokaryotes on Singy Island, Antarctica. Soil Biol Biogeochem 42:601–610

Christner BC, Mosley-Thompson E, Thompson LG, Zagorodnov V, Sandman K, Reeve JN (2000) Recovery and identification of viable bacteria immured in glacial ice. Icarus 144:479–485

Christner BC, Kvitko BH, Reeve JN (2003) Molecular identification of Bacteria and Eukarya inhabiting an Antarctic cryoconite hole. Extremophiles 7:177–183

Christner BC, Royston-Bishop G, Foreman CM, Arnold BR, Tranter M, Welch KA, Lyons WB, Tsapin AI, Studinger M, Priscu JC (2006) Limnological conditions in subglacial Lake Vostok, Antarctica. Limnol Oceanogr 51:2485–2501

Clocksin KM, Jung DO, Madigan MT (2007) Cold-active chemoorganotrophic bacteria from permanently ice-covered Lake Hoare, McMurdo Dry Valleys, Antarctica. Appl Environ Microbiol 73:3077–3083

Danks HV (2002) Modification of adverse conditions by insects. Oikos 99:10–24

de Pascale D, Cusano AM, Autore F, Parrilli E, di Prisco G, Marino G, Tutino ML (2008) The cold-active Lip1 lipase from the Antarctic bacterium *Pseudoalteromonas haloplanktis* TAC125 is a member of a new bacterial lipolytic enzyme family. Extremophiles 12:311–323

Deegenaars ML, Watson K (1998) Heat shock response in psychrophilic and psychrotrophic yeast from Antarctica. Extremophiles 2:41–49

DeLong EF (1998) Everything in moderation: Archaea as 'non-extremophiles'. Curr Opin Genet Dev 8:649–654

Devos N, Ingouff M, Loppes R, Matagne RF (1998) RUBISCO adaptation to low temperatures: a comparative study in psychrophilic and mesophilic unicellular algae. J Phycol 34:655–660

Dieser M, Greenwood M, Foreman CM (2010) Carotenoid pigmentation in Antarctic heterotrophic bacteria as a strategy to withstand environmental stresses. AAAR 42:396–405

Donachie SP, Bowman JP, Alam M (2004) *Psychroflexus tropicus* sp. nov., an obligately halophilic Cytophaga – Flavobacterium – Bacteroides group bacterium from an Hawaiian hypersaline lake. Int J Syst Evol Microbiol 54:935–940

Duarte CM, Agustí S, Vaqué D, Agawin NSR, Felipe J, Casamayor EO, Gasol JM (2005) Experimental test of bacteria – phytoplankton coupling in the Southern Ocean. Limnol Oceanogr 50:1844–1854

Feller G, Gerday C (2003) Psychrophilic enzymes: hot topics in cold adaptation. Nat Rev Microbiol 1:200–208

Feller G, Zekhnini Z, Lamotte-Brasseur J, Gerday C (1997) Enzymes from cold-adapted microorganisms. The class C β-lactamase from the Antarctic psychrophile *Psychrobacter immobilis* A5. Eur J Biochem 244:186–191

Fenchel T, Finlay BJ (2004) The ubiquity of small species: patterns of local and global diversity. Bioscience 54:777–784

Ferrer M, Golyshina O, Beloqui A, Golyshin PN (2007) Mining enzymes from extreme environments. Curr Opin Microbiol 10:207–214

Fields PA (2001) Protein function at thermal extremes: balancing stability and flexibility. Comp Biochem Physiol A 129:417–431

Finlay BJ (2002) Global dispersal of free-living microbial eukaryote species. Science 296:1061–1063

Finlay BJ, Clarke KJ (1999) Ubiquitous dispersal of microbial species. Nature 400:828

Finlay BJ, Esteban GF, Olmo JL, Tyler PA (1999) Global distribution of free-living microbial species. Ecography 22:138–144

Franzmann PD, Liu Y, Balkwill DL, Aldrich HC, De Macario EC, Boone DR (1997) *Methanogenium frigidum* sp. nov., a psychrophilic, H_2-using methanogen from Ace Lake, Antarctica. Int J Syst Bacteriol 47:1068–1072

Giaquinto L, Curmi PMG, Siddiqui KS, Poljak A, DeLong F, DasSarma S, Cavicchioli R (2007) The structure and function of cold shock proteins in archaea. J Bacteriol 189:5738–5748

Gilichinsky D, Wagener S, Vishnevetskaya T (1995) Permafrost microbiology. Permafrost Periglacial Process 6:281–291

Gocheva YG, Tosi S, Krumova ET, Slokoska LS, Miteva JG, Vassilev SV, Angelova MB (2009) Temperature downshift induces antioxidant response in fungi isolated from Antarctica. Extremophiles 13:273–281

Herbert RA (1986) The ecology and physiology of psychrophilic microorganisms. In: Herbert RA, Codd GA (eds) Microbes in extreme environments. The Society for General Microbiology, Academic Press, London, pp 1–24

Herbert RA (1989) Microbial growth at low temperature. In: Gould GW (ed) Mechanisms of action of food preservation procedures. Elsevier Applied Science, London, pp 71–96

Hirsch P, Gallikowski CA, Siebert J, Peissl K, Kroppenstedt R, Schumann P, Stackebrandt E, Anderson R (2004a) *Deinococcus frigens* sp. nov., *Deinococcus saxicola* sp. nov., and *Deinococcus marmoris* sp. nov., low temperature and drought-tolerating, UV resistant bacteria from continental Antarctica. Syst Appl Microbiol 27:636–645

Hirsch P, Mevs U, Kroppenstedt RM, Schumann P, Stackebrandt E (2004b) Cryptoendolithic actinomycetes from Antarctic sandstone rock samples: *Micromonospora endolithica* sp. nov. and two isolates related to *Micromonospora coerulea* Jensen 1932. Syst Appl Microbiol 27:166–174

Hochachka PW, Somero GN (2002) Biochemical adaptation: mechanism and process in physiological evolution. Oxford University Press, New York, USA

Hua MX, Chi Z, Liu GL, Buzdar MA, Chi ZM (2010) Production of a novel and cold-active killer toxin by *Mrakia frigida* 2E00797 isolated from sea sediment in Antarctica. Extremophiles 14:515–521

Hughes J, Smith HG (1989) Temperature relations of *Heteromita globosa* Stein in Signy Island fellfields. In: Heywood RB (ed) Proceedings of British Antarctic Survey Antarctic Special Topic Award Scheme Symposium, 9–10 November 1988, University Research in Antarctica. British Antarctic Survey, Natural Environment Research Council, Cambridge, pp 117–122

Hughes KA, McCartney HA, Lachlan-Cope TA, Pearce DA (2004) A preliminary study of airborne biodiversity over peninsular Antarctica. Cell Mol Biol 50:537–542

Jung DO, Achenbach LA, Karr EA, Takaichi S, Madigan MT (2004) A gas vesiculate planktonic strain of the purple non-sulfur bacterium *Rhodoferax antarcticus* isolated from Lake Fryxell, Dry Valleys, Antarctica. Arch Microbiol 182:236–243

Karl DM, Bird DF, Björkman K, Houlihan T, Shackelford R, Tupas L (1999) Microorganisms in the accreted ice of Lake Vostok, Antarctica. Science 286:2144–2147

Kawamoto J, Kurihara T, Kitagawa M, Kato I, Esaki N (2007) Proteomic studies of an Antarctic cold-adapted bacterium, *Shewanella livingstonensis* Ac10, for global identification of cold-inducible proteins. Extremophiles 11:819–826

Labrenz M, Lawson PA, Tindall BJ, Hirsch P (2009) *Roseibaca ekhonensis* gen. nov., sp. nov., an alkalitolerant and aerobic bacteriochlorophyll a-producing alphaproteobacterium from hypersaline Ekho Lake. Int J Syst Evol Microbiol 59:1935–1940

Lanoil B, Skidmore M, Priscu JC, Han S, Foo W, Vogel SW, Tulaczyk S, Engelhardt H (2009) Bacteria beneath the West Antarctic Ice Sheet. Environ Microbiol 11:609–615

Lawley B, Ripley S, Bridge P, Convey P (2004) Molecular analysis of geographic patterns of eukaryotic diversity in Antarctic soils. Appl Environ Microbiol 79:5963–5972

Laybourn-Parry J (2002) Survival mechanisms in Antarctic lakes. Philos Trans R Soc Lond B Biol Sci 357(1423):863–869

Laybourn-Parry J (2009) No place too cold. Science 324:1521–1522

Laybourn-Parry J, Pearce DA (2007) The biodiversity and ecology of Antarctic lakes – models for evolution. Philos Trans R Soc Lond B Biol Sci 362(1488):2273–2289

Lee CC, Fenchel T (1972) Studies on ciliates associated with sea ice from Antarctica. II. Temperature responses and tolerances in ciliates from Antarctic, temperate and tropical habitats. Arch Protistenk 114:237–244

Li S, Xiao X, Yin X, Wang F (2006) Bacterial community along a historic lake sediment core of Ardley Island, west Antarctica. Extremophiles 10:461–467

Liu S, Zhang P, Cong B, Liu C, Lin X, Shen J, Huang X (2010) Molecular cloning and expression analysis of a cytosolic Hsp70 gene from Antarctic ice algae *Chlamydomonas* sp. ICE-L. Extremophiles 14:329–337

Mavromatis K, Feller G, Kokkinidis M, Bouriotis V (2003) Cold adaptation of a psychrophilic chitinase: a mutagenesis study. Protein Eng 16:497–503

Médigue C, Krin E, Pascal G, Barbe V, Bernsel A, Bertin PN, Cheung F, Cruveiller S, D'Amico S, Duilio A, Fang G, Feller G, Ho C, Mangenot S, Marino G, Nilsson J, Parrilli E, Rocha EP, Rouy Z, Sekowska A, Tutino ML, Vallenet D, von Heijne G, Danchin A (2005) Coping with cold: the genome of the versatile marine Antarctica bacterium *Pseudoalteromonas haloplanktis* TAC125. Genome Res 15:1325–1335

Mikucki JA, Priscu JC (2007) Bacterial diversity associated with Blood Falls, a subglacial outflow from the Taylor Glacier, Antarctica. Appl Environ Microbiol 73:4029–4039

Möhlmann DTF (2003) Unfrozen subsurface water on Mars: presence and implications. Int J Astrobiol 2:213–216

Montes MJ, Mercade E, Bozal N, Guinea J (2004) *Paenibacillus antarcticus* sp. nov., a novel psychrotolerant organism from the Antarctic environment. Int J Syst Evol Microbiol 54:1521–1526

Morgan-Kiss RM, Priscu JC, Pocock T, Gudynaite-Savitch L, Huner NP (2006) Adaptation and acclimation of photosynthetic microorganisms to permanently cold environments. Microbiol Mol Biol Rev 70:222–252

Naganuma T, Hua PN, Okamoto T, Ban S, Imura S, Kanda H (2005) Depth distribution of euryhaline halophilic bacteria in Suribati Ike, a meromictic lake in East Antarctica. Polar Biol 28:964–970

Newsham KK, Pearce DA, Bridge PD (2010) Minimal influence of water and nutrient content on the bacterial community composition of a maritime Antarctic soil. Microbiol Res 165:523–530

Nogi Y, Kato C, Horikoshi K (1998) Taxonomic studies of deep-sea barophilic *Shewanella* strains and description of *Shewanella violacea* sp. nov. Arch Microbiol 170:331–338

Olson JB, Steppe TF, Litaker RW, Paerl HW (1998) N_2 fixing microbial consortia associated with the ice cover of Lake Bonney, Antarctica. Microb Ecol 36:231–238

Pearce DA (2005) The structure and stability of the bacterioplankton community in Antarctic freshwater lakes, subject to extremely rapid environmental change. FEMS Microbiol Ecol 53:61–72

Pearce DA (2008) Climate change and the microbiology of the Antarctic Peninsula region. Sci Prog 91:203–217

Pearce DA (2009) Antarctic subglacial lakes – a new frontier in microbial ecology. ISME J 3:877–880

Pearce DA, Wilson WH (2003) Viruses in Antarctic ecosystems. Antarctic Sci 15:319–331

Pearce DA, Bridge PD, Hughes K, Sattler B, Psenner R, Russell NJ (2009) Microorganisms in the atmosphere over Antarctica. FEMS Microbiol Ecol 69:143–157

Pearce DA, Hughes KA, Harangozo SA, Lachlan-Cope TA, Jones AE (2010) Biodiversity of air-borne microorganisms at Halley station, Antarctica. Extremophiles 14:145–159

Pocock T, Lachance MA, Proschold T, Priscu JC, Kim SS, Huner NPA (2004) Identification of a psychrophilic green alga from Lake Bonney Antarctica: *Chlamydomonas raudensis* Ettl. (UWO 241) *Chlorophyceae*. J Phycol 40:1138–1148

Poglazova MN, Mitskevich IN, Abyzov SS, Ivanov MV (2001) Microbiological characterization of the accreted ice of subglacial Lake Vostok, Antarctica. Microbiology 70:723–730

Pointing SB, Chan Y Lacap DC, Lau MCY, Jurgens JA, Farrell RL (2009) Highly specialized microbial diversity in hyper-arid polar desert. Proc Nat Acad Sci USA 106:19964–19969

Poli A, Esposito E, Orlando P, Lama L, Giordano A, de Appolonia F, Nicolaus B, Gambacorta A (2006) *Halomonas alkaliantarctica* sp. nov., isolated from saline lake Cape Russell in Antarctica, an alkalophilic moderately halophilic, exopolysaccharide-producing bacterium. Syst Appl Microbiol 30:31–38

Pommier T, Canback B, Riemann L, Bostrom KH, Simu K, Lundberg P, Unlid A, Hagström (2007) Global patterns of diversity and community structure in marine bacterioplankton. Mol Ecol 16:867–880

Prescott GW (1978) How to know the freshwater algae, 3rd edn. Wm C Brown, Dubuque, IA, USA

Price PB (2000) A habitat for psychrophiles in deep Antarctic ice. Proc Nat Acad Sci USA 97:1247–1251

Price PB (2006) Microbial life in glacial ice and implications for a cold origin of life. FEMS Microbiol Ecol 59:217–231

Price PB, Nagornov OV, Bay R, Chirkin D, He Y, Miocinovic P, Richards A, Woschnagg K, Koci B, Zagorodnov V (2002) Temperature profile for glacial ice at the South Pole: implications for life in a nearby subglacial lake. Proc Natl Acad Sci USA 99:7844–7847

Priscu JC, Adams EE, Lyons WB, Voytek MA, Mogk DW, Brown RL, McKay CP, Takacs CD, Welch KA, Wolf CF, Kirshtein JD, Avci R (1999) Geomicrobiology of subglacial ice above Lake Vostok, Antarctica. Science 286:2141–2144

Priscu JC, Kennicutt II, MC, Bell RE, Bulat SA, Ellis-Evans JC, Lukin VV, Petit J-R, Powell RD, Siegert MJ, Tabacco I (2005) Exploring subglacial Antarctic Lake environments. Am Geophys Union EOS Trans 86:193–200

Rakusa-Suszczewski S (1980) The role of near-shore research in gaining and understanding of the functioning of the Antarctic ecosystem. Pol Arch Hydrobiol 27:229–233

Ravenschlag K, Sahm K, Pernthaler J, Amann R (1999) High bacterial diversity in permanently cold marine sediments. Appl Environ Microbiol 65:3982–3989

Reddy GS, Prakash JS, Vairamani M, Prabhakar S, Matsumoto GI, Shivaji S (2002) *Planococcus antarcticus* and *Planococcus psychrophilus* spp. nov. isolated from cyanobacterial mat samples collected from ponds in Antarctica. Extremophiles 6:253–261

Reddy GS, Raghavan PU, Sarita NB, Prakash JS, Nagesh N, Delille D, Shivaji S (2003) *Halomonas glaciei* sp. nov. isolated from fast ice of Adelie Land, Antarctica. Extremophiles 7:55–61

Reid IN, Sparks WB, Lubow S, McGrath M, Livio M, Valenti J, Sowers KR, Shukla HD, MacAuley S, Miller T, Suvanasuthi R, Belas R, Colman A, Robb FT, DasSarma P, Müller JA, Coker JA, Cavicchioli R, Chen F, DasSarma S (2006) Terrestrial models for extraterrestrial life: methanogens and halophiles at Martian temperatures. Int J Astrobiol 5:89–97

Sattler B, Puxbaum H, Psenner R (2001) Bacterial growth in supercooled cloud droplets. Geophys Res Lett 28:239–242

Saunders NF, Thomas T, Curmi PM, Mattick JS, Kuczek E, Slade R, Davis J, Franzmann PD, Boone D, Rusterholtz K, Feldman R, Gates C, Bench S, Sowers K, Kadner K, Aerts A, Dehal P, Detter C, Glavina T, Lucas S, Richardson P, Larimer F, Hauser L, Land M, Cavicchioli R (2003) Mechanisms of thermal adaptation revealed from the genomes of the Antarctic Archaea *Methanogenium frigidum* and *Methanococcoides burtonii*. Genome Res 13:1580–1588

Säwström C, Anesio MA, Granéli W, Laybourn-Parry J (2007) Seasonal viral loop dynamics in two large ultraoligotrophic Antarctic freshwater lakes. Microb Ecol 53:1–11

Schiraldi C, De Rosa M (2002) The production of biocatalysts and biomolecules from extremophiles. Trends Biotechnol 20:515–521

Sharma A, Scott JH, Cody GD, Fogel ML, Hazen RM, Hemley RJ, Huntress WT (2002) Microbial activity at gigapascal pressures. Science 295:1514–1516

Sharp M, Parks J, Cragg B, Fairchild I, Lamb H, Tranter M (1999) Widespread bacterial population at glacier beds and their relationship to rock weathering and carbon cycling. Geology 27:107–110

Sheridan PP, Loveland-Curtze J, Miteva VI, Brenchley JE (2003) *Rhodoglobus vestalii* gen. nov. sp. nov., a novel psychrophilic organism isolated from an Antarctic Dry Valley Lake. Int J Syst Evol Microbiol 53:985–994

Shivaji S, Reddy GSN, Raghavan PUM, Sarita NB, Delille D (2004) *Psychrobacter salsus* sp. nov. and *Psychrobacter adeliensis* sp. nov. isolated from fast ice from Adelie Land, Antarctica. Syst Appl Microbiol 27:628–635

Siddiqui KS, Cavicchioli R (2006) Cold adapted enzymes. Annu Rev Biochem 75:403–433

Siegert MJ, Tranter M, Ellis-Evans JC, Priscu JC, Berry Lyons W (2003) The hydrochemistry of Lake Vostok and the potential for life in Antarctic subglacial lakes. Hydrol Process 17:795–814

Sjöling S, Cowan DA (2003) High 16S rDNA bacteria diversity in glacial meltwater lake sediment, Bratina island, Antarctica. Extremophiles 7:275–282

Smith JJ, Ah Tow L, Stafford W, Cary C, Cowan DA (2006) Bacterial diversity in three different Antarctic cold desert mineral soils. Microb Ecol 51:413–421

Sonjak S, Frisvad JC, Gunde-Cimerman N (2006) *Penicillium* mycobiota in Arctic subglacial ice. Microb Ecol 52:207–216

Spring S, Merkhoffer B, Weiss N, Kroppenstedt RM, Hippe H, Stackebrandt E (2003) Characterization of novel psychrophilic clostridia from an Antarctic microbial mat: description of *Clostridium frigoris* sp. nov., *Clostridium lacusfryxellense* sp. nov., *Clostridium bowmanii* sp. nov. and *Clostridium psychrophilum* sp. nov. and reclassification of *Clostridium laramiense* as *Clostridium estertheticum* subsp. *laramiense* subsp. nov. Int J Syst Evol Microbiol 53:1019–1029

Stingl U, Cho J-C, Foo W, Vergin KL, Lanoil B, Giovannoni SJ (2008) Dilution-to-extinction culturing of psychrotolerant planktonic bacteria from permanently ice-covered lakes in the McMurdo Dry Valleys, Antarctica. Microb Ecol 55:395–405

Tatur A (2002) Ornithogenic ecosystems in the Maritime Antarctic – formation, development and disintegration. In: Beyer L, Bölter M (eds) Geoecology of Antarctic ice-free coastal landscapes. Series ecological studies, vol 154. Springer, Berlin, Heidelberg, pp 161–184

Thomas DN, Dieckmann GS (2002) Antarctic sea ice – a habitat for extremophiles. Science 295:641–644

Tindall B (2004) Prokaryotic diversity in the Antarctic: the tip of the iceberg. Microb Ecol 47:271–283

Ting L, Williams TJ, Cowley MJ, Lauro FM, Guilhaus M, Raftery MJ, Cavicchioli R (2010) Cold adaptation in the marine bacterium, *Sphingopyxis alaskensis* assessed using quantitative proteomics. Environ Microbiol 12:2658–2676

Turner J, King JC, Lachlan-Cope TA, Jones PD (2002) Recent temperature trends in the Antarctic. Nature 418:291–292

Van Trappen S, Mergaert J, Eygen SV, Dawyndt P, Cnockaert MC, Swings J (2002) Diversity of 746 heterotrophic bacteria isolated from microbial mats from ten Antarctic lakes. Syst Appl Microbiol 25:603–610

Van Trappen S, Mergaert J, Swings J (2003) *Flavobacterium gelidilacus* sp. nov., isolated from microbial mats in Antarctic lakes. Int J Syst Evol Microbiol 53:1241–1245

Van Trappen S, Vandecandelaere I, Mergaert J, Swings J (2004a) *Algoriphagus antarcticus* sp. nov., a novel psychrophile from microbial mats in Antarctic lakes. Int J Syst Evol Microbiol 54:1969–1973

Van Trappen S, Mergaert J, Swings J (2004b) *Lokanella salsilacus* gen. nov., sp. nov., *Lokanella fryxellensis* sp. nov. and *Lokanella vestfoldensis* sp. nov., new members of the *Rhodobacter* group, isolated from microbial mats in Antarctic lakes. Int J Syst Evol Microbiol 54:1263–1269

Van Trappen S, Vandecandelaere I, Mergaert J, Swings J (2004c) *Flavobacterium degerlachei* sp. nov., *Flavobacterium frigoris* sp. nov. and *Flavobacterium micromati* sp. nov., novel psychrophilic bacteria isolated from microbial mats in Antarctic lakes. Int J Syst Evol Microbiol 54:85–92

Van Trappen S, Vandecandelaere I, Mergaert J, Swings J (2005) *Flavobacterium fryxellicola* sp. nov. and *Flavobacterium psychrolimnae* sp. nov., novel psychrophilic bacteria isolated from microbial mats in Antarctic lakes. Int J Syst Evol Microbiol 55:769–772

Vishnivetskaya TA, Siletzky R, Jefferies N, Tiedje JM, Kathariou S (2007) Effect of low temperature and culture media on the growth and freeze – thawing tolerance of *Exiguobacterium* strains. Cryobiology 54:234–240

Willerslev E, Hansen A, Christensen B, Steffensen J, Arctander P (1999) Diversity of Holocene life forms in fossil glacier ice. Proc Natl Acad Sci USA 96:8017–8021

Wilson WH, Lane D, Pearce DA, Ellis-Evans JC (2000) Transmission electron microscope analysis of virus-like particles in the freshwater lakes of Signy Island, Antarctica. Polar Biol 23:657–660

Yergeau E, Schoondermark-Stolk SA, Brodie EL, Déjean S, DeSantis TZ, Gonçalves O, Piceno YM, Andersen GL, Kowalchuk GA (2008) Environmental microarray analyses of Antarctic soil microbial communities. ISME J 3:340–351

Yi H, Oh HM, Lee JH, Kim SJ, Chun J (2005a) *Flavobacterium antarcticum* sp. nov., a novel psychrotolerant bacterium isolated from the Antarctic. Int J Syst Evol Microbiol 55:637–641

Yi H, Yoon HI, Chun J (2005b) *Sejongia antarctica* gen. nov., sp. nov. and *Sejongia jeonii* sp. nov., isolated from the Antarctic. Int J Syst Evol Microbiol 55:409–416

Yoshimune K, Galkin A, Kulakova L, Yoshimura T, Esaki N (2005) Cold-active DnaK of an Antarctic psychrotroph *Shewanella* sp. Ac10 supporting the growth of dnaK-null mutant of *Escherichia coli* at cold temperatures. Extremophiles 9:145–150

Yu Y, Xin YH, Liu HC, Chen B, Sheng J, Chi ZM, Zhou PJ, Zhang DC (2008) *Sporosarcina antarctica* sp. nov., a psychrophilic bacterium isolated from the Antarctic. Int J Syst Evol Microbiol 58:2114–2117

Anhydrobiotic rock-inhabiting cyanobacteria: potential for astrobiology and biotechnology*

Daniela Billi

Department of Biology, University of Rome "Tor Vergata", Via della Ricerca Scientifica, Rome, Italy

1 Introduction

Deciphering how microorganisms can adapt to what we consider, in an anthropocentric way, extreme, is not only challenging intellectually, but also an issue of intense social and commercial interest. The metabolism and physiology of extremophiles have such peculiar features as to be fascinating per se; however, their commercial potential, albeit long recognized, is far from being fully realized. Discovering the extremes at which life can occur has made more plausible the search for life on other planets, with many more discoveries likely to come due to improvements in exploration and analytical technology (Rothschild and Mancinelli 2001). The International Space Station provides a unique opportunity in establishing the limits of endurance of life as we know it; results of ongoing research will provide insights into the potential of life to survive beyond Earth (Rabbow et al. 2009).

In cold and hot deserts, such as the McMurdo Dry Valleys in Antarctica and the Atacama Desert in Chile, both considered the Earth's nearest equivalent to the Martian environment, life is pushed to its physical limits due to extreme water deficit and/or freezing temperatures. In these places organisms escaped from prohibitive external conditions by colonizing rocks, the last refuge for life: Such are the photosynthesis-based lithic communities. The discovery of these communities sheds light on the possible history of Martian microbial life, if it ever existed, and has led to identifying rock-inhabiting cyanobacteria as pioneers in the colonization of the ultimate desert – Mars (Grilli Caiola and Billi 2007).

---

*In memoriam of Imre Friedmann and Roseli Ocampo-Friedmann.

2 Cyanobacteria in hot and cold desert rocks

Lithobionts (from the Greek, lithos: rock; bios: life) are mainly microbionts colonizing rock surfaces and rock interiors in a wide variety of environments. In dry environments, such as hot and cold deserts, lithic ecosystems harbor much of the extant life. Lithobionts are distinguished according to the location in respect to substrate and functional criteria: epilithics dwelling on the rock surface; hypolithics forming biofilms at the stone–soil interface; endolithics colonizing microscopic fissures (chasmoendoliths) and structural cavities (cryptoendoliths) of rocks; and finally euendoliths actively boring into rocks (Golubic et al. 1981). In hot and cold deserts, where life is pushed to its limits, members of the genus *Chroococcidiopsis* colonize porous rock or the stone–soil interface (Fig. 1a, b). The endurance of *Chroococcidiopsis* in the Dry Valleys in Antarctica as well as in the Atacama Desert in Chile, which are both considered terrestrial analogs of the two environmental extremes on Mars – cold and dryness

Fig. 1. Examples of photosynthesis-based lithic communities. (**a**) Hypolithics; (**b**) endolithics (photos: E. Imre Friedmann); (**c**) *Chroococcidiopsis* sp. isolated from endolithic growth in Nubian sandstone; (**d**) endoevaporitic growth (photo: Nunzia Stivaletta and Roberto Barbieri)

(Warren-Rhodes et al. 2006) – is remarkable: The photosynthesis-based lithic communities found offer unique model systems for microbial ecology, geobiology, and astrobiology (Walker and Norman 2007).

Research into photosynthesis-based lithic communities was pioneered by E. Imre Friedmann and Roseli Ocampo-Friedmann, who first described the *Chroococcidiopsis*-dominated lithic communities in the Negev Desert, Israel (Friedmann et al. 1967). Later research was extended to hot and cold deserts worldwide, leading to the establishment of the Culture Collection of Microorganisms from Extreme Environments (CCMEE); currently about 250 desert strains of *Chroococcidiopsis* and a few related genera are maintained by this author at the University of Rome, Tor Vergata.

Members of the genus *Chroococcidiopsis* have a developmental cycle in which a spherical cell (called baeocyte) undergoes repeated binary fissions, yielding aggregates with only a few cells; thereafter, multiple fission occurs in almost all of the cells within an aggregate, followed by the release of numerous baeocytes (Fig. 1c). The phylogenetic relationships within the *Chroococcidiopsis* lineage remain to be resolved. Due to the high sequence divergences of 16S rRNA genes it has been suggested that some forms could be even regarded as different species or genera (Fewer et al. 2002). Based on the sequencing of 16S rRNA genes it was reported that four desert isolates of *Chroococcidiopsis* were divergent from other pleurocapsalean representatives (Billi et al. 2001). The extension of the phylogenetic analysis to 12 desert strains of *Chroococcidiopsis* revealed their clustering into different groups (Billi and Wilmotte, unpublished).

The Dry Valleys are the largest ice-free region on the Antarctic continent and were considered to be virtually sterile (Horowitz et al. 1972). Then came the discovery of the "lichen-dominated communities," formed by fungi, algae, bacteria, and cyanobacteria of the genera *Chroococcidiopsis* and *Gloeocapsa*, in sedimentary rocks (Friedmann and Ocampo-Friedmann 1976). As a consequence of culture-independent analysis based on molecular methods it is now well recognized that the bacterial diversity and abundance in the Dry Valleys is not as low as first assumed (Cary et al. 2010). Recently, a culture-independent survey of the microbial biodiversity in the McKelvey Valley showed that the greatest diversity occurred in endolithic and chasmolithic communities in sandstone, where *Chroococcidiopsis* was dominant. The soil, on the other hand, was relatively impoverished and lacked a significant photoautotrophic component, except for isolated islands of hypolithic cyanobacterial colonization on quartz rocks (Pointing et al. 2009). However, the finding of lichen-dominated communities in the Dry Valleys raised new hopes of finding life on Mars at a time when the Viking missions had shown the Martian soil to be lifeless and depleted in organic material. Since then considerable interest has been aroused concerning rock-inhabiting communities, in the contest of identifying signature for past or present life forms Mars and to investigate the survival potential of terrestrial microbial life on Mars of present days (Cockell et al. 2005). It has been speculated that before their extinction Martian microbes, if they ever existed, could have escaped the adverse environmental conditions during the cooling of the planet by withdrawing into the rocks (Friedmannn 1986).

In hot deserts life must adapt to severe stress due to sudden changes between warm/humid and hot/dry, and this is thought to be the reason for the exclusion of eukaryotic organisms from endolithic communities in extremely dry deserts. It was reported that along an environmental gradient of hot and cold deserts in China the diversity of hypolithic communities was affected by the availability of liquid water rather than by temperature, *Chroococcidiopsis* being ubiquitous in different thermal and moisture conditions, and dominant in the most arid sites (Pointing et al. 2007). In the effort to identify the absolute dry limit of life on Earth, the Atacama Desert in Chile has become one of the most relevant hot places. This desert is so dry that a hyperarid core has been identified: decades without rain were recorded, along with the virtual absence of heterotrophic bacteria (Navarro-González et al. 2003). In the Atacama Desert, hypolithic cyanobacterial communities occur along the aridity gradient, which reaches its limit at the hyperarid core. Here rare *Chroococcidiopsis*-based communities exist in small spatially isolated islands amidst a microbially impoverished soil (Warren-Rhodes et al. 2006). When applied to Mars, the Atacama model predicts that if microhabitats exist (or have existed) on Mars, they will be difficult to detect, being dispersed in virtually lifeless surroundings (Warren-Rhodes et al. 2006). Remarkably, in the hyperarid zone of the Atacama, cyanobacteria of the genus *Chroococcidiopsis* colonize a peculiar habitat provided by halite deposits (Wierzchos et al. 2006). Such a finding suggests this cyanobacterium can achieve metabolic activity during periods of moisture availability, by adsorbing moisture from halite deliquescence, e.g., the absorption of moisture from the atmosphere (Davila et al. 2008). Recently, environmental 16S rRNA gene sequences obtained from endolithic growth in halite rock collected from the Atacama Desert (Fig. 1d) identified a cyanobacterial sequence closely related to hot desert strains of *Chroococcidiopsis* isolated from geographically distinct desert areas (Billi, Stivaletta and Barbieri, unpublished). Photosynthetic microorganisms within dry evaporate rocks have been considered as relevant models in the search for life within our Solar System (Wierzchos et al. 2006). Mars has been suggested as an intriguing location to search for halophiles (or their remnants) outside Earth, as it may have been a wetter and warmer place in the past and recent data suggest the presence of halite on Mars (Leuko et al. 2010).

3 Cyanobacterial adaptation to Earth's deserts

In extremely dry environments, such as the ice-free Ross desert or hyperarid hot deserts, rock-inhabiting communities of *Chroococcidiopsis* come into contact with water for only a few hours per year; thus they persist in an ametabolic dry and/or frozen state for the greater part of their life (Friedmann et al. 1993; Warren-Rhodes et al. 2006). How they manage to survive is still a mystery.

Water is essential for life: Only a small but taxonomically diverse group of organisms can withstand desiccation by entering into a state of suspended animation, a phenomenon known as anhydrobiosis (from the Greek "life without water") (Van Leeuwenhoek 1702). Anhydrobiotic cyanobacteria can withstand desiccation without differentiating into any specialized cell types, as other anhydrobiotes do, e.g., certain bacteria, lichens, mosses, ferns, certain angiosperm genera (known as "resurrection plants"), representatives of rotifers, tardigrades, and nematode taxa. Other anhydrobiotes produce stage-specific anhydrobiotic forms, e.g., the spores of some bacteria and fungi, the embryonic cysts of brine shrimps, larvae of certain insects (chironomids), and some plant seeds and pollen (Alpert 2006). Anhydrobiotic cyanobacteria such as *Chroococcidiopsis* spp. and *Nostoc commune* are of particular significance since they need to prevent oxidative damage exacerbated by oxygenic photosynthesis (Billi and Potts 2000, 2002). Despite the interest in comprehending such a peculiar phenomenon, the mechanisms underlying anhydrobiosis have not been fully understood in any organism (Alpert 2006). Adaptation to desiccation has the singular distinction that dried cells enter full metabolic arrest. This raises intriguing questions as to the role of this function opposed to adaptation to extremes of pH, temperature, or pressure, once it is assumed that evolution is driven toward optimum function rather than maximum stability (Potts et al. 2005). In desiccation-sensitive cells the complete removal of water is lethal due to the damage induced at every level of cellular organization. The removal of the hydration shell from phospholipids of membrane bilayers increases the van der Waal's interactions between adjacent lipids, causing an increase in the phase transition temperature of membranes, and their transition to the gel phase at environmentally relevant temperatures (Crowe 2007). Oxidative damage to proteins, lipids, and nucleic acids is induced when the production of reactive oxygen species (ROS) due to dysfunction in enzymes and/or electron transport chains exceeds the antioxidant system (França et al. 2007). Other damage to proteins and nucleic acids results from metal-catalyzed Haber–Weiss and Fenton reactions as well as via the Maillard (browning) reaction (Potts et al. 2005).

The occurrence of live and dead cells in dried multicellular aggregates of *Chroococcidiopsis* is a feature of its response to desiccation that warrants emphasis. Cytological and ultrastructural studies carried out on samples dried for 5 years identified cells retaining typical ultrastructural features and others with varying degrees of degeneration (Grilli Caiola et al. 1993, 1996a). Subsequent investigations based on molecular probes used with fluorescence microscopy highlighted the fact that desiccation surviving *Chroococcidiopsis* avoids and/or limits genomic fragmentation and covalent modifications, preserves intact plasma membranes and autofluorescence of photosynthetic pigments, and undergoes a spatially reduced ROS accumulation (Billi 2009a). The co-occurrence of live and dead cells within a given dried aggregate of *Chroococcidiopsis* poses intriguing questions: Is cell death the outcome of a passive externally driven process? Or does it result from programmed cell death, thus further corroborating the idea that desiccation resistance is not a simple process.

The capability of dried cells of *Chroococcidiopsis* to avoid and/or reduce subcellular damage indicates that protection mechanisms are relevant in the process of desiccation resistance. However, other evidence indicates that an efficient repair DNA systems is relevant to the ability of *Chroococcidiopsis* to survive prolonged desiccation, a condition in which oxidative damage continues even in the absence of metabolic activity, or when dried cells experience additional environmental stressors. Actively growing *Chroococcidiopsis* cells can repair the genomic fragmentation induced by 5 kGy of ionizing radiation (Billi et al. 2000a) while DNA damage caused in dried cells by a simulated unattenuated Martian UV flux is repaired upon rehydration (Cockell et al. 2005).

The capability of desert strains of *Chroococcidiopsis* to repair genome fragmentation resembles that reported for *Deinococcus radiodurans* (Cox and Battista 2005). Since naturally occurring environments result in exposures exceeding 400 mGy per year, it has been proposed that radioresistance is a consequence of the adaptation to DNA-damaging conditions, such as desiccation (Cox and Battista 2005). However, unlike *D. radiodurans*, *Chroococcidiopsis* does not undergo genome fragmentation upon desiccation (Billi 2009a) as reported for *N. commune* (Shirkey et al. 2003). In dried cells of *Chroococcidiopsis* at least a subset of proteins, namely phycobilisomes and esterase, were protected against oxidative damage (Billi 2009a). This might have a bearing on the desiccation tolerance of *Chroococcidiopsis*, given that oxidative protein damage, but not DNA damage, has been proposed to determine bacterial survival of DNA-damaging conditions. Indeed, protein damage is thought to kill irradiation-sensitive bacteria after exposure to low doses of ionizing radiation which cause less than one DNA double-strand break per genome (Daly 2009).

How *Chroococcidiopsis* can manage to resist oxidative stress remains largely unknown. The accumulation of an iron superoxide dismutase has been observed in dried cells of *Chroococcidiopsis* (Grilli Caiola et al. 1996b) while an abundant, active iron superoxide dismutase has been described in dried *N. commune* (Shirkey et al.

2000). Recently, high Mn (II) contents have been correlated with the resistance of *D. radiodurans* to both ionizing radiation and desiccation (Daly 2009). A close correlation between high Mn (II) concentration, high levels of resistance to ionizing irradiation and low susceptibility to desiccation-induced protein oxidation have also been reported in bacteria isolated from desert environments (Fredrickson et al. 2008).

In the effort to decipher the structural, physiological, and molecular mechanisms of anhydrobiosis it is becoming clear that they are both numerous and highly diverse. A crucial structural mechanism in the adaptation of *Chroococcidiopsis* to anhydrobiosis is the production of abundant polysaccharide-rich envelopes (Grilli Caiola et al. 1996a). It has been proposed that extracellular polysaccharide (EPS) is a key component in cyanobacterial desiccation tolerance by providing a repository for water as well as a matrix which stabilizes desiccation-related enzymes and molecules (Wright et al. 2005). In anhydrobiotic cyanobacteria the EPS production might act in synergy with trehalose accumulation in the cytoplasm. This non-reducing disaccharide is produced in large amounts by most anhydrobiotes and, by substituting water molecules, prevents the phase transition of cellular membranes and stabilizes dried proteins (Crowe 2007). It is not yet known whether *Chroococcidiopsis* accumulates trehalose upon drying, although the involvement of this disaccharide in the desiccation resistance of *N. commune* has been reported (Shirkey et al. 2003).

Although hints on desiccation tolerance of *Chroococcidiopsis* suggest an interplay between protection and repair mechanisms, it is necessary to employ DNA microarray and proteomic tools to decipher the genetic basis of its anhydrobiotic adaptation. Remarkably *Chroococcidiopsis* is the only anhydrobiotic cyanobacterium suitable for genetic manipulation (Billi et al. 2001), for which gene activation was attempted (Billi 2009b). When the genome sequences of hot and cold desert strains of *Chroococcidiopsis*, namely CCMEE 029 from the Negev desert and CCMEE 134 from Beacon Valley (Antarctica), along with a Tibetan and Taklimakan isolate, will be completed (Billi and Pointing, unpublished), insights into the molecular aspects of desiccation tolerance are likely.

4 Survival of desert cyanobacteria beyond Earth

It is widely recognized that the capability of anhydrobiotes to withstand desiccation is often associated with an extraordinary resistance to other environmental stressors. Such a feature makes them ideal for investigating the survival potential of terrestrial organisms for outer space conditions or Mars, both of which require extreme tolerance to vacuum (imposing extreme dehydration), cold, and radiation (Baglioni et al. 2007).

Notably, spores of *Bacillus subtilis* survived space conditions in low Earth orbit for 6 years on the Long Duration Exposure Facility (Horneck 1993), while dried lichens and tardigrades survived in space for 2 weeks during the Biopan experiments (Sancho et al. 2007; Jönsson et al. 2008). Ground-based simulations of Martian and space conditions demonstrated the potential endurance of Antarctic rock-inhabiting fungi and desert strains of *Chroococcidiopsis* in extraterrestrial environments (Onofri et al. 2008; Billi et al. 2008). The mechanisms behind the tolerance of anhydrobiotes to outer space have not yet been revealed; however, they are considered to be a consequence of the resistance to desiccation and radiation (Jönsson et al. 2008).

In recognition of its environmental flexibility *Chroococcidiopsis* has long been identified as a photosynthetic model organism for space research (Friedmann and Ocampo-Friedmann 1995), at a time when NASA established the Astrobiology Program (Cockell 2002). It was proposed that *Chroococcidiopsis* could be used as a photosynthetic pioneer for Mars terraforming, if inoculated in proper desert pavements and periodically wetted (Friedmann and Ocampo-Friedmann 1995). The expected capability of *Chroococcidiopsis* to withstand environmental stressors not currently met in nature was corroborated by the finding that strains from hot and cold deserts survive doses of ionizing radiation as high as 15 kGy (Billi et al. 2000a). Remarkably, a monolayer of dried cells of *Chroococcidiopsis* survived 15-min exposure to an attenuated Martian UV flux (Cockell et al. 2005), thus proving more resistant than *B. subtilis* spores (Schuerger et al. 2003) and akinetes of *Anabaena cylindrica* (Olsson-Francis et al. 2009). It was also reported that the survival of dried cells of *Chroococcidiopsis* shielded under 1 mm of Martian soil simulant or gneiss was unaffected by 4 h of exposure to Martian UV radiation. The endurance of dried *Chroococcidiopsis* under simulated Martian UV radiation further supported its use in future approaches to mimic endolithic Martian exposure and allowed the speculation that it could survive and perhaps grow within lithic habitats in the presence of a source of liquid water and essential nutrients (Cockell et al. 2005). Dried cells of *Chroococcidiopsis* under a few millimeters of Antarctic sandstone were reported to withstand UV radiation corresponding to 1.5 years permanence in space (Billi et al. 2008; Billi et al. 2011). This corroborates the importance of lithic communities in the context of lithopanspermia, the transfer of living material inside rocks between planets (Nicholson 2009). Lithopanspermia is currently divided into three stages: (i) launch of microbe-bearing rocks from a donor planet into space; (ii) transit through space to a recipient planet; and (iii) entry into a recipient planet. Survival of launch pressures simulating the estimated values of Martian meteorites during escape was reported for spores of *B. subtilis* as well as the photobiont and mycobiont partners of the lichen *Xanthoria elegans* which, embedded in a Martian analog rock, encompassed shock pressures up to 40 GPa

(Horneck et al. 2008). By contrast, *Chroococcidiopsis* survived shock pressures ranging from 5 to 10 GPa. Therefore, given the low frequency of weakly shocked meteorites, the chances for interplanetary transport of cyanobacteria-type organisms seem reduced (Horneck et al. 2008). *Chroococcidiopsis* cells inoculated into rock to the depth at which they occur in nature, and mounted on the heat shield of a FOTON-M2 recoverable orbital capsule, did not survive the re-entry into the atmosphere (Cockell et al. 2007). In fact, during the atmospheric re-entry, the photosynthetic lithic community located at a depth at which light is available for photosynthesis, was heated to well above the upper temperature limit for life; this suggests that nonphotosynthetic organisms living deep within rocks have a better chance of surviving the exit and entry process (Cockell 2007).

5 Biotechnological exploitation of anhydrobiosis

Since the air-dried state is characterized by enhanced biostability there is considerable interest in the industrial applications of conferring desiccation tolerance to otherwise desiccation-sensitive cells. Logistical problems and costs associated with preservation and storage at ultra-low temperatures necessitate the development of novel technologies for air-dried pharmaceuticals, dried cell-based biosensors, and cell and tissue banks (Potts et al. 2005). With the understanding of the role played by the accumulation of trehalose and sucrose in the majority of anhydrobiotes, a whole field of research has been opened up. Embedding in trehalose or sucrose has been used successfully to stabilize dry membranes and enzymes, as well as intact bacterial cells (Crowe 2007). In view of the simplicity of the biosynthetic pathway of trehalose and sucrose the metabolic engineering of desiccation-sensitive cells has been attempted. The expression of a cyanobacterial *sps*A gene encoding for a sucrose-6-phosphate synthase in *Escherichia coli* led to a marked increase in survival after air drying, freeze-drying and chemical desiccation over phosphorus pentoxide (Billi et al. 2000b). However, when as an extension of this principle the *sps*A gene was expressed in human kidney cells, only a low percentage of the cells underwent growth; moreover, this occurred with significantly less vigor than in cells that had never been desiccated (Bloom et al. 2001). Thus, even if it is in principle possible to stabilize air-dried human cells, additional studies are required to fully optimize such a process. In striking contrast to the possibility of enabling prokaryotes and isolated cells to tolerate desiccation, there is a lack of success in achieving desiccation tolerance of whole, desiccation-sensitive, multicellular animals and plants (Alpert 2005). While trehalose loading could be used to usefully stabilize human blood platelets, the survival of human embryonic kidney cells to air drying was enhanced only by the synergetic action of trehalose loading and expression of a gene codifying for the stress protein p26 obtained from an anhydrobiotic organism (Crowe et al. 2005).

Mechanisms identified as relevant in the adaptation to desiccation of the cyanobacterium *N. commune* were exploited to stabilize desiccation-sensitive cells as well. It was first proved that the extracellular glycan of *N. commune* can be used to inhibit fusion of membrane vesicles during desiccation and freeze-drying (Hill et al. 1997). Indeed cyanobacteria produce EPS that are so varied in their gel and sol properties that they may actually have properties similar to those described for glass-forming polymers and thus allow the stabilization of cell membranes during periods of desiccation (Pereira et al. 2009). In fact, in the amorphous state molecular diffusion is reduced and uncontrolled reactions that would be disastrous over the prolonged desiccation are avoided (Crowe et al. 1998). Unlike bacterial EPS which contain less than four different monomers, cyanobacterial EPS are complex heteropolysaccharides composed of more than six different monosaccharides, whose alternative composition is relevant to a variety of industrial applications (Pereira et al. 2009). It was also reported that the overexpression in the aquatic desiccation-sensitive cyanobacterium *Anabaena* sp. PCC 7120 of the group 3 sigma factor *sigJ* gene obtained from the desiccation-tolerant cyanobacterium *Nostoc* sp. HK-01 resulted in an increased ESP production and acquisition of desiccation tolerance (Yoshimura et al. 2007). When subsequently the exploitation of the extracellular glycan of *N. commune* was attempted in order to dry and revive nucleated mammalian cells, the cells proved to be nonviable despite structural preservation after several weeks of desiccation (Bloom et al. 2001). The use of exogenous addition of water stress proteins, abound in the extracellular matrix of dried cells of *N. commune*, for air-dry stabilization of *E. coli* proved evidence that a two-step rehydration protocol must be followed to avoid deleterious effects of rapid rewetting (Potts et al. 2005).

Until now, the composition and properties of EPS produced by desert strains of *Chroococcidiopsis* remain unknown as do any changes in their proteome following desiccation. However, unlike *N. commune*, which is refractory to gene transfer, desert strains of *Chroococcidiopsis* have been identified as suitable to genetic manipulation by means of electroporation and conjugation (Billi et al. 2001). The identification of a pDU1-based plasmid capable of autonomous replication in *Chroococcidiopsis* along with the possibility of driving gene expression by using an inducible promoter of *E. coli* (Billi et al. 2001), will contribute to the biotechnological exploitation of this anhydrobiotic cyanobacterium. Air-dried cells of *Chroococcidiopsis* could be used to develop novel biosensors for detecting DNA-damaging conditions, which only requires water for re-activation. The ability to withstand prolonged desiccation and high doses of ionizing radiation make *Chroococcidiopsis* an effective candidate for biosensor fabrication, important performance criteria in this regard being robustness, resistance to environmental extremes, and portability. The exploitation of genetically modified cells of *Chroococcidiopsis* has real potential in space research:

Dried cells of *Chroococcidiopsis* could be used to develop an air-dried bank of a wide varieties of metabolites with useful applications within the framework of human space exploration.

Acknowledgments

This research was funded by the Italian Ministry of Foreign Affairs, Directorate General for Development Cooperation and by the Italian Space Agency.

References

Alpert P (2005) The limits and frontiers of desiccation-tolerant life. Integr Comp Biol 45:685–695
Alpert P (2006) Constraints of tolerance: why are desiccation-tolerant organisms so small or rare? J Exp Biol 209:1575–1584
Baglioni P, Sabbatini M, Horneck G (2007) Astrobiology experiments in low earth orbit – facilities, instrumentation and results. In: Rettberg P, Horneck G (eds) Complete course in astrobiology. Wiley-VCH, Berlin, pp 273–319
Billi D (2009a) Subcellular integrities in *Chroococcidiopsis* sp. CCMEE 029 survivors after prolonged desiccation revealed by molecular probes and genome stability assays. Extremophiles 13:49–57
Billi D (2009b) Loss of topological relationships in a Pleurocapsalean cyanobacterium (*Chroococcidiopsis* sp.) with partially inactivated *ftsZ*. Ann Microbiol 59:235–238
Billi D, Potts M (2000) Life without water: responses of prokaryotes to desiccation. In: Storey KB, Storey JM (eds) Cell and molecular response to stress: environmental stressors and gene responses. Elsevier Science, Amsterdam, pp 181–192
Billi D, Potts M (2002) Life and death of dried prokaryotes. Res Microbiol 153:7–12
Billi D, Friedmann EI, Hofer KG, Grilli Caiola M, Ocampo-Friedmann R (2000a) Ionizing-radiation resistance in the desiccation-tolerant cyanobacterium *Chroococcidiopsis*. Appl Environ Microbiol 66:1489–1492
Billi D, Wright DJ, Helm RF, Prickett T, Potts M, Crowe JH (2000b) Engineering desiccation tolerance in *Escherichia coli*. Appl Environ Microbiol 66:1680–1684
Billi D, Friedmann EI, Helm RF, Potts M (2001) Gene transfer to the desiccation-tolerant cyanobacterium *Chroococcidiopsis*. J Bacteriol 183:2298–2305
Billi D, Ghelardini P, Onofri S, Cockell CS, Rabbow E, Horneck G (2008) Desert Cyanobacteria under simulated space and Martian conditions. In: European Planetary Science Congress, Abstr EPSC2008-A-00474
Billi D, Viaggiu E, Cockell CS, Rabbow E, Horneck G, Onofri S (2011) Damage escape and repair in dried *Chroococcidiopsis* spp. from hot and cold deserts exposed to simulated space and Martian conditions. Astrobiology 11:65–73.
Bloom F, Price RP, Lao GF, Xia JL, Crowe JH, Battista JR, Helm RF, Slaughter S, Potts M (2001) Engineering mammalian cells for solid-state sensor applications. Biosens Bioelectron 16:603–608
Cary SC, McDonald IR, Barrett JE, Cowan DA (2010) On the rocks: the microbiology of Antarctic Dry Valley soils. Nat Rev Microbiol 8:129–138
Cockell CS (2002) Astrobiology – a new opportunity for interdisciplinary thinking. Space Policy 18:263–266

Cockell CS (2007) The interplanetary exchange of photosynthesis. Orig Life Evol Biosph 38:87–104

Cockell CS, Schuerger AC, Billi D, Friedmann EI, Panitz C (2005) Effects of a simulated Martian UV flux on the cyanobacterium, *Chroococcidiopsis* sp. 029. Astrobiology 5:127–140

Cockell CS, Brack A, Wynn-Williams DD, Baglioni P, Brandstatter F, Demets R, Edwards HG, Gronstal AL, Kurat G, Lee P, Osinski GR, Pearce DA, Pillinger JM, Roten CA, Sancisi-Frey S (2007) Interplanetary transfer of photosynthesis: an experimental demonstration of a selective dispersal filter in planetary island biogeography. Astrobiology 7:1–9

Cox MM, Battista JR (2005) *Deinococcus radiodurans* – the consummate survivor. Nat Rev Microbiol 3:882–892

Crowe JH (2007) Trehalose as a "chemical chaperone": fact and fantasy. Adv Exp Med Biol 94:143–158

Crowe JH, Carpenter JF, Crowe LM (1998) The role of vitrification in anhydrobiosis. Annu Rev Physiol 60:73–103

Crowe JH, Crowe LM, Wolkers WF, Oliver AE, Ma X, Auh J-H, Tang M, Zhu S, Norris J, Tablin F (2005) Stabilization of dry mammalian cells: lessons from nature. Integr Comp Biol 45:810–820

Daly MJ (2009) A new perspective on radiation resistance based on *Deinococcus radiodurans*. Nat Rev Microbiol 7:237–245

Davila AF, Gómez-Silva B, de los Rios A, Ascaso C, Olivares H, McKay CP, Wierzchos J (2008) Facilitation of endolithic microbial survival in the hyperarid core of the Atacama Desert by mineral deliquescence. J Geophys Res 113. DOI: 10.1029=2007JG000561

Fewer D, Friedl T, Büdel B (2002) *Chroococcidiopsis* and heterocyst-differentiating cyanobacteria are each other's closest living relatives. Mol Phylogenet Evol 23:82–90

França MB, Panek AD, Eleutherio EC (2007) Oxidative stress and its effects during dehydration. Comp Biochem Physiol A Mol Integr Physiol 146:621–631

Fredrickson JK, Li SMW, Gaidamakova EK, Matrosova VY, Zhai M, Sulloway HM, Scholten JC, Brown MG, Balkwill DL, Daly MJ (2008) Protein oxidation: key to bacterial desiccation resistance? Int J Syst Evol Microbiol 2:393–403

Friedmann EI (1986) The Antarctic cold desert and the search for traces of life on Mars. Adv Space Res 6:265–268

Friedmann EI, Ocampo-Friedmann R (1976) Endolithic blue-green-algae in dry valleys – primary producers in Antarctic desert ecosystem. Science 193:1247–1249

Friedmann EI, Ocampo-Friedmann R (1995) A primitive cyanobacterium as pioneer microorganism for terraforming Mars. Adv Space Res 15:243–246

Friedmann EI, Lipkin Y, Ocampo-Paus R (1967) Desert algae of the Negev (Israel). Phycologia 6:185–199

Friedmann EI, Kappen L, Meyer MA, Nienow JA (1993) Long-term productivity in the cryptoendolithic microbial community of the Ross Desert, Antarctica. Microb Ecol 25:51–69

Golubic S, Friedmann EI, Schneider J (1981) The lithobiontic ecological niche, with special reference to microorganisms. J Sediment Petrol 51:475–478

Grilli Caiola M, Billi D (2007) *Chroococcidiopsis* from desert to Mars. In: Seckbach J (ed) Algae and cyanobacteria in extreme environments, vol 11. Cellular origin, life in extreme habitats and astrobiology. Springer-Verlag, Berlin, pp 553–568

Grilli Caiola M, Ocampo-Friedmann R, Friedmann EI (1993) Cytology of long-term desiccation in the cyanobacterium *Chroococcidiopsis* (Chroococcales). Phycologia 32:315–322

Grilli Caiola M, Billi D, Friedmann EI (1996a) Effect of desiccation on envelopes of the cyanobacterium *Chroococcidiopsis* sp. (Chroococcales). Eur J Phycol 31:97–105

Grilli Caiola M, Canini A, Ocampo-Friedmann R (1996b) Iron superoxide dismutase (Fe-SOD) localization in *Chroococcidiopsis* (Chroococcales, Cyanobacteria). Phycologia 35:90–94

Hill DR, Keenan TW, Helm RF, Potts M, Crowe LM, Crowe JH (1997) Extracellular polysaccharide of *Nostoc commune* (Cyanobacteria) inhibits fusion of membrane vesicles during desiccation. J Appl Phycol 9:237–248

Horneck G (1993) Responses of *Bacillus subtilis* spores to the space environment: results from experiments in space. Orig Life Evol Biosph 23:37–52

Horneck G, Stöffler D, Ott S, Hornemann U, Cockell CS, Moeller R, Meyer C, de Vera JP, Fritz J, Schade S, Artemieva NA (2008) Microbial rock inhabitants survive hypervelocity impacts on Mars-like host planets: first phase of lithopanspermia experimentally tested. Astrobiology 8:17–44

Horowitz NH, Cameron RE, Hubbard JS (1972) Microbiology of the Dry Valleys of Antarctica. Science 176:242–245

Jönsson KI, Rabbow E, Schill RO, Harms-Ringdahl M, Rettberg P (2008) Tardigrades survive exposure to space in low Earth orbit. Curr Biol 18:R729–R731

Leuko S, Rothschild LJ, Burns BP (2010) Halophilic archaea and the search for extinct and extant life on Mars J. Cosmology 5:940–950

Navarro-González R, Rainey FA, Molina P, Bagaley DR, Hollen BJ, de la Rosa J, Small AM, Quinn RC, Grunthaner FJ, Cáceres L, Gomez-Silva B, McKay CP (2003) Mars-like soils in the Atacama Desert, Chile, and the dry limit of microbial life. Science 302:1018–1021

Nicholson WL (2009) Ancient micronauts: interplanetary transport of microbes by cosmic impacts. Trends Microbiol 17:243–250

Olsson-Francis K, de la Torre R, Towner MC, Cockell CS (2009) Survival of akinetes (resting-state cells of Cyanobacteria) in Low Earth Orbit and simulated extraterrestrial conditions. Orig Life Evol Biosph 39:565–579

Onofri S, Barreca D, Selbmann L, Isola D, Rabbow E, Horneck G, de Vera JP, Hatton J, Zucconi L. 2008. Resistance of Antarctic black fungi and cryptoendolithic communities to simulated space and Martian conditions. Stud Mycol 61:99–109

Pereira S, Zille A, Micheletti E, Moradas-Ferreira P, De Philippis R, Tamagnini P (2009) Complexity of cyanobacterial exopolysaccharides: composition, structures, inducing factors and putative genes involved in their biosynthesis and assembly. FEMS Microbiol Rev 33:917–941

Pointing SB, Warren-Rhodes KA, Lacap DC, Rhodes KL, McKay CP (2007) Hypolithic community shifts occur as a result of liquid water availability along environmental gradients in China's hot and cold hyperarid deserts. Environ Microbiol 9:414–424

Pointing SB, Chan Y, Lacap DC, Lau MCY, Jurgens JA, Roberta L, Farrell RL (2009) Highly specialized microbial diversity in hyper-arid polar desert. Proc Natl Acad Sci USA 106:19964–19969

Potts M, Slaughter SM, Hunneke FU, Garst JF, Helm RF (2005) Desiccation tolerance of prokaryotes: application of principles to human cells. Integr Comp Biol 45:800–809

Rabbow E, Horneck G, Rettberg P, Schott J-U, Panitz C, L'Afflitto A, von Heise-Rotenburg R, Willnecker R, Baglioni P, Hatton J, Dettmann J, Demets R, Reitz G (2009) EXPOSE, an astrobiological exposure facility on the International Space Station – from proposal to flight. Orig Life Evol Biosph 39:581–598

Rothschild LJ, Mancinelli RL (2001) Life in extreme environments. Nature 409:1092–1101

Sancho LG, de la Torre R, Horneck G, Ascaso C, de los Rios A, Pintado A, Wierzchos J, Schuster M (2007) Lichens survive in space: results from the 2005 LICHENS experiment. Astrobiology 7:443–454

Schuerger AC, Mancinelli RL, Kern RG, Rothschild LJ, McKay CP (2003) Survival of *Bacillus subtilis* on spacecraft surfaces under simulated Martian environments: implications for the forward contamination of Mars. Icarus 165:253–27

Shirkey B, Kovarcik DP, Wtight DJ, Wilmoth G, Prickett TF, Helm RF, Gregory EM, Potts M (2000) Active Fe-containing superoxide dismutase and abundant *sodF* mRNA in *Nostoc commune* (Cyanobacteria) after years of desiccation. J Bacteriol 182:189–197

Shirkey B, McMaster NJ, Smith SC, Wright DJ, Rodriguez H, Jaruga P, Birincioglu M, Helm RF, Potts M (2003) Genomic DNA of *Nostoc commune* (Cyanobacteria) becomes covalently modified during long-term (decades) desiccation but is protected from oxidative damage and degradation. Nucl Acids Res 31:2995–3005

Van Leeuwenhoek A (1702) On certain animalcules found in the sediments in gutter of the roofs of houses. Letter 144 to Hendrik van Bleyswijk, dated 9 February 1702. In: The select works of Anton van Leeuwenhoek, vol 2. London, pp 207–213

Walker JJ, Norman RP (2007) Endolithic microbial ecosystems. Annu Rev Microbiol 61:331–47

Warren-Rhodes KA, Rhodes KL, Pointing SB, Ewing SA, Lacap DC, Gómez-Silva B, Amundson R, Friedmann EI, McKay CP (2006) Hypolithic cyanobacteria, dry limit of photosynthesis, and microbial ecology in the hyperarid Atacama Desert. Microb Ecol 52:389–398.

Wierzchos J, Ascaso C, McKay CP (2006) Endolithic cyanobacteria in halite rocks from the hyperarid core of the Atacama Desert. Astrobiology 6:415–422

Wright DJ, Smith SC, Joardar V, Scherer S, Jervis J, Warren A, Helm RF, Potts M (2005) UV irradiation and desiccation modulate the three-dimensional extracellular matrix of *Nostoc commune* (Cyanobacteria). J Biol Chem 280:40271–40281

Yoshimura H, Okamoto S, Tsumuraya Y, Ohmori M (2007) Group 3 sigma factor gene, *sigJ*, a key regulator of desiccation tolerance, regulates the synthesis of extracellular polysaccharide in cyanobacterium *Anabaena* sp. strain PCC 7120. DNA Res 14:13–24

Psychrophilic microorganisms as important source for biotechnological processes

Sergiu Fendrihan[1] and Teodor G. Negoiță[2]

[1]Romanian Bioresource Centre and Advanced Research, Bucharest, Romania
[2]Romanian Institute of Polar Research, Bucharest, Romania

1 Introduction

The major parts of Earth's environments are cold and have temperatures below 5°C (Gounot 1999; Russell and Cowan 2005). About 70% of the freshwater is ice and about 14% from the Earth's biosphere is represented by terrestrial and aquatic polar areas (Priscu and Christner 2004). The depth of the oceans, the poles, and high mountains are the most important cold regions on Earth (Russell and Cowan 2005). Global ice, for example, covers 6.5 million km^2 which increases to 14.4 million km^2 in wintertime (Perovich et al. 2002). Here we can meet representatives from all domains of the living world. Two categories of microorganisms were discovered in such cold environments. First, the psychrophiles with an optimum growth temperature of about 15°C or even less, which cannot grow above 20°C (Moyer and Morita 2007); second, the psychrotolerants with an optimum growth temperature of 20–30°C, which are able to grow and exhibit activity at temperatures close to the freezing point of water (Madigan and Jung 2003). The lowest temperature for life's activities is −20°C under certain defined conditions (Rivkina et al. 2000; Gilichinsky 2002; D'Amico et al. 2006); others consider the temperature limits for reproduction as −12°C and for metabolism as −20°C (Bakermans 2008). *Colwellia psychrerythraea* strain 34H is motile at −10°C, as observed by transmitted light microscopy (Junge et al. 2003). Psychrophilic microorganisms are dominant in permanently cold environments such as Antarctic waters and have important roles in the biogeochemical cycles in the polar zones (Helmke and Weyland 2004). Not only are prokaryotes adapted to the cold, but also are many eukaryotes such as algae (Takeuchi and Kohshima 2004) and macroorganisms from crustaceans to fishes. The present work will focus on prokaryotes and some microscopic eukaryotes of biotechnological importance.

2 Diversity of cold-adapted microorganisms

The psychrophilic and psychrotolerant microorganisms belong to all three of life's principal domains, Archaea, Bacteria and Eukarya. It is interesting to note that viruses are omnipresent and even so in those inhospitable places. Viruses from the families *Podoviridae*, *Siphoviridae*, and *Myoviridae* were identified in cold environments (Wells 2008). Bacteriophages were identified in inner polar waters and in ice (Säwström et al. 2007) infecting psychrophilic microorganisms; for example, phage 9A of *Colwellia psychrerythrea* strain 34 is capable of forming plaques at low temperatures, but not at 13°C (Wells 2008).

Archaea found in cold environments are methanogens for example, from genera *Methanogenium*, *Methanococcoides* and *Methanosarcina*, but halophilic (*Halorubrum*) and other strains can also occur (Cavicchioli 2006).

Bacteria. The majority of isolates from polar areas belong to the groups of Beta-, Gamma-, Delta-Proteobacteria, Actinobacteria, Acidobacteria, the Cytophaga–Flexibacter–Bacteroides group, and green nonsulfur bacteria. Many strains of Bacteria as well as Archaea and Eukarya were revealed by 16S rRNA and 18S rRNA gene clone libraries (Tian et al. 2009). Soils of the McMurdo Dry Valleys host species of *Pseudonocardia*, *Nocardioides*, *Geodermatophilus*, *Modestobacter*, *Sporichthya* and *Streptomyces* (Babalola et al. 2008). Cyanobacteria as photoautotrophs were retrieved from ice, soils, rocks, lakes, ponds, marine ecosystems, and alpine areas (Zakhia et al. 2008). *Chamaesiphon* sp., *Chroococcidiopsis* (from sandstone) and *Synechococcus* sp. (from lakes, marine water, and others) are examples of cyanobacterial genera with cold-adapted strains.

Algae. Species of *Chlamydomonas* were retrieved from water derived from melting glacier ice and from some layer species of *Rhodomonas* and *Chromulina*. Species of *Tribonemataceae* were found in Antarctic terrestrial environments (Rybalka et al. 2008). Several microalgae can be found in all known cold environments as in snow (*Chlamydomonas* and *Chloromonas*), seawater (diatoms), sea ice (diatoms and dinoflagellates), on rocks as endoliths (*Hemichloris antarctica*), ice-covered lakes (*Chloromonas* sp., *Chlamydomonas intermedia*, and *Chlamydomonas raudensis*) and at high altitudes (reviewed by Mock and Thomas 2008). Samples from the Tyndall Glacier in Patagonia, Chile contained algal species of the genera *Mesotaenium*, *Cylindrocystis*, *Ancylonema*, *Closterium*, *Chloromonas*, and some cyanobacteria (Takeuchi and Kohshima 2004).

Yeasts. Yeast strains such as *Sporobolomyces*, *Cryptococcus*, and *Rhodotorula* sp. were isolated from Lake Vanda (Goto et al. 1969) and from other Antarctic and alpine environments, including psychrophilic yeasts such as the novel species *Mrakia robertii*, *M. blollopis*, and *M. niccombsi* (Thomas-Hall et al. 2010). Several yeasts, which are producers of lipases and proteases, were isolated from cold marine water and freshwater (Rashidah et al. 2007), such as *Cryptococcus antarcticus* and *Cryptococcus albidosimilis, Basidioblastomycetes* (Vishniac and Kurtzman 1992), *Cryptococcus nyarrowii* (Thomas-Hall and Watson 2002), *Cryptococcus watticus* (Guffogg et al. 2004), and *Leucosporidium antarcticum* – the latter from Antarctic waters (Turkiewicz et al. 2005) – and *Mrakia* strains (Thomas-Hall et al. 2010).

Fungi were isolated from many cold environments. For example, *Penicillium, Aspergillus, Paecilomyces, Cladosporium, Mortierella, Candida,* and *Rhodotorula* were isolated from soils of Terra Nova Bay and Edmonson Point, Antarctica (Gesheva 2009). Some authors described isolates from soils, such as *Chrysosporium* sp., *Phoma exigua, Heterocephalum aurantiacum, Aureobasidium pullulans, Fusarium oxysporum, Trichoderma viride,* and *Penicillium antarcticum* (Negoiță et al. 2001a). From the soils of Schirmacher Oasis, Antarctica, fungi such as *Acremonium, Aspergillus,* and *Penicillium* were isolated, the majority surviving as spores in those harsh environments, and some species possess unique features of their mycelia (Singh et al. 2006). Frisvad (2008b) reviewed the fungi from cold ecosystems and indicated their isolation from soils and permafrost, caves, rocks, mosses and lichens, glacier ice, freshwater, as well as from frozen foods. The fungi belong to the Ascomycetes (*Acremonium antarcticum, A. psychrophilum,* and *Penicillium antarcticum*), Zygomycetes (*Mortierella alpina* and *Absidia psychrophila*) and basidiomycetous yeasts, which are very rare in cold areas. Endolithic fungi resistant to low temperature and low water activity were isolated by Onofri et al. (2007).

3 Ecology and biology

Some of the microorganisms are polyextremophiles, for example halo-psychrophiles, or piezo-psychrophiles, which tolerate high pressure (Nogi 2008) and cannot grow at atmospheric pressure and at temperatures above 20°C, such as strains of *Shewanella, Colwellia, Moritella,* and *Psychromonas*. In these categories all the physiological and metabolic types can be found – anaerobes and aerobes, methanogens, methanotrophs, chemolithotrophs, sulfate reducers, and organotrophs. Anaerobic cold-adapted *Clostridium* sp. (e.g., *C. frigoris, C. bowmannii,* and *C. psychrophilum*) were isolated from Antarctic microbial mats (Spring et al. 2003) or some psychrotolerants, such as *C. frigidicarnis* and *C. algidixylanolyticum,* from frozen products (Finster 2008). Sulfate-reducing psychrophiles *Desulfotalea, Desulfofaba,* and *Desulfofrigus* (Knoblauch et al. 1999), sulfur-oxidizing bacteria (SOB), occurring in such organic carbon depleted environments as subglacial waters (Sattley and Madigan 2006), as well as denitrifying microorganisms in sea ice (Rysgaard et al. 2008) were found. Ammonia oxidizers were identified by genetic methods in all of the samples taken from lakes Fryxell, Bonney, Hoare, Joyce, and Vanda in Antarctica, belonging to the *Proteobacteria* (Voytek et al. 1999). Acetogenic bacterial sequences originating from *Acetobacterium tundrae* and others related to *Acetobacterium bakii* (Sattley and Madigan 2007) were isolated from sediments of Lake Fryxell. From the same lake different phototrophic purple bacteria were identified with molecular methods (Karr et al. 2003) as well as methanogenic and other Archaea (Karr et al. 2006). Biological methane oxidation and sulfate reduction by Archaea occur in alpine lakes (such as Lake Lugano deeps) in anoxic zones (Blees et al. 2010). Methanogens were detected in soils, water sediments, sea and lake waters from cold environments (Cavicchioli 2006). Methanotrophy was detected indirectly in Lake Untersee (Antarctica) by identification of hopanoids and two steroids (4-methyl steroid and 4,4-dimethyl steroid), one hopanoid (diplopterol) having a specific low isotopic ^{13}C content, and originating from the aerobic methylotroph *Methylococcus* sp. (Niemann et al. 2010). Some *Shewanella* and *Pseudomonas* strains from Antarctic lakes are able to mediate redox reactions of manganese under stimulation by Co and Ni (Krishnan et al. 2009).

4 Cold environments

Cold deserts. There are cold deserts in Antarctica where the precipitation is very low, the temperatures range between $-55°C$ and $10°C$, UV radiation is high and water activity is low; these are some of the most extreme environments on Earth. Many microorganisms can be found in endolithic communities composed of cyanobacteria such as *Acaryochloris marina* and *Gloeocapsa* species (de los Ríos et al. 2007). Endolithic bacteria, fungi, archaea, green algae, yeasts, and lichens were found in McMurdo Dry Valley (Gounot 1999), analyzed by staining with the BacLight LIVE/DEAD kit and observed with confocal laser scanning microscopy to demonstrate their survival (Wierzchos et al. 2004).

Soils covered with snow. From Arctic wetland soil methanotrophic bacteria were retrieved such as *Methylocystis rosea* (Wartiainen et al. 2006). In soils of Lapland microbial communities were discovered, similarly as in soils from alpine zones, where the temperatures can reach $-25°C$ in wintertime. In addition, soils from Spitsbergen contained many fungi such as *Mucor, Mortierella, Alternaria, Fusarium*, and *Zygorrhinchus* (Negoiță et al. 2001b), genera which are very probably psychrotolerants.

Permafrost. Permafrost soils in the geological sense stay below $0°C$ for two consecutive years or more and are specific for arctic areas covering about 26% of the surface of the Northern Hemisphere. The average temperature is $-16°C$; in Siberia $-11°C$ and in Antarctica $-18°C$ to $-27°C$ were measured (Vorobyova et al. 1997). From those soils over 100 bacterial strains were isolated, also some methanogenic archaea from the families *Methanomicrobiaceae*, *Methanosarcinaceae*, and *Methanosetaceae* (Ganzert et al. 2007), methane oxydizing bacteria (Liebner and Wagner 2007), sulfate-reducing bacteria, aerobic and anaerobic heterotrophs (Gilichinsky 2002), denitrifiers, and iron and sulfate reducers (Rivkina et al. 1998). The majority of strains included species of *Micrococcus*, *Bacillus*, *Paenibacillus*, *Rhodococcus*, *Arthrobacter*, *Haloarcula*, and *Halobaculum* (Steven et al. 2007), which were isolated in quantities of 10^7-10^9 cells per gram of dry soil. From layers of permafrost which were demonstrated to be about 3–5 million years old, viable cells of bacteria were isolated (Rodrigues-Diaz et al. 2008), which have to face low temperatures and natural irradiation by radionuclides (Gilichinsky et al. 2008). From a layer of an arctic permafrost ice wedge from Canada (temperature $-17.5°C$, pH 6.5, salt concentration 14.6 g/l, age about 25,000 years) bacteria were isolated (Katayama et al. 2007) belonging to the classes *Actinobacteria* and *Gamma-Proteobacteria*.

Snow, ice, and glaciers. The ice glaciers in Antarctica contain approximately 90% of the ice of our planet according to the National Snow and Ice Data Centre of USA (http://nsidc.org/, cited by Christner et al. 2008). Some aspects of the soils covered partially with ice on the shore of the Antarctic sea are shown in Figs. 1 and 2. Sea and lake ice glaciers are hosting considerable quantities of biological material, consisting of microorganisms, bacteria, spores, and pollen grains, the majority being transported there by air. The microbiota can survive in crevices and capillary tunnels containing concentrated ionic solutions with a lower freezing point (Price 2006). The number of viable microorganisms decreases with the depth of the ice layers; there is a supraglacial community (bacteria, viruses, diatoms, tardigrades, and

Fig. 1. Larsemann Hills Coast, Law-Racovita Base area, East Antarctica, 69°23'0"S; 76°23'0"E (photo T.G. Negoita, 2007)

Fig. 2. Antarctic ice cap in the Schirmacher Oasis, 70°46'S; 11°50'E, 100 km inside the continent (photo T.G. Negoita, 2007)

rotifers), a subglacial community (aerobic and anaerobic) and an endoglacial community (Hodson et al. 2008). Hollibaugh et al. (2007) studied the Sea Ice Microbial Community (SIMCO) formed of bacteria, algae, and fungi of various metabolic types: sulfate reducers, chemolithotrophs, methanogens, anaerobic nitrate reducers (Skidmore et al. 2000), and viruses (Deming 2007). In ice there is an incredible diversity of *Proteobacteria*, of the phylum *Cytophaga–Flavobacterium–Bacteriodes*, high GC Gram positives and low GC Gram positives (Miteva 2008), and a large metabolic and physiological diversity. Some of them were entrapped for very long periods of time, such as the strain *Herminimonas glaciei*, a Gram-negative ultramicrobacterium, which was isolated from a 3042 m deep drilling core from a Greenland glacier of about 120,000-year-old ice (Loveland-Curtze et al. 2009), or *Chryseobacterium greenlandense* (Loveland-Curtze et al. 2010) and sequences from *Pseudomonas* and *Acinetobacter*, which stem from 750,000-year-old ice from the Qinghan-Tibetan plateau in Western China (Christner et al. 2003a). Many prokaryotes isolated from snow melt water belong to the *Beta-Proteobacteria* (21.3%), *Sphingobacteria* (16.4%), *Flavobacteria* (9.0%), *Acidobacteria* (7.7%), and *Alpha-Proteobacteria* (6.5%) and other groups (Larose et al. 2010). The cryoconite holes form another microhabitat containing various forms of life, such as diatoms, algae, prokaryotes, fungi, rotifers, and tardigrades (Wharton et al. 1985; Christner et al. 2003b).

Cold caves. Caves represent a constant temperature environment with low organic content, sometimes only 1 mg of organic matter per liter. Some strains are chemolithotrophs such as *Galionella*. In many cases there are more psychrotolerants than psychrotrophs. Some stenothermic bacterial strains were isolated which can grow at 10–20°C and only few which grow at 2°C or 28°C (Gounot 1999). The strain *Arthrobacter psychrophenolicum* was isolated from an Austrian ice cave (Margesin et al. 2004).

Cold lakes. The cold lakes in the polar and alpine zones can be covered with an ice layer (Antarctic lakes), which practically isolates the lake from the rest of the environment, and their content of organic carbon and oxygen is rather low. Christner et al. (2008) pointed out in their comprehensive review that there are 141 subglacial Antarctic lakes having a total volume of about 10,000 km^3. One of the largest lakes is Lake Vostok of 14,000 km^2, covered by a 4000 m thick ice sheet and being 400–800 m deep. The bottom is covered with a thick sediment layer. The temperature of the ice layer is about $-55°C$, but the lake has a constant temperature of $-2.65°C$ (Di Prisco 2007); the ice layer is about 15 million years old. Alpine lakes are only temporarily covered by ice and the microbiota there are subject to seasonal fluctuations (Pernthaler et al. 1998). The cold Antarctic lake environment is chemically driven, with reactions such as sulfide and iron oxidation (Christner et al. 2008), and contains methanogens such as *Methanosarcina*, *Methanoculleus*, and anoxic methanotrophs (Karr et al. 2006). The saline lakes host euryhalophiles related to *Halomonas* and *Marinobacter* (Naganuma et al. 2005).

Cold marine waters. From marine waters of Ushuaia, a sub-Antarctic town in Argentina, many sequences were identified belonging to the *Alpha-* and *Gamma-Proteobacteria*, *Cytophaga–Flavobacterium–Bacteroidetes* group, the genera *Marinomonas*, *Colwellia*, *Cytophaga*, *Glacieola*, *Cellulophaga*, *Roseobacter*, *Staleya*, *Sulfitobacter*, *Psychrobacter*, *Polaribacter*, *Ulvibacter*, *Tenacibacter*, *Arcobacter*, and *Formosa* (Prabagaran et al. 2007). In the depth of the ocean the temperature is about $3°C$, and a considerable pressure exists (the pressure increases by 1 atm per each 10 m of depth). Here a very diverse bacterial community can be retrieved, for example, from the sediments of the Japanese Trench (Hamamoto 1993). From the deep sediments were, all the domains of life are represented, an important microbiota was identified by molecular methods (Tian et al. 2009). The archaeal sequences can reach 17% from total microbiota in marine coastal waters (Murray et al. 1998). Sulfate-reducing bacteria form a large community in sediments of the Arctic ocean, being active at about $2.6°C$ with a sulfate reduction rate similar to that under mesophilic conditions (Knoblauch et al. 1999).

Anthropic cold environments. Artificial cooling and freezing systems can be visualized as man-made environments. *Pseudomonas fluorescens* is one of the lipolytic food spoiling bacteria which is active in the cold, and its hydrolytic activity at low temperatures was studied as a function of water activity (Andersson et al. 1979). The lower temperatures and lower water activity did not affect the enzymatic activity since the substrates were hydrophobic. In cooling devices the bacterium *Pseudomonas fragi* is frequently found, which is supported by temperatures between 2°C and 35°C; it possesses some cold shock proteins (Csps) and degrades frozen foods. Another bacterium from water-cooling systems is *Chryseobacterium aquifrigidense* (Park et al. 2008). The psychrophilic strain *Lactobacillus algidus* was isolated from refrigerated, packed beef meat (Kato et al. 2000).

Air. From the Antarctic continent air several psychrotolerant microorganisms were isolated, such as *Sphingomonas aurantiaca*, *Sphingomonas aerolata*, and *Sphingomonas faeni* sp. nov. (Busse et al. 2003).

5 Adaptation to cold environments

Growth and activity. The temperature has a direct influence on microbial growth and the relationship between growth and temperature generally conforms to the Arrhenius law (Gounot 1999). Christner (2002) reported the incorporation of DNA and protein precursors by *Arthrobacter* and *Psychrobacter* at $-15°C$. *Polaromonas hydrogenivorans* has a lower temperature limit of 0°C for growth (Sizova and Panikov 2007), and psychrophilic methanotrophs can grow at about 2°C (Liebner and Wagner 2007). The psychrophilic strain *Psychromonas ingrahami* showed growth at $-12°C$ with a slow rate of 10 days of generation time (Breezee et al. 2004). The activity of microorganisms was proven by measurement of ATP as a result of biomass activity in soils and permafrost (Cowan and Casanueva 2007); truly psychrophilic microorganisms showed an increase of the ATP content at lower temperatures, which is the opposite reaction of mesophiles (Napolitano and Shain 2004). Other information can be obtained by determination of the Indicator of Enzymatic Soil Activity Potential, the Indicator of Vital Activity Potential, and Biologic Synthetic Indicator (Negoiţă et al. 2001b). These indicators were introduced by Ştefanic (1994) in order to obtain comprehensive information about the biological activity of soils and to compare them for agricultural uses.

Membrane polar lipids. There are differences regarding the composition of membrane lipids and there are clear contributions to cold adaptation, depending also on bacterial taxonomy. The cytoplasmic membrane contains lipids with fatty acids of lengths ranging mainly between C_{14} and C_{18} Gram negatives possess in addition an outer membrane containing lipopolysaccharides, Archaea contain ether-linked glycerol alkyl lipids instead of fatty acids, and eukaryotes contain sterols (Russell 2008). Membrane fluidity depends on the degree of saturation of the polar lipids; the membranes from psychrophiles contain a higher amount of unsaturated and/or polyunsaturated and branched fatty acids, with methyl groups and a larger percentage of double bonds of the *cis* type (Chintalapati et al. 2004). The changes in amount and type of methyl-branched fatty acids of Gram-positive bacteria are a possibility for increasing membrane fluidity at low temperatures. The amount of unsaturated fatty acids contributes to the flexibility of the membrane structure in cold-adapted microorganisms, including eukaryotic photobionts such as diatoms and algae (Morgan-Kiss et al. 2006). The presence of polyunsaturated fatty acids (PUFAs) does not completely explain the adaptation to cold environments, because there are many marine strains without them (Russell and Nichols 1999). Archaeal adaptation to the cold shows a similar increase in desaturation of their isoprenoids containing lipids; *Methanococcoides burtoni* for example generates unsaturated lipids during growth at low temperatures by selective saturation and not by using a desaturase such as bacteria (Cavicchioli 2006).

The proteome. Cold-adapted bacterial proteins have a reduced amount of arginine, glutamic acids, and proline (salt bridge forming residues) and reduced amounts of hydrophobic clusters (Grzymski et al. 2006). A comparison of the contents of amino acids of psychrophilic enzymes was made by Gianese et al. (2001); they found that generally Arg and Glu residues in the exposed sites of alpha helices were replaced by Lys and Ala in psychrophiles. Studying the crystal structure of the β-lactamase from several psychrophilic strains (*P. fluorescens* and others) some authors found that the enzymes from psychrophiles have a lower content of arginine in comparison with lysine and a lower proline content than mesophilic enzymes (Michaux et al. 2008). The lysine residues are of great importance for the cold adaptation mechanism in enzymes, for example in α-amylase from *Pseudoalteromonas haloplanktis* (Siddiqui et al. 2006). A similar replacement is observed with Archaea having a higher content of noncharged amino acids (as glutamine and threonine) and lower contents of hydrophobic amino acids such as leucine (Cavicchioli 2006). At the same time the number of hydrogen bonds (Michaux et al. 2008) and the number of disulfide bridges are reduced (Sælensminde et al. 2009). The cellulase Cel5G from *P. haloplanktis* possesses a catalytic domain and a carbohydrate-binding domain which are joined by a long-linker region containing three loops closed by disulfide bridges. By experimental shortening of this linker region, the enzyme became less flexible approaching the activity of its mesphilic counterpart, which suggested that a long-linker region is an appropriate adaptation of this enzyme to low temperates (Sonan et al. 2007). Studying the thermal adaptations of psychrophilic, mesophilic and thermophilic DNA ligases, the conclusion was that "the active site of the cold enzyme is destabilized by an excess of hydrophobic surfaces and contains a decreased number of charged residues compared with its thermophilic counterpart" (Georlette et al. 2003). The proteins must keep a balance between their stability and flexibility, especially enzymes, which are to be active at lower temperatures than their mesophilic counterparts (Georlette et al. 2003). An intensive study of the proteomics of psychrophilic microorganisms (Kurihara and Esaki 2008) showed that there are various proteins involved in transcription, folding of RNA and proteins, modulation of gene expression, and others, which are inducibly produced at low temperatures.

The following three main types of proteins are of interest for mechanisms of adaptation:

Csps, which are induced by exposure to low temperatures, are involved in several cellular processes (fluidity of membranes, transcription, translation). Proteins from psychrophiles (Caps, cold acclimation proteins) are similar to the Csps. The regulation of the CspA protein takes place at the transcriptional level at the level of stabilization of mRNA (*cspA* mRNA) and at the level of translation (Phadtare and Inouye 2008). Numerous Csps and proteins helping in the adaptation to low temperatures were isolated and characterized (Russell 2008). They play a role in the cold adaptation during stress response and also act as RNA chaperones. Similar proteins can be found in Archaea, such as *Methanogenium frigidum*, which are bound to nucleic acids (Giaquinto et al. 2007).

Antifreeze proteins (AFPs). AFPs and antifreeze glycoproteins (AFGPs) can lower the temperature of the freezing point of water (D'Amico et al. 2006; Kawahara 2008). They can inhibit the formation of ice crystals and prevent the penetration of ice into cells (Zachariassen and Lundheim 1999). One of the examples is the protein Hsc25 produced by the bacterium *Pantoea ananatis* KUIN 3, which helps to refold denatured proteins in the cold (Kawahara 2008).

Antinucleating proteins (ANPs). ANPs and other compounds inhibit ice nucleation and formation of intracellular ice crystals, avoiding thereby the damage of cells. *Acinetobacter aceticus* can release such antinucleating proteins with a mass of 550 kDa. The proteins can be used in the preservation of livers (in a concentration of 20 µg/ml) at subzero temperatures without freezing, with addition of an antioxidant such as ascorbic acid (Kawahara 2008). An ice-binding protein of a mass of 54 kDa, isolated from a bacterial strain from an ice core of over 3000 m depth, was able to inhibit the recrystallization of ice (Raymond et al. 2007).

Enzymes. The rate of a chemical reaction is temperature dependent, according to the Arrhenius equation $K = A \exp(-E_a/RT)$, where K is the reaction rate, E_a is the activation energy, R is the gas constant, T is absolute temperature in Kelvin, and A is a constant. It is well known that biological reactions showed a 16- to 80-fold reduction when the temperature is reduced from 37°C to 0°C (Collins et al. 2008).

While psychrophiles exhibit a high metabolic rate at cold temperatures, they are usually inactivated at mesophilic temperatures because of the flexibility and lower stability as a consequence of the plasticity of catalytic zones of their molecules, due to a reduction of hydrophobic and hydrogen bonds and a lower content of arginine and proline (D'Amico et al. 2006). Shifting to optimum activation energy allows them to keep normal reaction rates at low temperatures (Siddiqui and Cavicchioli 2006). At the same time the 3D structure is also important (Tkaczuk et al. 2005), such as intramolecular bond modifications (Feller and Gerday 1997; Bae and Phillips 2004), modification of amino acids in or near catalytic domains of enzymes (Papaleo et al. 2008), and a lower content of hydrogen bonds (Michaux et al. 2008). An intensive search for cold-adapted enzymes was performed by Morita et al. (1997) who isolated more than 130 bacterial strains and tested the properties of amylases, proteases and lipases, showing that they were easily inactivated at above optimum temperatures.

Other substances which can play a role in cold adaptation and cryoprotection are carotenoids, which contribute to the stability of cellular membranes (Russell 2008); extracellular polymeric substances (EPSs), some of them of high molecular weight or heteropolysaccharides (with additions of proteins), which are released by some microorganism into the neighboring environment and form a kind of gel with cryoprotective effects (Krembs and Deming 2008); polyhydroxyalkanoates (PHAs), which can reduce oxidative stress in the cold, maintaining the redox state (Ayub et al. 2009); trehalose, which is able to protect cells under conditions of shock exposure to high and low temperatures and osmotic stress (Phadtare and Inouye 2008) by stabilizing the cell membrane and removing free radicals, thus preventing denaturation of proteins. Generally speaking, when comparing with thermophiles and mesophiles, it appears that cold-adapted microorganisms adopted the strategy of more entropy by molecular mechanisms, which are allowing an enhanced flexibility for maintaining dynamics and functions of the molecules (Feller 2007).

Genetic features as adaptation mechanisms. So far, the following genomes from psychrophiles and psychrotolerants have been sequenced: *Methanococcoides burtonii* (Allen et al. 2009); *Methanogenium frigidum* (Saunders et al. 2003); *Colwellia psychrerythrea* 34H (Methé et al. 2005); *Desulphotalea psychrophila* (Rabus et al. 2004); *Idiomarina loihiensis* L2TR (Hou et al. 2004), *Pseudoalteromonas haloplanktis* TAC125 (Médigue et al. 2005); *Shewanella frigidimarina* (Copeland et al. 2006); *Psychrobacter arcticus* 253-4 (Ayala-del-Río et al. 2010), *Psychromonas ingrahamii* (Riley et al. 2008); 14 *Shewanella* strains (Hau and Gralnick 2007); and several others, which are partially sequenced. The analysis of the genes showed some principal features of the mechanisms for cold adaptation (Bowman 2008): a lower content in arginine and proline, which influences the flexibility of proteins, was observed, especially in sequences related to growth and development (Ayala-del-Río et al. 2010). Nucleic acids of psychrophiles showed a different proportion of uracil in 16S rRNA sequences, such as an inverse proportional relation to their optimum growth temperature (Khachane et al. 2005).

6 Applications of psychrophilic microorganisms

Bioprospecting and bioscreening of psychrophilic microbial resources (Nichols et al. 2002) have become real challenges and opportunities for biotechnology. Cold-adapted bioactive substances provide advantages in different areas, such as activity at low temperatures; the possibility of challenging reactions with a sufficiently high reaction rate; energy savings; efficient production with lower processing costs; thermal protection of the products; and better quality of products. Presently the market for bioactive products and industrial enzymes is growing. Archaea, Bacteria, and Eukarya can be sources of valuable products. Huston (2008) reviewed the enzymes from cold-adapted microorganisms, identifying compounds and enzymes for the food and cosmetic industry, pharmaceuticals, biofuels, substances for molecular biology studies, and even for nanobiotechnology. An extensive compilation of applications of psychrophilic and psychrotolerant microorganisms is presented in Table 1.

Bioscreening for valuable products is generally not made anymore in the classical way by isolation and cultivation of microorganisms. Now high-throughput culturing technologies enable the isolation of a major proportion of the microbiota in environmental samples; combined with metagenomics and gene expression studies, genome data mining permits an efficient search for bioproducts (see Huston 2008). Psychrophilic proteins, for example, can have some interesting applications, and their production can be achieved directly or expressed in an adequate host such as *Escherichia coli*, which was used for the α-amylase from *P. haloplanktis* (Feller et al. 1998). It can be difficult to obtain a stable production, due to autolytic deterioration. A possible solution is overexpression using a plasmid vector from *P. haloplanktis* pMTBL and the plasmid of *E. coli* pJB3 (Tutino et al. 2001). This type of expression technology promises a wide application for the problem of efficient gene expression systems and rapid purification steps. Recombinant proteins can be obtained by expressing them in prokaryotic cells of cold-adapted (*P. haloplanktis* TAC125) and eukaryotic cells (*Saccharomyces cerevisiae*; Parrilli et al. 2008).

Antibiotics. The isolates from the Antarctic Ocean, Ross Bay, were shown to have antibiotic activities which were tested with the terrestrial bacteria *E. coli*,

Table 1. Applications of psychrophilic and psychrotolerant microoganisms isolated from cold environments

Microorganism	Enzymes and other metabolites	Applications	References
Bacteria			
Acinetobacter sp.	Lipases	Lipid hydrolysis, detergent additives	Ramteke et al. (2005) and Joseph et al. (2007)
Achromobacter sp.	Lipases	Lipid hydrolysis	Ramteke et al. (2005) and Joseph et al. (2007)
Aeromonas sp.	Lipases	Lipid hydrolysis	Lee et al. (2003)
Arthrobacter strains	Antibiotics	Pharma industry	Lo Giudice et al. (2007) and Benešova et al. (2005)
Arthrobacter sp.	Alkaline phosphatases	Alkaline phosphatase (removal of 5′ phosphate groups from DNA and RNA)	De Prada et al. (1996)
Arthrobacter C2-2	α-Glucosidase, β-glucosidase	Cleavage of maltose at β-1,4 bonds; pharma industry, medicine	Benešova et al. (2005)
Arthrobacter strain 20B	β-Galactosidases	Lactose hydrolysis	Białkowska et al. (2009)
Arthrobacter psychrolactophilus strain F2	β-Galactosidase rBglAp	Produces trisaccharides from lactose; food industry	Nakagawa et al. (2007)
Arthrobacter psychrophenolicus		Degradation of phenol and phenolic compounds	Margesin et al. (2004)
Bacillus subtilis strain MIUG 6150	α-, β-Amylases	Starch hydrolysis; food industry	Bahrim and Negoiţă (2004)
Bacillus subtilis strain MIUG 6150	Proteases	Protein hydrolysis	Bahrim and Negoiţă (2004)
Brevibacterium antarcticum		Bioremediation; resistant to metals in soils (Cu, Cr, Hg, and others)	Tashyrev (2009)
Colwellia demingiae	Protease (azocasein)		Nichols et al. (1999)
Colwellia demingiae	Protease (azoalbumin)		Nichols et al. (1999)
Colwellia demingiae	Trypsin-like enzyme	Protein hydrolysis	Nichols et al. (1999)
Colwellia demingiae	Phosphatase		Nichols et al. (1999)
Colwellia-like strain	Trypsin-like enzyme		Nichols et al. (1999)
Colwellia-like strain	Phosphatase		Nichols et al. (1999)
Colwellia-like strain	β-Galactosidase	Lactose hydrolysis	Nichols et al. (1999)
Colwellia-like strain	Protease (azocasein)		Nichols et al. (1999)
Colwellia-like strain	Protease (azoalbumin)	Protein hydrolysis	Nichols et al. (1999)

(*continued*)

Table 1 (*continued*)

Microorganism	Enzymes and other metabolites	Applications	References
Colwellia-like strain	Trypsin		Nichols et al. (1999)
Colwellia-like strain	β-Galactosidase	Removal of lactose	Nichols et al. (1999)
Colwellia-like strain	α-Amylase	Starch hydrolysis	Nichols et al. (1999)
Colwellia-like strain	Alkaline phosphatase		Nichols et al. (1999)
Colwellia demingiae	Synthesizes docosahexaenoic acid	PUFAs as precursors for prostaglandins, thromboxanes, leucotrienes; medicine, pharma industry	Bowman et al. (1998) and Lees (1990)
Colwellia hornerae	Synthesizes docosahexaenoic acid	Pharma industry	Bowman et al. (1998)
Colwellia maris	Malate synthase, *iso*-citrate lyase	Bioethanol and biomethane production; wastewater treatment, bioremediation	Brenchley (1996) and Cavicchioli et al. (2002)
Colwellia rossensis	synthesizes docosahexaenoic acid	Pharma industry	Bowman et al. (1998)
Colwellia psychrotropica	Synthesizes docosahexaenoic acid	Pharma industry	Bowman et al. (1998)
Dactylsporangium roseum	Antibiotics	Pharma industry, medicine	Nguyen et al. (2010)
Erythrobacter litoralis HTCC2594	Epoxide hydrolase	Epoxide hydrolase, for enantio-selective hydrolysis of styrene oxide	Woo et al. (2007)
Fibrobacter succinogenes S85	Cellulase	Animal food industry, detergents, textile industry	Cavicchioli et al. (2002)
Flavobacerium sp.	β-Mannanase	Decreases viscosity in food products	Zakaria et al. (1998)
Flavobacterium frigidarium	Xylanolytic and laminarinolytic	Xylane degradation	Humphry et al. (2001)
Flavobacterium frigidimaris	Malate dehydrogenase		Oikawa et al. (2005)
Flavobacterium hibernum sp. nov.	β-Galactosidase	Lactose degradation	McCammon et al. (1998)
Flavobacterium limicola		Organic polymer degradation	Tamaki et al. (2003)
Glaciecola chathamensis	Exopolysaccharides	Food processing industry; medical and industrial uses	Matsuyama et al. (2006)
Glaciecola chathamensis	Polysaccharide-producing strain	Exopolysaccharides, industrial applications	Matsuyama et al. (2006)
Instrasporangium sp.	Antibiotics	Pharma industry	Nguyen et al. (2010)

(*continued*)

Table 1 (*continued*)

Microorganism	Enzymes and other metabolites	Applications	References
Janibacter sp.	Antibiotics	Pharma industry, medicine	Lo Giudice et al. (2007)
Kordiimonas gwangyangensis	Cold-adapted enzymes	Capable of degrading polycyclic aromatic hydrocarbons (PAHs)	Kwon et al. (2005)
Micromonospora sp.	Antibiotics	Pharma industry	Nguyen et al. (2010)
Moraxella sp.	Lipases	Pharma industry, medicine, food additives	Ramteke et al. (2005) and Joseph et al. (2007)
Oceanibulbus indolifex	Indole and several indole derivatives	Cosmetics industry, pharma industry, cancer prevention	Wagner-Döbler et al. (2004) and Auborn et al. (2003)
Oceanibulbus indolifex	Cyclic dipeptides cyclo-(Leu,Pro), cyclo-(Phe,Pro), and cyclo-(Tyr,Pro)	Compounds with antiviral, antibiotic, and antitumor activity	Wagner-Döbler et al. (2004) and Milne et al. (1998)
Oceanibulbus indolifex	Tryptanthrin	Activity against some Gram-positive bacteria and fungi	Wagner-Döbler et al. (2004)
Oleispira antarctica	Cold-adapted enzymes	Hydrocarbonoclastic; for bioremediation	Yakimov et al. (2003)
Photobacterium frigidiphilum	Lipases	Lipid hydrolysis	Seo et al. (2005)
Planococcus sp.	β-Galactosidase	Lactose hydrolysis	Sheridan and Brenchley (2000)
Polaromonas naphthalenivorans	Enzymes	Degrades naphthalene	Jeon et al. (2004)
Polaromonas sp. strain JS666	Enzymes	*cis*-1,2-Dichloroethene as carbon source; for bioremediation	Mattes et al. (2008)
Pseudoalteromonas sp.	Protease (azocasein)		Nichols et al. (1999)
Pseudoalteromonas sp.	Trypsin-like enzyme		Nichols et al. (1999)
Pseudoalteromonas	Antibiotics	Pharma industry, medicine	Lo Giudice et al. (2007)
Pseudoalteromonas sp.	Lipases		Ramteke et al. (2005)
Pseudoalteromonas haloplanktis TAE 47	β-Galactosidase	Lactose hydrolysis	Hoyoux et al. (2001)
Pseudomonas sp. strain B11-1	Lipases, esterases		Suzuki et al. (2001)
Pseudoalteromonas sp. SM9913	Subtilase		Yan et al. (2009)
Psychrobacter okhotskensis	Lipase-producing strain		Yumoto et al. (2003)

(*continued*)

Table 1 (*continued*)

Microorganism	Enzymes and other metabolites	Applications	References
Psychrobacter sp.			Ramteke et al. (2005)
Rhodococcus	Antibiotics	Pharma industry, medicine	Lo Giudice et al. (2007)
Rhodococcus sp.		Biodegradation of phenol and phenolic compounds	Margesin and Schinner (1999)
Rhodococcus sp. strain N774	Nitrile hydratase	Acrylamide production	Kobayashi et al. (1992)
Rhodococus sp. Q15 i		Degrades short- and long-chain aliphatic alkanes from diesel fuel	Whyte et al. (1998)
Rhodococcus ruber			Murygina et al. (2000)
Rhodococcus erythrococcus		Product "Rhoder" for bioremediation of oil polluted environments	Murygina et al. (2000)
Serratia proteamaculans	Trypsin-like protease		Mikhailova et al. (2006)
Shewanella sp.	Produces omega 3 fatty acids	Essential fatty acid for humans	Hau and Gralnick (2007)
Shewanella sp.		Waste removal of radionuclides (uranium, technetium)	Hau and Gralnick (2007)
Shewanella sp.		Reduction of organic chlorine compounds	Hau and Gralnick (2007)
Shewanella donghaensis	High levels of polyunsaturated fatty acid	Medicine, food supplements	Yang et al. (2007)
Shewanella gelidimarina	β-Galactosidase	Lactose hydrolysis	Nichols et al. (1999)
Shewanella frigidimarina	Eicosapentaenoic acid (20:w503)	Food additives	Bowman et al. (1997) and Bozal et al. (2002)
Shewanella pacifica	Produces polyunsaturated fatty acids	Food additives, pharma industry	Ivanova et al. (2004)
Serratia proteamaculans	Trypsin-like protease	Protein hydrolysis	Mikhailova et al. (2006)
Serratia sp.	Lipases	Lipid hydrolysis	Ramteke et al. (2005)
Sphingmonas paucimobilis	Proteases	Meat industry, detergent industry, molecular biology	Cavicchioli et al. (2002)
Streptomyces sp.	Amylases, proteases, cellulases, lipases, antibiotics, other bioactive compounds	Detergent additives, starch industry, bread industry, antibiotics, immuno-suppressants, anticancer agents, extracellular hydrolytic enzymes, degradation of ligno-cellulosic materials	Cavicchioli et al. (2002), Galante and Formantici (2003), and Morita et al. (1997)

(*continued*)

Table 1 (continued)

Microorganism	Enzymes and other metabolites	Applications	References
Streptomyces fradiae	Antibiotics, amylase, protease, cellullase, lipases	Pharma industry, medicine, food industry, detergent additives	Nguyen et al. (2010)
Streptomyces anulatus	Dextranase	Dextrane hydrolysis; sugar industry	Doaa Mahmoud and Wafaa Helmy (2009)
Streptoverticillium	Antibiotics	Pharma industry	Nguyen et al. (2010)
Fungi			
Candida antarctica	Lipases		Joseph et al. (2007)
Candida antarctica		Conversion of n-alkanes into glycolipid; biosurfactants	Kitamoto et al. (2001)
Cryptococcus albidus	Xylanase	Hydrolyzing xylane for improvement of food, waste treatment, food industry	Amoresano et al. (2000)
Cryptococcus laurentii	Phytase	animal feeding	Pavlova et al. (2008)
Cryptococcus laurentii	β-Galactosidases	Dairy industry	Law and Goodenough (1995)
Cryptococcus cylindricus	Pectinases	Clarification of fruit juices; improving filterability, and extractability of juices	Nakagawa et al. (2004)
Cystofilobasidium capitatum	Pectinases	Clarification of fruit juices; improving filterability, and extractability of juices	Nakagawa et al. (2004)
Mrakia frigida	Pectinases	Clarification of fruit juices; improving filterability, and extractability of juices	Nakagawa et al. (2004)
Pichia lynferdii strain Y-7723	Lipase		Kim et al. (2010)
Rhodotorula psychrophenolica		degradation of phenolic compounds	Margesin et al. (2007)
Algae			
Porphyridium cruentum	Eicosapentaenoic acid, arachidonic acid	Pharma industry	Cohen (1990)

Pseudomonas aeruginosa, Staphylococcus aureus, Micrococcus luteus, Bacillus subtilis, Proteus mirabilis, Salmonella enterica, and the yeast *Candida albicans*, following incubation at 37°C on nutrient agar (Lo Giudice et al. 2007). The isolates were identified by 16S rRNA and characterized by biochemical tests. From 580 isolated strains belonging to *Arthrobacter, Rhodococcus, Pseudoalteromonas*, and *Janibacter*, some were able to inhibit the test strains. The actinomycetes *Intrasporangium* sp., *Micromonospora* sp., *Streptoverticillium* sp., *Streptomyces* sp., and *Dactylsporangium roseum*, isolated from the soils from India at over 4000 m altitude, showed different antibiotic activities against strains of *Streptococcus* isolated from dental plaque (Raja et al. 2010). From cold environments, strains close to *Serratia* and *Pseudomonas* were isolated; both are producers of antimicrobials, probably class II microcins acting in the cold (Sanchez et al. 2009).

Enzymes. Many psychrophilic and psychrotolerant bacteria possess the capacity to produce extracellular enzymes such as lipases, proteases, amylases, cellulases, chitinases, and β-galactosidase, when induced by the presence of specific substrates. Producers were strains of sea ice microorganisms, for example, for lipases *Colwellia psychroerythrea, Shewanella livingstonensis,* and *Marinomonas prymoriensis*; for hydrolysis of polysaccharides *Colwellia, Marinomonas, Pseudoalteromonas, Pseudomonas,* and *Shewanella*; for hydrolysis of chitin *Pseudoalteromonas tetraodonis, Pseudoalteromonas elyakovii, Bacillus firmus,* and *Janibacter melonis*, which degrade the organic matter from phyto- and zooplankton (Yu et al. 2009). A simple screening of 137 cold water isolates belonging to *Moraxella, Pseudomonas, Aeromonas, Chromobacterium, Vibrio,* and others showed that about 62% can produce gelatinase, 71% proteases, 31% lipases, 47% amylases, 36% chitinases, 36% β-galactosidases, 47% cellulases, and 25% alginate lyases (Ramaiah 1994). Groudieva et al. (2004) found that from 116 strains isolated from the Spitsbergen sea ice 40% possessed the ability to degrade skim milk, casein, lipids, starch and proteins. Enzymes such as dehydrogenase from cold-adapted microorganisms can also be used as biosensors or in biotransformations (Gomes and Steiner 2004). Protease-producing strains from the genera *Pseudoalteromonas, Shewanella, Colwellia,* and *Planococcus* were isolated from the sub-Antarctic marine sediments of Isla de Los Estados (Olivera et al. 2007).

The enzymes from psychrophilic and psychrotolerant strains can be divided into three categories: (1) heat sensitive, but similar to mesophilic enzymes; (2) heat sensitive and more active at low temperatures than mesophilic enzymes; and (3) heat sensitive exactly as mesophilic enzymes, but more active at lower temperatures (Ohgiya et al. 1999). Another example is a complex of enzymes generated by an Antarctic isolate, *B. subtilis* strain MIUG 6150, which produces α- and β-amylases and proteases (Bahrim and Negoiță 2004). The productivity of the microorganisms showed a strong dependency on the culture media used for growth and other conditions (Bahrim et al. 2007). The cold-adapted strains of *Streptomyces* can be a source of valuable enzymes (Cotârleț et al. 2008).

Proteases. Strains producing proteases belong to the genera *Bacillus* and *Pseudomonas* and were isolated from Antarctic cyanobacterial mats. They showed good production at a temperature of about 20°C, with glucose and maltose as carbon sources (*Pseudomonas* sp.) and soybean meal and peptone as nitrogen sources (Singh and Ramana Venkata 1998). A trypsin-like protease was identified and characterized which is produced by *Serratia proteamaculans* (Mikhailova et al. 2006). These proteases are used in the dairy industry to enhance the flavour development in cheese. In the chemical industry, the enzymes from psychrophilic strains are used in detergents, food industry, and leather manufacturing (Cavicchioli et al. 2002).

Alkaline phosphatase was isolated and purified from the strain *Shewanella* sp. (Ishida et al. 1998). Interestingly, the enzyme showed a maximum activity at 40°C, 39% of that activity at 0°C, and a tendency to loose activity at 20°C. Two different extracellular alkaline phosphatases were identified from an *Arthrobacter* sp. strain (De Prada et al. 1996). Purified recombinant serine alkaline protease (in *E. coli*) from another *Shewanella* strain showed activity between 5°C and 15°C (Kulakova et al. 1999). The enzymes, such as the microorganisms themselves, face sometimes diverse polyextreme conditions; e.g., the cold-active protease MCP-03 from *Pseudoalteromonas* sp. SM9913 was active and stable also in a high salt environment of about 3 M NaCl/KCl (Yan et al. 2009).

β-Galactosidases. One possible applications of this enzyme is obtaining an ice cream with reduced lactose content for lactose intolerant peoples (Phadtare and Inouye 2008). The removal of lactose from milk is very important for persons with lactose intolerance. The cold-active β-galactosidases (EC3.2.1.23) isolated from psychrophilic yeasts (e.g., *Cryptococus laurentis*; Law and Goodenough 1995) and fungi can be used to hydrolyze lactose and therefore a new method of supplementation of milk with dormant cultures was proposed (Somkutl and Holsinger 1997). The utilization of cold-active β-galactosidase (optimum activity at 10°C) from *Arthrobacter psychrolactophilus* strain F2, which was overexpressed in *E. coli*, for the production of trisaccharides from lactose was also tested for applications in the food industry (Tomoyuki et al. 2007). Some strains contain isoenzymes (C2-2-1 and C2-2-2) such as an *Arthrobacter* strain (Karasová et al. 2002), which was found – as a first example – being able to catalyze transglycosylation reactions in the cold.

Lipases. Cold-active lipases (triacylglycerol acylhydrolases, EC3.1.1.3) were isolated from many psychrophilic strains and can have many industrial applications such as additives for detergents, additives in food products, in bioremediation, and in molecular biology (Joseph et al. 2007, 2008). Strains of *Acinetobacter, Achromobacter, Moraxella, Psychrobacter, Pseudoalteromonas, Serratia*, and others are lipase producers. An *Aeromonas* strain produces a cold-active lipase (Lee et al. 2003), and fungal lipase producers such as *Candida antarctica, Geotrichum*, and *Aspergillus* sp. were described. From 137 anaerobic strains isolated from soils in Schirmacher Oasis, Antarctica, 49 isolates showed lipolytic activity on Tween-agar medium (Ramteke et al. 2005). *Psychrobacter okhotskensis* was isolated from Okhotsk seawater (Yumoto et al. 2003) and is a producer of cold-active lipases. These cold-active enzymes have a larger K coefficient and high efficiency down to temperatures of zero degree; they are inactivated by raising the temperature. They are used in the food industry and chemical industry; the latter utilizes them for catalyzing reactions with compounds which are unstable at higher temperatures (Suzuki et al. 2001), for example, lipases and esterases from *Pseudomonas* sp. strain B11-1. Many detergents contain a mixture of proteases, lipases, and amylases (Ohgiya et al. 1999). Some lipases can be used to remove fatty stains from various textiles (Araújo et al. 2008). Lipases can also be used for the synthesis of biopolymers, biodiesel, pharmaceuticals, and certain aromatic products (Joseph et al. 2007). The authors reported also that lipases from psychrophilic and psychrotolerant strains can be used in cold environments for the bioremediation areas contaminated by oil and grease, as well as in the detergent industry. From deep-sea sediments of the Pacific ocean, *Photobacterium frigidiphilum*, a lipolytic psychrophilic bacterium, was isolated (Seo et al. 2005). Some yeasts such as *Pichia lynferdii* strain Y-7723 produce a cold-adapted lipase (Kim et al. 2010). Lipases have a wide range of applications reviewed by Joseph et al. (2007, 2008), such as aryl aliphatic glycolipids, synthesis of fine chemicals, production of fatty acids, interesterification of fats, detergent additives, synthesis of biodiesel, removal of hydrocarbons, oils, and lipidic pollutants. Other uses are in the food industry and concern the improvement of food structure and gelling of fish meat (Cavicchioli and Siddiqui 2004). Nielsen et al. (1999) isolated two rather thermotolerant lipases A and B, with uses in the textile industry for the removal of waxes and lipids from fibers. The lipases obtained from microorganisms, which are used in different detergent formulations, are covered by many patents issued for industrial companies (Hasan et al. 2010). Lipase B from *C. antarctica* was immobilized onto epoxy-activated macroporous poly(methyl methacrylate) Amberzyme beads and on nanoparticles, in order to improve contact with the substrate and the reaction activity for polycondensation (Chen et al. 2008a). The enzyme was quickly adsorbed on the polystyrene porous particles (Chen et al. 2008b).

Pectinases. Several cold-adapted yeasts strains were isolated from the soil of Hokkaido Island (Japan), which were taxonomically affiliated with *Cryptococcus cylindricus, Mrakia frigida,* and *Cystofilobasidium capitatum*. The strains showed pectinolytic activity at temperatures less than 5°C and can be used for the production of pectinolytic enzymes (pectin methylesterase EC3.1.1.11; endopolygalacturonase EC3.2.1.15) for the clarification of fruit juices at low temperatures (Nakagawa et al. 2004), improving at the same time the filterability and extractability of the juice.

Malate dehydrogenases. A malate dehydrogenase was purified from *Flavobacterium frigidimaris* KUC1 and characterized by Oikawa et al. (2005). It contains lower amounts of proline and arginine residues compared to other malate dehydrogenases and is dependent on NAD(P)+. The enzyme looses its activity at 55°C within 30 min of incubation. The enzyme can be used for producing malate at low temperatures.

Dextranases. An important problem in the sugar industry is the removal of dextran, a high-molecular-weight polymer of D-glucose, which can lower the recovery of sugar, interfere with material processing and lead to a poor quality of the final product. Bacterial cold-active dextranases can resolve this problem at low temperatures of about 4°C (Doaa Mahmoud and Wafaa Helmy 2009), such as the dextranase from psychrophilic strain *Streptomyces anulatus*.

β-Amylases. The need for cold-active amylases (EC3.2.1.1) and related starch hydrolyzing enzymes, especially for obtaining sweeteners such as palatinose, a disaccharide of glucose and fructose, and cyclodextrin, was reported (Rendleman 1996).

Phytases. A cold-active phytase is produced very efficiently by the Antarctic strain *Cryptococcus laurentii* AL 27 (Pavlova et al. 2008). Phytase is an enzyme which catalyzes the conversion of undigestible phytate to phosphorylated myo-inositol derivatives and inorganic phosphate, which are digestible. Its applications are in the fields of animal food additives and the pharmaceutical industry.

Xylanases. Xylanase from the yeast *Cryptococcus albidus*, isolated from Antarctica, is a glycoprotein; its structure was investigated by mass spectroscopy (Amoresano et al. 2000). The xylanases hydrolyze the heteropolysaccharide xylane (a hemicellulose containing a backbone chain of β-1,4-linked xylanopyranoside residues) and have found wide applications, e.g. improvement of maceration processes, clarification of juices, improvement of filtration efficiency, maceration of grape skins in wine technology, reducing viscosity of coffee extracts, improve drying and lyophilization processes, improving the elasticity of dough and bread textures. Xylanases can also be used to degrade xylane from agricultural wastes in order to obtain energy from biomass. Hydrolyzing xylane from the cell walls of plants at low temperatures will allow energy savings and the production of more accessible feedstock (Lee et al. 2006). Furthermore, xylanases are used for the pulping process in the paper industry and for biobleaching (Beg et al. 2001), thereby reducing the use of alkali. They also improve energy consumption in the textile industry, being used in the microbiological retting of textile materials, which replaces chemical retting. They are useful for obtaining fermentation products, bioethanol, and other chemicals as well as improving the separation of starch and gluten in the starch industry. The glycoside hydrolase family 8 xylanases can be used in baking processes in order to improve the flexibility of dough and product quality (Collins et al. 2006).

β-Mannanase. β-Mannanase (EC3.2.1.78) was isolated from *Flavobacterium* sp. and showed good activity at 4°C; it can be used to decrease the viscosity in food products (Zakaria et al. 1998).

Nitrile hydratase. Companies such as Nitrochemicals developed many years ago the production of acrylamide with the help of *Rhodoccus* sp. strain N774 (Kobayashi et al. 1992), which produces the enzyme nitrile hydratase (EC4.2.1.84).

Cellulases. Some cellulases are used in bleaching and bio-stoning of textile material (Gomes and Steiner 2004), and alkaline cellulases used in detergents are active toward amorphous cellulose (Ito et al. 1989).

Trehalose. In agriculture, trehalose-producing systems can be used for reducing crop losses due to the lower temperatures with the help of genetic engineering (Phadtare and Inouye 2008).

EPSs. Extracellular polymeric substances released by microorganisms, which promote the formation of biofilms and have presumably protective roles. They can be used in the chemical industry to produce biodegradable plastic materials. Several strains with potential for this type of production were investigated as well as the conditions for production, such as temperature, pressure and pH (Marx et al. 2009). The authors reviewed useful microorganisms isolated from cold Antarctic environments, e.g., *Morixella, Psychrobacter,* and *Aeromonas* from polar waters, and psychrotolerants such as *Pseudomonas* and *Photobacterium.* The strains showed a good production of EPS at $-4°C$ to $-10°C$ and resistance under high-pressure conditions between 1 and 200 atm.

Medicinal uses. Besides improving the quality of foods, antifreeze proteins can also improve the preservability of human organs for transplants, for example livers (Kawahara 2008). Frisvad (2008a) reviewed bioactive products from cold-adapted fungi, such as griseofulvin and cycloaspeptide A from *Penicillium soppi* and *P. lanosum* from cold soils; the latter compound can be used as an antimalarial product. Cycloaspeptides were found so far only in cold-adapted fungi.

PUFAs. PUFAs are produced by many different organisms, for instance by strains such as *S. frigidimarina* (Bowman et al. 1997). They can be used as food supplements and medicinal products. Russell and Nichols (1999) showed that the bacterial PUFA-producing strain cannot compete with the fungal PUFA-producing strain, but can be an alternative for feedstock in the food chains used in aquaculture.

Biomining. The biomining industry is developing processes at low temperatures in three-phase systems: the solid phase, which is represented by the mineral ore; a liquid phase, which contains the microorganisms and nutrients; and the gaseous phase (Rossi 1999). Such a process can be performed in stirred tank reactors or in airlift reactors. The process was used for the release and recovery of copper from sulfide minerals, of uranium and for the pretreatment of gold ores (Ovalle 1987).

Bioremediation. Psychrophilic strains can be used to degrade the organic pollutants from soils and waters at low temperatures. Many strains possessing biodegrading properties were isolated from polar and alpine areas, from soils and waters, but more research of some aspects is required, such as the stability of the bacterial community, the accessibility of the pollutant for microorganisms, and the low removal rate (Margesin and Schinner 2001). Petroleum spills can produce catastrophic damages and their cleanup is an important goal. Petroleum is a complex mixture of water-soluble and -insoluble compounds (linear cyclic alkanes, aromatic hydrocarbons, paraffin, asphalt, and waxy oils; Brakstad 2008), being very hard to degrade. About 200 bacterial, cyanobacterial, fungal, and algal genera possess the capacity to do it (Prince 2005). The main bacterial genera able to degrade petroleum are *Acinetobacter, Arthrobacter, Colwellia, Cytophaga, Halomonas, Marinobacter, Marinomonas, Pseudoalteromonas, Oleispira, Rhodococcus,* and *Shewanella* (Brakstad 2008). Both anaerobic and aerobic degradation are possible and bioremediation can occur by stimulation of the local hydrocarbon degraders (using dispersants and nutrients) and less by bioaugmentation (inoculation of cultures of hydrocarbonoclastic bacteria; Margesin and Schinner 1999). *Dietzia psychralkaliphila* can grow on defined culture media containing *n*-alkenes as sole carbon source (Yumoto et al. 2002). The strain *Rhodocous* sp. Q15 is able to degrade short- and long-chain aliphatic alkenes from diesel fuel at low temperatures of about 5°C (Whyte et al. 1998). Two strains of *Rhodococcus, R. ruber,* and *R. erythrococcus,* were used by Russian researchers for the product "Rhoder," which is applied for the removal of oil pollution (Murygina et al. 2000). Some strains isolated from alpine soils are able to degrade phenol and phenolic compounds (bacteria such as *Rhodococcus* spp., *Arthrobacter psychrophenolicus,* and *Pseudomonas*; yeasts such as *Rhodotorula psychrophenolica, Trichosporon dulcitum,* and *Leucosporidium watsoni*) and hydrocarbons from oil at low temperatures (Margesin 2007; Margesin et al. 2007), even though a complete biodegradation cannot be obtained. Polychlorophenols are toxic and persistent pollutants which are used as biocidal wood preservatives (Langwaldt et al. 2008). The authors list several genera such as *Ralstonia, Burkholderia, Arthrobacter, Rhodococcus, Mycobacteria,* and anaerobes such as *Desulfomonile* and *Desulfitobacterium,* which are able to degrade polychlorophenol compounds at low temperatures.

Polaromonas sp. strain JS666 is an isolate which can grow on *cis*-1,2-dichloroethene as carbon source; the investigation of its genome showed genes for the metabolism of aromatic compounds, alkanes, alcohols and others (Mattes et al. 2008). Another strain, *Polaromonas naphthalenivorans*, was isolated from a contaminated freshwater environment and is capable of degrading naphthalene (Jeon et al. 2004). The isolation of the genes for naphthalene dehydrogenase from cold environments was reported (Flocco et al. 2009). For bioremediation, the genus *Shewanella*, which can use a wide range of electron acceptors, is important, since many members of this group show capabilities for degrading several pollutants. The genus showed the possibility to be used in bioremediation of radionuclide and reduction of elements such as Co, Hg, Cr, and As, as well as for the removal of organics such as halogenated compounds, e.g., tetrachloromethane or nitramine (an explosive contaminant), as reported in the review by Hau and Gralnick (2007). Many strains such as *Brevibacterium antarcticum* have demonstrated a polyresistance to heavy metals in high concentrations, resisting concentrations of Cu^{2+}, Hg^{2+}, and CrO_4, up to 6000 ppm (Tashyrev 2009). Several psychrophilic microorganisms are able to degrade natural organic polymers (starch, agar, and gelatin) such as *Flavobacterium limicola*, which was isolated from freshwater sediments (Tamaki et al. 2003).

Wastes and wastewater treatments. The anaerobic treatment of wastewaters in treatment plants, using expanded granular sludge bed reactors at temperatures of 5–10°C, looks very promising (Lettinga et al. 2001). A mixture of microorganisms such as *Methanobrevibacter* sp., *Methanosarcina* sp., and *Methanosaeta* sp. has been explored (Lettinga et al. 1999). The psychrophilic treatment of landfill leachates using anoxic/oxic biofilters appears to be a good solution to prevent water and soil pollution with such leachates containing organic matter and also heavy metals (Kalyuzhnyi et al. 2004). *Oleispira antarctica* is able to degrade hydrocarbons in cold marine waters (Yakimov et al. 2003). Aerobic treatment of wastewater in cold lagoons has been performed in Canada's cold areas with success (Smith and Emde 1999). The cold-adapted xylanases can be used for the hydrolysis of agricultural and food industry wastes. A selection of cold-active degrading microorganism for wastewater treatment was performed (Gratia et al. 2009), where *A. psychrolactophilus* Sp 31.3 was isolated, which had the desired characteristics and was used further.

Acid mine drainage is the result of oxidation of certain sulfide minerals by exposure to environmental conditions and the activity of microorganisms. For example, ores containing pyrite and chalcopyrites are oxidized in the presence of water and oxygen and form highly acidic, sulfate-rich drainage. Ferrous iron (Fe^{2+}) develops in the process, which can be re-oxidized by acidophilic bacteria and archaea to ferric iron (Fe^{3+}), and the sulfur is oxidized to sulfate. These oxidations and the concomitant dissolution of sulfide minerals can take place in cold conditions, too. Sulfate reduction at low temperatures occurs with bacteria such as *Desulfofrigus*, *Desulfofaba*, *Desulfotalea*, and *Desulfovibrio* (Kaksonen et al. 2008). *Acidithiobacillus ferrooxidans* is also able to oxidize iron and sulfur compounds at low temperatures (Kaksonen et al. 2008).

Astrobiological models. A special theoretical application concerns astrobiology, since some scientists are considering the Antarctic a model of planet Mars or other planets, with respect to the low temperatures and water activity (Abyzov et al. 1998), and also a model of cold environments where certain microorganisms could possibly live (Deming 2007). At the same time, the protocols for sampling of ice and permafrost and their analysis can be used for the exploration of Mars, and could be relevant and helpful in the isolation of potential Martian microbiota and for the development of future protocols for the decontamination of extraterrestrial samples (Christner et al. 2005).

7 Conclusions

1. Psychrophilic and psychrotolerant microorganisms can be retrieved from very diverse environments – oceans and freshwater, hypersaline cold waters, sediments, soils, permafrost, ice, glaciers, cold deserts, alpine soils, lakes and snow, cold man-made environments, and some microecosystems. The microorganisms in ice layers constitute not real ecosystems, even if some activity at subzero temperatures was proven; instead, most of them are only opportunistic assemblages of mixtures of microorganisms brought together by air and water currents from other environments. The diversity of so-called cold environments is much greater than was thought initially, and many microenvironments can be distinguished. In addition, the cold-adapted members of microbiota can have different other adaptations to extreme conditions – resistance to high radiation, oligotrophy, adaptation to high pressures, and perhaps others.
2. Psychrophiles are found in all the three domains of life and have a very diverse taxonomic origin. The most frequent taxonomic groups in cold environments are *Alpha-*, *Beta-*, *Delta-*, and *Gamma-Proteobacteria*, the phylum *Cytophaga–Flavobacterium–Bacteriodetes*, and Actinobacteria. Together with prokaryotes (Archaea and Bacteria) numerous eukaryotes are present such as algae, yeasts, and fungi.
3. Special adaptations allowing life in the cold include membrane lipids with branched unsaturated fatty acids, proteins and enzymes with a more flexible 3D structure due to the reduction of the number of weak intramolecular bonds, reduction of salt bridges, reduction of aromatic interactions, density of charged surface residues, increased surface hydrophobicity, and increased clustering of

glycine residues. Special proteins are Csps, ice nucleation proteins and antifreeze proteins, which protect the structure of cells from the cold and from formation of ice crystals.
4. The molecular biology of psychrophiles showed that their enzymes have low activation energy requirements due to their structure discussed in the text and to the flexibility of near active sites domains, and that they are easily inactivated by higher temperatures.
5. The different possibilities of adaptation of the microorganisms, either psychrophiles or psychrotolerants, showed complicated mechanisms, combining adaptation features and environmental opportunities. Their adaptation mechanisms are thus much more flexible than we have thought, providing possibilities and strategies of survival in extreme conditions. There is evidence for survival of psychrophiles in such conditions for very long periods of time which suggests the possibility of survival of similar microorganisms on other planets.
6. Psychrophiles produce bioactive and useful compounds, especially enzymes, pharmaceuticals, biodegradable plastics, substances for medical care, agriculture, biomining and bioremediation of wastes; all being usable in low temperature conditions, which entails important energy savings.
7. More laboratory and field bioprospecting should be envisaged for the isolation and identification of appropriate microorganisms for psychrophilic biotechnology.

Acknowledgments

We thank Dr. D.S. Nichols and his co-workers for the kind permission to integrate their data on enzymes from psychrophilic microorganisms in the table of this chapter.

This work benefitted from the research funded by the Executive Unit for Funding of High Level Education and of Universities Scientific Research, Romania (U.E.F.I.S.C.S.U.) in the frame of contract nr 1, Europolar 2010.

References

Abyzov SS, Mitskevich IN, Poglazova MN, Barkov NI, Lipenkov VY, Bobin NE, Koudryashov BB, Pashkevich VM (1998) Antarctic ice sheet as a model in search of life on other planets. Adv Space Res 22:363–368

Allen MA, Lauro FM, Williams TJ, Burg D, Siddiqui KS, De Francisci D, Chong KW, Pilak O, Chew HH, De Maere MZ, Ting L, Katrib M, Ng C, Sowers KR, Galperin MY, Anderson IJ, Ivanova N, Dalin E, Martinez M, Lapidus A, Hauser L, Land M, Thomas T, Cavicchioli R (2009) The genome sequence of the psychrophilic archaeon, *Methanococcoides burtonii*: the role of genome evolution in cold adaptation. ISME J 3:1012–1035

Amoresano A, Andolfo A, Corsaro MM, Zocchi I, Petrescu I, Gerday I, Marino G (2000) Structural characterization of a xylanase from psychrophilic yeast by mass spectrometry. Glycobiology 10:451–458

Andersson RE, Hedlund CB, Jonsson U (1979) Thermal inactivation of a heat-resistant lipase produced by the psychrotrophic bacterium *Pseudomonas fluorescens*. J Dairy Sci 62:361–367

Araújo R, Casal M, Cavaco-Paulo A (2008) Application of enzymes for textile fibres processing. Biocatal Biotransform 26:332–349

Auborn KJ, Fan S, Rosen EM, Goodwin L, Chandraskaren A, Williams DE, Chen D, Carter TH (2003) Indole-3-carbinol is a negative regulator of estrogen. J Nutr 133(Suppl):2470S–2475S

Ayala-del-Río HL, Chain PS, Grzymski JJ, Ponder MA, Ivanova N, Bergholz PW, Di Bartolo G, Hauser L, Land M, Bakermans C, Rodrigues D, Klappenbach J, Zarka D, Larimer F, Richardson P, Murray A, Thomashow M, Tiedje JM (2010) The genome sequence of *Psychrobacter arcticus* 273-4, a psychroactive Siberian permafrost bacterium, reveals mechanisms for adaptation to low-temperature growth. Appl Environ Microbiol 76:2304–2312

Ayub ND, Tribelli PM, Lopez N (2009) Polyhydroxyalkanoates are essential for maintenance of redox state in the Antarctic bacterium *Pseudomonas* sp. 14-3 during low temperature adaptation. Extremophiles 13:59–66

Babalola OO, Kirby BM, Le Roes-Hill M, Cook AE, Craig CS, Burton SG, Cowan DA (2008) Phylogenetic analysis of actinobacterial populations associated with Antarctic Dry Valley mineral soils. Environ Microbiol 11:566–576

Bae E, Phillips GN Jr (2004) Structures and analysis of highly homologous psychrophilic, mesophilic, and thermophilic adenylate kinases. J Biol Chem 279:28202–28208

Bahrim G, Negoiță TG (2004) Effects of inorganic nitrogen and phosphorous sources on hydrolase complex production by the selected *Bacillus subtilis* Antarctic strain. Rom Biotechnol Lett 9:1925–1932

Bahrim GE, Scântee M, Negoiță TG (2007) Biotechnological conditions of amylase and protease complex production and utilization involving filamentous bacteria. In: Annals Univers "Dunărea de Jos" Galați, Fasc VI Food Technol, pp 76–81

Bakermans C (2008) Limits for microbial life at subzero temperatures. In: Margesin R, Schinner R, Marx JC, Gerday C (eds) Psychrophiles: from biodiversity to biotechnology. Springer, Berlin, Heidelberg, pp 17–28

Beg QK, Kapoor M, Mhajan L, Hoondal GS (2001) Microbial xylanases and their industrial applications – a review. Appl Microbiol Biotechnol 56:326–338

Benešova E, Markova M, Kralova B (2005) Alpha glucosidase and beta glucosidase from psychrophilic strain *Arthrobacter* sp. C2-2. Czech J Food Sci 23:116–120

Białkowska AM, Cieslinski H, Niowakowska KM, Kur J, Turkievich M (2009) A new beta-galactosidase with a low temperature optimum isolated from the Antarctic *Arthrobacter* sp. 20B: gene cloning, purification, characterization. Arch Microbiol 191:825–835

Blees J, Wenk C, Niemann C, Schubert C, Zopfi J, Veronesi M, Simona M, Lehmann M (2010) The isotopic signature of methane oxidation in a deep south-alpine lake Geophysical Research Abstracts 12, EGU2010-788

Bowman JP (2008) Genomic analysis of psychrophilic prokaryotes. In: Margesin R, Schinner F, Marx JC, Gerday C (eds) Psychrophiles: from biodiversity to biotechnology. Springer, Berlin, Heidelberg, pp 265–284

Bowman JP, McCammon SA, Nichols DS, Skerratt JH, Rea SM, Nichols PD, McMeekin TA (1997) *Shewanella gelidimarina* sp. nov. and *Shewanella frigidimarina* sp. nov., novel Antarctic species with the ability to produce eicosapentaenoic acid (20:5 omega 3) and grow anaerobically by dissimilatory Fe(III) reduction. Int J Syst Bacteriol 47:1040–1047

Bowman JP, Gosink JJ, McCammon SA, Lewis TE, Nichols DS, Nichols PD, Skeratt JH, Staley JT, McMeekin TA (1998) *Colwellia demingiae* sp. nov., *Colwellia hornerae* sp. nov., *Colwellia rossensis* sp. nov. and *Colwellia psychrotropica* sp. nov.: psychrophilic Antarctic species with the ability to synthesize docosahexaenoic acid (22:6ω3) Int J Syst Bacteriol 48:1171–1180

Bozal N, Montes MJ, Tudela E, Jimenez F, Guinea J (2002) *Shewanella frigidimarina* and *Shewanella livingstonensis* sp. nov. isolated from Antarctic coastal areas. Int J Syst Evol Microbiol 52:195–205

Brakstad OG (2008) Natural and stimulated biodegradation of petroleum in cold marine environments. In: Margesin R, Schinner F, Marx JC, Gerday C (eds) Psychrophiles: from biodiversity to biotechnology. Springer, Berlin, Heidelberg, pp 389–428

Breezee J, Cady N, Staley JT (2004) Subfreezing growth of the sea ice bacterium *Psychromonas ingrahamii*. Microbial Ecol 47:300–304

Brenchley JE (1996) Psychrophilic microorganisms and their cold-active enzymes. J Ind Microbiol Biotechnol 17:432–437

Busse H-J, Denner EBM, Buczolits S, Salkinoja-Salonen M, Bennasar A, Kämpfer P (2003) *Sphingomonas aurantiaca* sp. nov., *Sphingomonas aerolata* sp. nov. and *Sphingomonas faeni* sp. nov., air- and dustborne and Antarctic, orange pigmented, psychrotolerant bacteria, and emended description of the genus *Sphingomonas*. Int J Syst Evol Microbiol 53:1253–1260

Cavicchioli R (2006) Cold adapted Archaea. Nat Rev Microbiol 4:331–343

Cavicchioli R, Siddiqui KS (2004) Cold-adapted enzymes. In: Paney A, Webb C, Socol CR, Larroche C (eds) Enzyme technology. Asia Tech Publishers, New Delhi, pp 615–638

Cavicchioli R, Siddiqui KS, Andrews D, Sowers KR (2002) Low temperature extremophiles and their applications. Curr Opin Biotechnol 13:253–261

Chen B, Hu J, Miller EM, Xie W, Cai M, Gross RA (2008a) *Candida antarctica* lipase B chemically immobilized on epoxy-activated micro- and nanobeads: catalysts for polyester synthesis. Biomacromolecules 9:463–470

Chen B, Miller ME, Gross RA (2008b) Immobilization of *Candida antarctica* lipase B on porous polystyrene resins: protein distribution and activity. In: Polymer biocatalysis and biomaterials II, ACS symposium series, vol 999. American Chemical Society, Washington, DC, pp 165–177

Chintalapati S, Kiran MD, Shivaji S (2004) Role of membrane lipid fatty acids in cold adaptation. Cell Mol Biol 50:631–642

Christner BC (2002) Incorporation of DNA and protein precursors into macromolecules by bacteria at $-15°$C. Appl Environ Microbiol 68:6435–6438

Christner BC, Mosley-Thompson E, Thompson LG, Reeve JN (2003a) Bacterial recovery from ancient glacial ice. Environ Microbiol 5:433–436

Christner BC, Kvitko BH, Reeve JN (2003b) Molecular identification of bacteria and eukarya inhabiting an Antarctic cryoconite hole. Extremophiles 7:177–183

Christner BC, Mikucki JA, Foreman CM, Denson J, Priscu JC (2005) Glacial ice cores: a model system for developing extraterrestrial decontamination protocols. Icarus 174:572–584

Christner BC, Skidmore ML, Priscu JC, Tranter M, Foreman C (2008) Bacteria in subglacial environments. In: Margesin R, Schinner F, Marx JC, Gerday C (eds) Psychrophiles: from biodiversity to biotechnology. Springer, Berlin, Heidelberg, pp 51–71

Cohen Z (1990) The production potential of eicosapentaenoic acid and arachidonic acid of the red algae *Porphyridium cruentum*. J Am Oil Chem Soc 67:916–920

Collins T, Dutron A, Georis J, Genot B, Dauvrin T, Arnaut F, Gerday C, Feller G (2006) Use of glycoside hydrolase family 8 xylanases in baking. J Cereal Sci 43:79–84

Collins T, Roulling F, Piette F, Marx JC, Feller G, Gerday C, D'Amico S (2008) Fundamentals of cold adapted enzymes. In: Margesin R, Schinner F, Marx JC, Gerday C (eds) Psychrophiles: from biodiversity to biotechnology. Springer, Berlin, Heidelberg, pp 211–228

Copeland A, Lucas S, Lapidus A, Barry K, Detter JC, Glavina del Rio T, Hammon N, Israni S, Dalin E, Tice H, Pitluck S, Fredrickson JK, Kolker E, McCuel LA, DiChristina T, Nealson KH, Newman D, Tiedje JM, Zhou J, Romine MF, Culley DE, Serres M, Chertkov O, Brettin T, Bruce D, Han C, Tapia R, Gilna P, Schmutz J, Larimer F, Land M, Hauser L, Kyrpides N, Mikailova N, Richardson P (2006) Complete sequence of *Shewanella frigidimarina* NCIMB 400. Submitted (Aug. Sept. 2006). Released 09/14/2006 by the DOE Joint Genome Institute

Cotârleț M, Negoiță TG, Bahrim G, Stougaard P (2008) Screening of polar streptomycetes able to produce cold-active hydrolytic enzymes using common and chromogenic substrates. Rom Biotechnol Lett 13:69–80 [special issue, edited for Int Conf Ind Microbiol Appl Biotechnol]

Cowan DA, Casanueva A (2007) Stability of ATP in Antarctic mineral soils. Polar Biol 30:1599–1603

D'Amico S, Collins T, Marx JC, Feller G, Gerday C (2006) Psychrophilic microorganisms: challenges for life. EMBO Rep 7:385–389

de los Ríos A, Grube M, Sancho LG, Ascaso C (2007) Ultrastructural and genetic characteristics of endolithic cyanobacterial biofilms colonizing Antarctic granite rocks. FEMS Microbiol Ecol 59:386–395

De Prada P, Loveland-Curtze J, Brenchley JE (1996) Production of two extracellular alkaline phosphatases by a psychrophilic *Arthrobacter* strain. Appl Environ Microbiol 62:3732–3738

Deming JW (2007) Life in ice formations at very cold temperatures. In: Gerday C, Glansdorff N (eds) Physiology and biochemistry of extremophiles. ASM Press, Washington, DC, pp 133–145

Di Prisco G (2007) Lake Vostok and subglacial lakes of Antarctica: do they host life? In: Gerday C, Glansdorff F (eds) Physiology and biochemistry of extremophiles. ASM Press, Washington, DC, pp 145–154

Doaa Mahmoud AR, Wafaa Helmy A (2009) Application of cold-active dextranase in dextran degradation and isomaltotriose synthesis by micro-reaction technology. Aust J Basic Appl Sci 3:3808–3817

Feller G (2007) Life at low temperatures: is disorder the driving force? Extremophiles 11:211–216

Feller G, Gerday C (1997) Psychrophilic enzymes: molecular basis of cold-adaptation. Cell Mol Life Sci 53:830–841

Feller G, Le Bussy O, Gerday C (1998) Expression of psychrophilic genes in mesophilic hosts: assessment of the folding state of a recombinant α-amylase. Appl Environ Microbiol 64:1163–1165

Finster K (2008) Anaerobic bacteria and archaea in cold ecosystems. In: Margesin R, Schinner F, Marx JC, Gerday C (eds) Psychrophiles: from biodiversity to biotechnology. Springer, Berlin, Heidelberg, pp 103–119

Flocco CG, Newton C, Gomes M, MacCormack W, Smalla K (2009) Occurrence and diversity of naphthalene dioxygenase genes in soil microbial communities from the maritime Antarctic. Environ Microbiol 11:700–714

Frisvad JC (2008a) Cold adapted fungi as source of valuable Metabolites. In: Margesin R, Schinner F, Marx JC, Gerday C (eds) Psychrophiles: from biodiversity to biotechnology. Springer, Berlin, Heidelberg, pp 381–388

Frisvad JC (2008b) Fungi in cold environment. In: Margesin R, Schinner F, Marx JC, Gerday C (eds) Psychrophiles: from biodiversity to biotechnology. Springer, Berlin, Heidelberg, pp 137–156

Galante YM, Formantici C (2003) Enzyme applications in detergency and in manufacturing industries. Curr Org Chem 7:1399–1422

Ganzert L, Jurgens G, Münster U, Wagner D (2007) Methanogenic communities in permafrost-affected soils of the Laptev Sea Coast, Siberian Arctic, characterized by 16S rRNA gene fingerprints. FEMS Microbiol Ecol 59:476–488

Georlette D, Damien B, Blaise V, Depiereux E, Uversky VN, Gerday C, Feller G (2003) Structural and functional adaptations to extreme temperatures in psychrophilic, mesophilic and thermophilic DNA ligases. J Biol Chem 278:37015–37023

Gesheva V (2009) Distribution of psychrophilic microorganisms in soils of Terra Nova Bay and Edmonson Point, Victoria Land and their biosynthetic capabilities. Polar Biol 32:1287–1291

Gianese G, Argos P, Pascarella S (2001) Structural adaptation of enzymes to low temperatures. Prot Eng 14:141–148

Giaquinto L, Curmi PMG, Siddiqui KS, Poljak A, DeLong E, DasSarma S, Cavicchioli R (2007) Structure and function of cold shock proteins in archaea. J Bacteriol 189:5738–5748

Gilichinsky DA (2002) Permafrost. In: Bitton G (ed) Encyclopedia of environmental microbiology. John Wiley & Sons, New York, pp 2367–2385

Gilichinsky D, Vishnivetskaya T, Petrova M, Spirina E, Mamkyn V, Rivkina E, (2008) Bacteria in Permafrost. In: Margesin R, Schinner F, Marx JC, Gerday C (eds) Psychrophiles: from biodiversity to biotechnology. Springer, Berlin, Heidelberg, pp 83–102

Gomes J, Steiner W (2004) The biocatalytic potential of extremophiles and extremozymes. Food Technol Biotechnol 42:223–235

Goto S, Sugiyama J, Iizuka HA (1969) A taxonomical study of Antarctic yeasts. Mycologia 61:748–774

Gounot AM (1999) Microbial life in permanently cold soils. In: Margesin R, Schinner F (eds) Cold-adapted organisms. Springer, Berlin, Heidelberg, New York, pp 3–15

Gratia E, Weekers F, Margesin R, D'Amico S, Thonart P, Feller G (2009) Selection of a cold-adapted bacterium for bioremediation of wastewater at low temperatures. Extremophiles 13:763–768

Groudieva T, Kambourova M, Yusef H, Royter M, Grote R, Trinks H, Antranikian G (2004) Diversity and cold-active hydrolytic enzymes of culturable bacteria associated with Arctic sea ice, Spitzbergen. Extremophiles 8:475–488

Grzymski JJ, Carter BJ, DeLong EF, Feldman RA (2006) Comparative genomics of DNA fragments from six Antarctic marine planktonic bacteria. Appl Environ Microbiol 72:1532–1541

Guffogg SP, Thomas-Hall S, Holloway P, Watson K (2004) A novel psychrotolerant member of the hymenomycetous yeasts from Antarctica: *Cryptococcus watticus* sp. nov. Int J Syst Evol Microbiol 54:275–277

Hamamoto T (1993) Psychrophilic microorganisms from deep sea environments. Riken Rev 3:9–10

Hasan F, Shah AA, Javed S, Hameed A (2010) Enzymes used in detergents: lipases. African J Biotechnol 9:4836–4844

Hau HH, Gralnick JA (2007) Ecology and biotechnology of the genus *Shewanella*. Ann Rev Microbiol 61:237–258

Helmke E, Weyland H (2004) Psychrophilic versus psychrotolerant bacteria – occurrence and significance in polar and temperate marine habitats. Cell Mol Biol 50:553–561

Hodson A, Anesio AM, Tranter M, Fountain A, Osborn M, Priscu J, Laybourn-Parry J, Sattler B (2008) Glacial ecosystems. Ecol Monogr 78:41–67

Hollibaugh JT, Lovejoy C, Murray AE (2007) Microbiology in polar ocean. Oceanography 20: 140–147

Hou S, Saw JH, Lee KS, Freitas TA, Belisle C, Kawarabayasi Y, Donachie SP, Pikina A, Galperin MY, Koonin EV, Makarova KS, Omelchenko MV, Sorokin A, Wolf YI, Li QX, Keum YS, Campbell S, Denery J, Aizawa S, Shibata S, Malahoff A, Alam M (2004) Genome sequence of the deep-sea gamma-proteobacterium *Idiomarina loihiensis* reveals amino acid fermentation as a source of carbon and energy. Proc Natl Acad Sci USA 101:18036–18041

Hoyoux, IJ, Dubois P, Genicot S, Dubail F, Franc JM, Baise Ois E, Feller G, Gerday C (2001) Cold-adapted beta-galactosidase from the Antarctic psychrophile *Pseudoalteromonas haloplanktis*. Appl Environ Microbiol 67:1529–1535

Humphry DR, George A, Black GW, Cummings SP (2001) *Flavobacterium frigidarium* sp. nov., an aerobic, psychrophilic, xylanolytic and laminarinolytic bacterium from Antarctica. Int J Syst Evol Microbiol 51:1235–1243

Huston AL (2008) Biotechnological aspects of cold adapted enzymes. In: Margesin R, Schinner F, Marx JC, Gerday C (eds) Psychrophiles: from biodiversity to biotechnology. Springer, Berlin, Heidelberg, pp 347–364

Ishida Y, Tsuruta H, Tsuneta ST, Uno T, Watanabe K, Aizono I (1998) Characteristics of psychrophilic alkaline phosphatase. Biosci Biotechnol Biochem 62:2246–2250

Ito S, Shikata S, Ozaki K, Kawai S, Okamoto K, Inoue S, Takei A, Ohta Y, Satoh T (1989) Alkaline cellulases for laundry detergents production by *Bacillus* sp. KSM 635 and enzymatic properties. Agric Biol Chem 53:1275–1281

Ivanova EP, Gorshkova NM, Bowman JP, Lysenko AM, Zhukova NV, Sergeev AF, Mikhailov VV, Nicolau DV (2004) *Shewanella pacifica* sp. nov., a polyunsaturated fatty acid-producing bacterium isolated from sea water. Int J Syst Evol Microbiol 54:1083–1087

Jeon CO, Park W, Ghiorse WC, Madsen EL (2004) *Polaromonas naphthalenivorans* sp. nov., a naphthalene-degrading bacterium from naphthalene-contaminated sediment. Int J Syst Evol Microbiol 54:93–97

Joseph B, Ramteke PW, Thomas G, Shrivastava N (2007) Standard review. Cold-active microbial lipases: a versatile tool for industrial applications. Biotechnol Mol Biol Rev 2:39–48

Joseph B, Ramteke PW, Thomas G (2008) Cold active microbial lipases: some hot issues and recent developments. Biotechnol Adv 26:457–470

Junge K, Eicken, H, Deming JW (2003) Motility of *Colwellia psychrerythraea* strain 34H at subzero temperatures. Appl Environ Microbiol 69:4282–4284

Kaksonen AH, Dopson M, Karnachuk O, Tuovinen OH, Puhakka JA (2008) Biological iron oxidation and sulfate reduction in the treatment of acid mine drainage at low temperatures. In: Margesin R, Schinner F, Marx JC, Gerday C (eds) Psychrophiles: from biodiversity to biotechnology. Springer, Berlin, Heidelberg, pp 429–454

Kalyuzhnyi SV, Gladchenko M, Epov A (2004) Combined anaerobic-aerobic treatment of landfill leachates under mesophilic, submesophilic and psychrophilic conditions. Water Sci Technol 48:311–318

Karasová P, Spiwok V, Malá Š, Králová B, Russel NJ (2002) Beta-galactosidase activity in psychrotrophic microorganisms and their potential use in food industry. Czech J Food Sci 20:43–47

Karr EA, Sattley WM, Jung DO, Madigan MT, Achenbach LA (2003) Remarkable diversity of phototrophic purple bacteria in permanently frozen Antarctic lake. Appl Environ Microbiol 69:4910–4914

Karr EA, Ng JM, Belchik SM, Sattley WM, Madigan MT, Achenbach LA (2006) Biodiversity of methanogenic and other archaea in the permanently frozen lake Fryxwell, Antarctica. Appl Environ Microbiol 72:1662–1666

Katayama T, Tanaka M, Moriizumi J, Nakamura T, Brouchkov A, Douglas TA, Fukuda M, Tomita F, Asano K (2007) Phylogenetic analysis of bacteria preserved in a permafrost ice wedge for 25,000 years. Appl Environ Microbiol 73:2360–2363

Kato Y, Sakala RM, Hayashidani H, Kiuchi A, Kaneuchi C, Ogawa M (2000) *Lactobacillus algidus* sp. nov., a psychrophilic lactic acid bacterium isolated from vacuum-packaged refrigerated beef. Int J Syst Evol Microbiol 50:1143–1149

Kawahara H (2008) Cryoprotectant and ice binding proteins. In: Margesin R, Schinner F, Marx JC, Gerday C (eds) Psychrophiles: from biodiversity to biotechnology. Springer, Berlin, Heidelberg, pp 229–246

Khachane AN, Timmis KN, Martins dos Santos VAP (2005) Uracil content of 16S rRNA of thermophilic and psychrophilic prokaryotes correlates inversely with their optimal growth temperatures. Nucleic Acids Res 33:4016–4022

Kim H-R, Kim I-H, Hou CT, Kwon K-Il, Shin B-S (2010) Production of a novel cold-active lipase from *Pichia lynferdii* Y-7723. J Agric Food Chem 58:1322–1326

Kitamoto D, Ikegami T, Suzuki GT Sasaki A, Takeyama Y, Idemoto Y, Koura N, Yanagishita H (2001) Microbial conversion of n-alkanes into glycolipid biosurfactants, annosylerythritol lipids by *Pseudozyma* (*Candida antarctica*). Biotechnol Lett 23:1709–1714

Knoblauch C, Sahm K, Jorgensen BB (1999) Psychrophilic sulfate-reducing bacteria isolated from permanently cold Arctic marine sediments: description of *Desulfofrigus oceanense* gen. nov., sp. nov., *Desulfofrigus fragile* sp. nov., *Desulfofaba gelida* gen. nov., sp. nov., *Desulfotalea psychrophila* gen. nov., sp. nov. and *Desulfotalea arctica* sp. nov. Int J Syst Bact 49:631–643

Kobayashi M, Nagasawa T, Yamada H (1992) Enzymatic synthesis of acrylamide manufacturing process using microorganisms. Trends Biotechnol 10:402–408

Krembs C, Deming JW (2008) The role of exopolymers in microbial adaptation to sea ice. In: Margesin R, Schinner F, Marx JC, Gerday C (eds) Psychrophiles: from biodiversity to biotechnology. Springer, Berlin, Heidelberg, pp 247–286

Krishnan KP, Sinha RK, Krishna K, Nair S, Singh SM (2009) Microbially mediated redox transformations of manganese (II) along with some other trace elements: a study from Antarctic lakes Polar Biol 32:1765–1778

Kulakova L, Galkin A, Kurihara T, Yoshimura T, Esaki N (1999) Cold-active serine alkaline protease from the psychrotrophic bacterium *Shewanella* strain Ac10: gene cloning and enzyme purification and characterization. Appl Environ Microbiol 65:611–617

Kurihara T, Esaki N (2008) Proteomic studies of psychrophilic microorganisms. In: Margesin R, Schinner F, Marx JC, Gerday C (eds) Psychrophiles: from biodiversity to biotechnology. Springer, Berlin, Heidelberg, pp 333–346

Kwon KK, Lee HS, Yang SH, Kim SJ (2005) *Kordiimonas gwangyangensis* gen. nov., sp. nov., a marine bacterium isolated from marine sediments that forms a distinct phyletic lineage (*Kordiimonadales* ord. nov.) in the '*Alphaproteobacteria*'. Int J Syst Evol Microbiol 55:2033–2037

Langwaldt JH, Tirola M, Puhakka JA (2008) Microbial adaptation to boreal saturated subsurface: implication in bioremediation of polychlorophenols. In: Margesin R, Schinner F, Marx JC, Gerday C (eds) Psychrophiles: from biodiversity to biotechnology. Springer, Berlin, Heidelberg, pp 409–428

Larose C, Berger S, Ferrari C, Navarro E, Dommergue A, Schneider D, Vogel TM (2010) Microbial sequences retrieved from environmental samples from seasonal Arctic snow and meltwater from Svalbard, Norway. Extremophiles 14:205–212

Law BA, Goodenough PW (1995) Enzymes in milk and cheese production. In: Tucker GA, Woods LFJ (eds) Enzymes in food processing, 2nd edn. Blackie Academic and Professional, Bishopbriggs, Glasgow, UK, pp 114–143

Lee HK, Ahn MJ, Kwak SH, Song WH, Jeong BC (2003) Purification and characterization of cold active lipase from psychrotrophic *Aeromonas* sp. LPB 4. J Microbiol 41:22–27

Lee CC, Smith MR, Accinelli R, Williams TG, Wagschal KC, Wong D, Robertson GH (2006) Isolation and characterization of a psychrophilic xylanase enzyme from *Flavobacterium* sp. Curr Microbiol 52:112–116

Lees RS (1990) Impact of dietary fats on human health. Food Sci Technol 37:1–38

Lettinga G, Rebac S, van Lier J, Zeman G (1999) The potential of sub-mesophilic and/or psychrophilic anaerobic treatment of low strength wastewaters. In: Margesin R, Schinner F (eds) Biotechnological applications of cold adapted organisms. Springer, Berlin, Heidelberg, pp 221–234

Lettinga G, Rebac S, Zeeman G (2001) Challenge of psychrophilic anaerobic wastewater treatment. Trends Biotechnol 19:363–370

Liebner S, Wagner D (2007) Abundance, distribution and potential activity of methane oxidizing bacteria in permafrost soils from the Lena Delta, Siberia. Environ Microbiol 9:107–117

Lo Giudice A, Bruni V, Michaud L (2007) Characterization of Antarctic psychrotrophic bacteria with antibacterial activities against terrestrial microorganisms. J Bas Microbiol 47:496–505

Loveland-Curtze J, Miteva V, Brenchley JE (2009) *Herminiimonas glaciei* sp. nov., a novel ultramicrobacterium from 3042 m deep Greenland glacial ice. Int J Syst Evol Microbiol 59:1272–1277

Loveland-Curtze J, Miteva V, Brenchley J (2010) Novel ultramicrobacterial isolates from a deep Greenland icecore represent a proposed new species, *Chryseobacterium greenlandense* sp. nov. Extremophiles 14:61–69

Madigan MT, Jung DO (2003). Extremophiles. Worldbook Encyclopedia, Science Year 2004 Annual, pp 74–89

Margesin R (2007) Alpine microorganisms: useful tools for low-temperature bioremediation. J Microbiol 45:281–285

Margesin R, Schinner F (1999) Biodegradation of organic pollutants at low temperature. In: Margesin R, Schinner F (eds) Biotechnological applications of cold adapted organisms. Springer, Berlin, Heidelberg, New York, pp 271–290

Margesin R, Schinner F (2001) Bioremediation (natural attenuation and biostimulation) of diesel-oil-contaminated soil in an alpine glacier skiing area. Appl Environ Microbiol 67:3127–3133

Margesin R, Schumann P, Spröer C, Gounot AM (2004) *Arthrobacter psychrophenolicus* sp. nov., isolated from an alpine ice cave. Int J Syst Evol Microbiol 54:2067–2072

Margesin R, Fonteyne PA, Schinner F, Sampaio JP (2007) *Rhodotorula psychrophila* sp. nov., *Rhodotorula psychrophenolica* sp. nov. and *Rhodotorula glacialis* sp. nov., novel psychrophilic basidiomycetous yeast species isolated from alpine environments. Int J Syst Evol Microbiol 57:2179–2184

Marx JG, Carpenter SD, Deming JW (2009) Production of cryoprotectant extracellular polysaccharide substances (EPS) by the marine psychrophilic bacterium *Colwellia psychrerythraea* strain 34H under extreme conditions. Can J Microbiol 55:63–72

Matsuyama H, Hirabayashi T, Kasahara H, Minami H, Hoshino T, Yumoto I (2006) *Glaciecola chathamensis* sp. nov., a novel marine polysaccharide-producing bacterium. Int J Syst Evol Microbiol 56:2883–2886

Mattes TE, Alexander AK, Richardson PM, Munk AC, Han CS, Stothard P, Coleman NV (2008) The Genome of *Polaromonas* sp. strain JS666: insights into the evolution of a hydrocarbon- and xenobiotic-degrading bacterium, and feature of relevance to biotechnology. Appl Environ Microbiol 74:6405–6416

McCammon SA, Innes BH, Bowman JP, Franzmann PD, Dobson SJ, Holloway PE, Skerratt JH, Nichols PD, Rankin LM (1998) *Flavobacterium hibernum* sp. nov., a lactose-utilizing bacterium from a freshwater Antarctic lake. Int J Syst Bacteriol 4:1405–1412

Médigue C, Krin E, Pascal G, Barbe V, Bernsel A, Bertin PN, Cheung F, Cruveiller S, D'Amico S, Duilio A, Fang G, Feller G, Ho C, Mangenot S, Marino G, Nilsson J, Parrilli E, Rocha EP, Rouy Z, Sekowska A, Tutino ML, Vallenet D, von Heijne G, Danchin A (2005) Coping with cold: the genome of the versatile marine Antarctica bacterium *Pseudoalteromonas haloplanktis* TAC125. Genome Res 15:1325–1335

Methé BA, Nelson KE, Deming JW, Momen B, Melamud E, Zhang X, Moult J, Madupu R, Nelson WC, Dodson RJ, Brinkac LM, Daugherty SC, Durkin AS, DeBoy RT, Kolonay JF, Sullivan SA, Zhou L, Davidsen TM, Wu M, Huston AL, Lewis M, Weaver B, Weidman JF, Khouri H, Utterback TR, Feldblyum TV, Fraser CM (2005) The psychrophilic lifestyle as revealed by the genome sequence of *Colwellia psychrerythraea* 34H through genomic and proteomic analyses. Proc Natl Acad Sci USA 102:10913–10918

Michaux C, Massant J, Kerff F, Frere J-M, Docquier J-D, Vandenberghe I, Samyn B, Pierrard A, Feller G, Charlier P, Van Beeumen J, Wouters J (2008) Crystal structure of a cold-adapted class C β-lactamase. FEBS J 275:1687–1697

Mikhailova G, Likhareva V, Khairullin RF, Lubenets NL, Rumsh LD, Demidyuk IV, Kostrov SV (2006) Psychrophilic trypsin-type protease from *Serratia proteamaculans*. Biochem Sci 71:563–570

Milne PJ, Hunt AL, Rostoll K, Van Der Walt JJ, Graz CJM (1998) The biological activity of selected cyclic dipeptides. J Pharm Pharmacol 50:1331–1337

Miteva V (2008) Bacteria in snow and glacier ice. In: Margesin R, Schinner F, Marx JC, Gerday C (eds) Psychrophiles: from biodiversity to biotechnology. Springer, Berlin, Heidelberg, pp 31–50

Mock T, Thomas DN (2008) Microalgae from Polar Regions: functional genomics and physiology. In: Margesin R, Schinner F, Marx JC, Gerday C (eds) Psychrophiles: from biodiversity to biotechnology. Springer, Berlin, Heidelberg, pp 347–368

Morgan-Kiss RM, Priscu JC, Pocock T, Gudynaite-Savitch L, Huner NPA (2006) Adaptation and acclimation of photosynthetic microorganisms to permanently cold environments. Microbiol Molec Biol Rev 70:222–252

Morita Y, Nakamura T, Hasan Q, Murakami Y, Yokoyama K, Tamiya E (1997) Cold-active enzymes from cold-adapted bacteria. J Am Oil Chem Soc 74:441–444

Moyer CL, Morita RY (2007) Psychrophiles and psychrotrophs. In: Encyclopedia of life sciences. John Wiley & Sons, Ltd. www.els.net

Murray E, Preston CM, Massana R, Taylor LT, Blakis A, Wu K, Delong EF (1998) Seasonal and spatial variability of bacterial and archaeal assemblages in the coastal waters near Anvers Island, Antarctica. Appl Environ Microbiol 64:2585–2595

Murygina V, Arinbasarov M, Kalyuzhnyi S (2000) Bioremediation of oil polluted aquatories and soils with novel preparation "Rhoder". Biodegradation 11:385–389

Naganuma T, Hua PN, Okamoto T, Ban S, Imura S, Kanda H (2005) Depth distribution of euryhaline halophilic bacteria in Suribati Ike, a meromictic lake in East Antarctica. Polar Biol 28:964–970

Nakagawa T, Nagaoka T, Taniguchi S, Miyaji T, Tomizuka N (2004) Isolation and characterization of psychrophilic yeasts producing cold-adapted pectinolytic enzymes. Lett Appl Microbiol 38:383–387

Nakagawa T, Ikehata R, Myoda T, Miyaji T, Tomizuka N (2007) Overexpression and functional analysis of cold-active β-galactosidase from *Arthrobacter psychrolactophilus* strain F2. Protein Expr Purif 54:295–299

Napolitano MJ, Shain DH (2004) Four kingdoms on glacier ice: convergent energetic processes boost energy levels as temperatures fall. Proc R Soc Lond B 271:S273–S276

Negoiță TG, Ştefanic G, Irimescu Orzan ME, Palanciuc V, Oprea G (2001a) Chemical and biological characterization of soils from the Antarctic East Coast. Polar Biol 24:565–571

Negoiță TG, Ştefanic G, Irimescu Orzan ME, Palanciuc V, Oprea G (2001b) Microbial chemical and enzymatic properties in Spitsbergen soils. Polar Forschung 71:41–46

Nguyen KT, Xiaowei H, Alexander DC, Li C, Gu J-Q, Mascio C, Van Praagh A, Mortin L, Chu M, Silverman JA, Brian P, Baltz RH (2010) Genetically engineered lipopeptide antibiotics related to A54145 and daptomycin with improved properties. Antimicrob Agents Chemother 54:1404–1413

Nichols DS, Bowman J, Sanderson C, Nichols CM, Lewis T, McMeekin T, Nichols PD (1999) Developments with Antarctic microorganisms: culture collections, bioactivity screening, taxonomy, PUFA production and cold-adapted enzymes. Curr Opin Biotechnol 10:240–246

Nichols DS, Sanderson K, Buia A, van de Kamp J, Holloway P, Bowman JP, Smith M, Mancuso C, Nichols PD, McMeekin T (2002) Bioprospecting and biotechnology in Antarctica. In: Jabour-Green J, Haward M (eds) The Antarctic: past, present and future. Antarctic CRC Research Report #28. Hobart, pp 85–103

Nielsen TB, Ishii M, Kirk O (1999) Lipases A and B from the yeast *Candida antarctica*. In: Margesin R, Schinner F (eds) Biotechnological applications of cold adapted organisms. Springer, Berlin, Heidelberg, New York, pp 48–61

Niemann H, Elvert M, Wand U, Samarkin VA, Lehmann MF (2010) First Lipid Biomarker evidence for aerobic methane oxidation in the water column of Lake Untersee (East Antarctica). Geophysical Research Abstracts Vol. 12, EGU2010-5921

Nogi Y (2008) Bacteria in deep sea: psychropiezophiles. In: Margesin R, Schinner F, Marx JC, Gerday C (eds) Psychrophiles: from biodiversity to biotechnology. Springer, Berlin, Heidelberg, pp 73–82

Ohgiya S, Hoshino T, Okuyama H, Tanaka S, Ishizaki K (1999) Biotechnology of enzymes from cold adapted microorganisms, In: Margesin R, Schinner F (eds) Biotechnological applications of cold adapted organisms. Springer, Berlin, Heidelberg, New York, pp 17–35

Oikawa T, Yamamoto N, Shimoke K, Uesato S, Ikeuki T, Fujioka T (2005) Purification, characterization and overexpression of psychrophilic and thermolabile malate dehydrogenase from a novel Antarctic psychrotolerant *Flavobacterium frigidimaris* KUC 1. Biosci Biotechnol Biochem 59:2146–2154

Olivera NL, Sequeiros C, Nievas ML (2007) Diversity and enzyme properties of protease-producing bacteria isolated from sub-Antarctic sediments of Isla de Los Estados, Argentina. Extremophiles 11:517–526

Onofri S, Zucconi L, Selbmann L, de Hoog S, de los Ríos A, Ruisi S, Grube M (2007) Fungal associations at the cold edge of life in algae and cyanobacteria in extreme environments. In: Seckbach J (ed) Cellular origins, life in extreme habitats and astrobiology, vol 11. Springer, Netherlands, pp 735–757

Ovalle AW (1987) In-place leaching of a block carving mine. In: Cooper WC, Lagos GE, Ugarte G (eds) Copper 87, vol 3: Hydrometallurgy and electrometallurgy of copper. Universidad de Chile, Santiago, Chile, pp 17–37

Papaleo E, Pasi M, Riccardi L, Sambi I, Fantucci P, De Gioia L (2008) Protein flexibility in psychrophilic and mesophilic trypsins. Evidence of evolutionary conservation of protein dynamics in trypsin-like serine-proteases. FEBS Lett 582:1008–1018

Park SC, Kim MS, Baik KS, Kim EM, Rhee MS, Seong CN (2008) *Chryseobacterium aquifrigidense* sp. nov., isolated from a water-cooling system. Int J Syst Evol Microbiol 58:607–611

Parrilli E, Duillio A, Tutino ML (2008) Heterologous protein expression on psychrophilic hosts. In: Margesin R, Schinner F, Marx JC, Gerday C (eds) Psychrophiles: from biodiversity to biotechnology. Springer, Berlin, Heidelberg, pp 365–380

Pavlova K, Gargova S. Hristozova T, Tankova Z (2008) Phytase from Antarctic yeast strain *Cryptococcus laurentiis* AL 27. Folia Microbiol 23:29–34

Pernthaler J, Glöckner FO, Unterholzner S, Alfreider A, Psenner R, Amann R (1998) Seasonal community and population dynamics of pelagic bacteria and archaea in a high mountain lake. Appl Environ Microbiol 64:4299–4306

Perovich DK, Grenfell TC, Light B, Hobbs PV (2002) Seasonal evolution of the albedo of multiyear Arctic sea-ice. J Geophys Res 107(C10):8044. DOI: 10.1029/2000JC000438

Phadtare S, Inouye M (2008) Cold shock proteins. In: Margesin R, Schinner F, Marx JC, Gerday G (eds) Psychrophiles: from biodiversity to biotechnology. Springer, Berlin, Heidelberg, pp 191–210

Prabagaran SR, Manorama R, Delile D, Shivaji S (2007) Predominance of *Roseobacter*, *Sulfitobacter*, *Glaciecola* and *Psychrobacter* in seawater collected from Ushuaia, Argentina, sub Antarctica. FEMS Microbiol Ecol 59:342–355

Price BP (2006) Microbial life in glacial ice and implications for a cold origin of life. FEMS Microbiol Ecol 59:217–231

Prince RC (2005) The microbiology of marine oil spills bioremediation. In: Olivier B, Margot M (eds) Petroleum microbiology. ASM Press, Washington, DC, pp 317–335

Priscu JC, Christner BC (2004) Earths icy biosphere. In: Bull AT (ed) Microbial diversity and bioprospecting. ASM Press, Washington, DC, pp 130–145

Rabus R, Ruepp A, Frickey T, Rattei T, Fartmann B, Stark M, Bauer M, Zibat A, Lombardot T, Amann J, Gellner K, Teeling H, Leuschner WD, Glockner F-O, Lupas AN, Amann R, Klenk H-P (2004) The genome of *Desulfotea psychrophila* a sulphate reducing bacterium from permanently cold Arctic sediment. Environ Microbiol 6:887–902

Raja A, Prabakaran P, Gjalakshmi P (2010) Isolation and screening of antibiotic producing actinomycetes and its nature from Rothang Hill soil against *Viridans Streptococcus* sp. Res J Microbiol 5:44–49

Ramaiah N (1994) Production of certain hydrolytic enzymes by psychrophilic bacteria from the Antarctic krill, zooplankton and seawater. In: Ninth Indian Expedition to Antarctica, Scientific Report, National Institute of Ocenography, Department of Ocean Development, Goa, India, Technical Publication 6:107–114

Ramteke PW, Joseph B, Kuddus M (2005) Extracellular lipases from anaerobic microorganism of Antarctic. Ind J Biotechnol 4:293–294

Rashidah A, Sabrina S, Ainihayati A, Shanmugapriya P, Nazalan N, Razip S (2007) Psychrophilic enzymes from the Antarctic isolates. In: Proceeedings of the international symposium Asian collaboration in IPY 2007–2008, Science Council of Japan, Tokyo 1st March 2007, National Institute of Polar Research, Tokio, Japan, pp 116–119

Raymond JA, Fritsen C, Shen K (2007) An ice-binding protein from an Antarctic sea ice bacterium. FEMS Microbiol Ecol 61:214–221

Rendleman JA Jr (1996) Enzymatic conversion of malto-oligosaccharides and maltodextrin into cyclodextrin at low temperature. Biotechnol Appl Biochem 24:129–137

Riley M, Staley JT, Danchin A, Wang TZ, Brettin TS, Hauser LJ, Land M, Thompson LS (2008) Genomics of an extreme psychrophile, *Psychromonas ingrahamii*. BMC Genomics 9:210

Rivkina E, Gilichinsky D, Wagener S, Tiedje J, McGrath J (1998) Biogeochemical activity of anaerobic microorganisms from buried permafrost sediments. Geomicrobiol J 15:187–193

Rivkina EM, Friedmann EI, McKay CP, Gilichinsky DA (2000) Metabolic activity of permafrost bacteria below the freezing point. Appl Environ Microbiol 66:3230–3233

Rodrigues-Diaz F, Ivanova N, He Z, Huebner M, Zhou J, Tiedje JM (2008) Architecture of thermal adaptation in an *Exiguobacterium sibiricum* strain isolated from 3 million year old permafrost: a genome and transcriptome approach. BMC Genomics 9:547

Rossi G (1999) Biohydrometallurgical processes and temperature. In: Margesin R, Schinner F (eds) Biotechnological applications of cold adapted organisms. Springer, Berlin, Heidelberg, New York, pp 291–308

Russell NJ (2008) Membrane components and cold sensing. In: Margesin R, Schinner F, Marx JC, Gerday C (eds) Psychrophiles: from biodiversity to biotechnology. Springer, Berlin, Heidelberg, pp 177–190

Russell NJ, Cowan DA (2005) Handling of psychrophilic microorganisms. In: Rainey FA, Oren A (eds) Extremophiles. Methods in Microbiology, vol 35. Elsevier, pp 371–393

Russell NJ, Nichols DS (1999) Polyunsaturated fatty acids in marine bacteria – a dogma rewritten. Microbiology 145:767–779

Rybalka N, Andersen RA, Kostikov I, Mohr KI, Massalski A, Olech M, Friedl T (2008) Testing for endemism, genotypic diversity and species concepts in Antarctic terrestrial microalgae of the *Tribonemataceae* (*Stramenopiles, Xanthophyceae*). Environ Microbiol Rep 11:554–565

Rysgaard S, Glud RN, Sejr MK, Blicher ME, Stahl HJ (2008) Denitrification activity and oxygen dynamics in Arctic sea ice. Polar Biol 31:527–537

Sanchez LA, Gomez FF, Delgado OD (2009) Cold-adapted microorganisms as a source of new antimicrobials. Extremophiles 13:111–120

Sattley WM, Madigan MT (2006) Isolation, characterization and ecology of cold-active, chemolithotrophic, sulfur-oxidizing bacteria from perennially ice-covered Lake Fryxell, Antarctica. Appl Environ Microbiol 72:5562–5568

Sattley WM, Madigan MT (2007) Cold-active acetogenic bacteria from surcial sediments of perennially ice-covered Lake Fryxell, Antarctica. FEMS Microbiol Lett 272:48–54

Saunders NFW, Thomas T, Curmi PMG, Mattick JS, Kuczek E, Slade R, Davis J, Franzmann PD, Boone D, Rusterholtz K, Feldman R, Gates C, Bench S, Sowers K, Kadner K, Aerts A, Dehal P, Detter C, Glavina T, Lucas S, Richardson P, Larimer F, Hauser L, Land M, Cavicchioli R (2003) Mechanisms of thermal adaptation revealed from the genomes of the Antarctic Archaea *Methanogenium frigidum* and *Methanococcoides burtonii*. Genome Res 13:1580–1588

Sælensminde G, Halskau Ø Jr, Jonassen I (2009) Amino acid contacts in proteins adapted to different temperatures: hydrophobic interactions and surface charges play a key role. Extremophiles 13:11–20

Säwström C, Laybourn-Parry J, Granéli W, Anesio AM (2007) Heterotrophic bacterial and viral dynamics in Arctic freshwaters: results from a field study and nutrient – temperature manipulation experiments. Polar Biol 30:1407–1415

Seo HJ, Bae SS, Lee J-H, Kim S-J (2005) *Photobacterium frigidiphilum* sp. nov., a psychrophilic, lipolytic bacterium isolated from deep-sea sediments of Edison Seamount. Int J Syst Evol Microbiol 55:1661–1666

Sheridan PP, Brenchley JE (2000) Characterization of a salt-tolerant family 42 beta-galactosidase from a psychrophilic Antarctic *Planococcus* isolate. Appl Environ Microbiol 66:2438–2444

Siddiqui KS, Cavicchioli R (2006) Cold-adapted enzymes. Annu Rev Biochem 75:403–433

Siddiqui KS, Poljak A, Guilhaus M, De Francisci D, Curmi PM, Feller G, D'Amico S, Gerday C, Uversky VN, Cavicchioli R (2006) Role of lysine versus arginine in enzyme cold-adaptation:

modifying lysine to homo-arginine stabilizes the cold-adapted alpha-amylase from *Pseudoalteromonas haloplanktis*. Proteins 1:486–501

Singh L, Ramana Venkata K (1998) Introduction. Isolation and characterization of psychrotrophic Antarctic bacteria from blue-green algal mats and their hydrolytic enzymes. In: Fourteenth Indian Expedition to Antarctica, Scientific Report, Depart Ocean Develop, Techn Public No 12, pp 199–206

Singh SM, Puja G, Bhat DJ (2006) Psychrophilic fungi from Schirmacher Oasis, East Antarctica. Curr Sci 90:1388–1392

Sizova M, Panikov N (2007) *Polaromonas hydrogenivorans* sp. nov., a psychrotolerant hydrogen-oxidizing bacterium from Alaskan soil. Int J Syst Evol Microbiol 57:616–619

Skidmore ML, Foght JM, Sharp MJ (2000) Microbial life beneath a high arctic glacier. Appl Environ Microbiol 66:3214–3220

Smith DW, Emde KME (1999) Effectiveness of wastewater lagoons in cold regions. In: Margesin R, Schinner F (eds) Biotechnological applications of cold adapted organisms. Springer, Berlin, Heidelberg, New York, pp 235–256

Somkutl GA, Holsinger VH (1997) Microbial technologies in the production of low-lactose dairy foods. Food Sci Technol Int J 3:163–169

Sonan GK, Receveur-Brechot V, Duez C, Aghari N, Czjzek M, Haser R, Gerday C (2007) The linker region plays a key role in the adaptation to cold of the cellulase from an Antarctic bacterium. Biochem J 407:293–302

Spring S, Merkhoffer B, Weiss N, Kroppenstedt RM, Hippe H, Stackebrandt E (2003) Characterization of novel psychrophilic clostridia from an Antarctic microbial mat: description of *Clostridium frigoris* sp. nov., *Clostridium lacusfryxellense* sp. nov., *Clostridium bowmanii* sp. nov. and *Clostridium psychrophilum* sp. nov. and reclassification of *Clostridium laramiense* as *Clostridium estertheticum* subsp. *laramiense* subsp. nov. Int J Syst Evol Microbiol 53:1019–1029

Ştefanic G (1994) Biological definition, quantifying method and agricultural interpretation of soil fertility. Rom Agric Res 2:107–116

Steven B, Briggs G, McKay CP, Pollard WH, Greer CW, Whyte LG (2007) Characterization of the microbial diversity in a permafrost sample from the Canadian high Arctic using culture-dependent and culture-independent methods. FEMS Microbiol Ecol 59:513–523

Suzuki T, Nakayama T, Kurihara T, Nishino T, Esaki N (2001) Cold active lipolytic activity of psychrotrophic *Acinetobacter* sp. strain no. 6. J Biosci Bioeng 92:144–148

Takeuchi N, Kohshima S (2004) A snow algal community on Tyndall glacier in the Southern Patagonia Icefield, Chile. Arct Antarct Alp Res 36:92–99

Tamaki H, Hanada S, Kamagata Y, Nakamura K, Nomura N, Nakano K, Matsumura M (2003) *Flavobacterium limicola* sp. nov., a psychrophilic, organic-polymer-degrading bacterium isolated from freshwater sediments. Int J Syst Evol Microbiol 53:519–526

Tashyrev OB (2009) The complex researches of structure and function of Antarctic terrestrial microbial communities. Ukr Antarctic J 8:343–357

Thomas-Hall S, Watson K (2002) *Cryptococcus nyarrowii* sp. nov., a basidiomycetous yeast from Antarctica. Int J Syst Evol Microbiol 52:1033–1038

Thomas-Hall S, Hall R, Turchetti B, Buzzini P, Branda E, Boekhout T, Theelen B, Watson K (2010) Cold-adapted yeasts from Antarctica and the Italian Alps – description of three novel species: *Mrakia robertii* sp. nov., *Mrakia blollopis* sp. nov. and *Mrakiella niccombsii* sp. nov. Extremophiles 14:47–59

Tian F, Yu Y, Chen B, Li H, Yao YF, Gua XK (2009) Bacterial, archaeal and eukaryotic diversity in Arctic sediment as revealed by 16S rRNA and 18S rRNA gene clone libraries analysis. Polar Biol 32:93–103

Tkaczuk KL, Buinicki JM, Bialkowska A, Bielecki S, Turkievicz M, Cieslinski H, Kur J (2005) Molecular modelling of a psychrophilic beta-galactosidase. Biocat Biotransf 23:201–209

Tomoyuki N, Ikehata R, Myoda T, Miyaji T, Tomizuka N (2007) Overexpression and functional analysis of cold-active beta-galactosidase from *Arthrobacter psychrolactophilus* strain F2. Prot Expr Purif 54:295–299

Turkiewicz M, Pazgier M, Donachie SP, Kalinowska H (2005) Invertase and glucosidase production by the endemic Antarctic marine yeast *Leucosporidium antarcticum*. Pol Polar Res 26:125–136

Tutino ML, Duilio A, Parrilli R, Remaut E, Sannia G, Marino G (2001) A novel replication element from an Antarctic plasmid as a tool for the expression of proteins at low temperature. Extremophiles 5:257–264

Vishniac HS, Kurtzman CP (1992) *Cryptococcus antarcticus* sp. nov., and *Cryptococcus albidosimilis* sp. nov., basidioblastomycetes from Antarctic soils. Int J Syst Bact 42:547–553

Vorobyova E, Soina V, Gorlenko M, Minkovskaya M, Zalinova N, Mamukelashvili A, Gilichinsky D, Rivkina E, Vishnivetskaya T (1997) The deep cold biosphere: facts and hypothesis. FEMS Microbiol Rev 20:277–290

Voytek MA, Priscu JC, Ward BB (1999) The dicversity and abundance of ammonia oxidizing bacteria in lake of the McMurdo Dry Valley, Antarctica. Hydrobiologia 401:113–130

Wagner-Döbler I, Rheims H, Felske A, El-Ghezal A, Flade-Schröder D, Laatsch H, Lang S, Pukall R Tindall BJ (2004) *Oceanibulbus indolifex* gen. nov., sp. nov., a North Sea alphaproteobacterium that produces bioactive metabolites. Int J Syst Evol Microbiol 54:1177–1184

Wartiainen I, Hestnes AG, McDonald IR, Svenning Mette M (2006) *Methylocystis rosea* sp. nov., a novel methanotrophic bacterium from Arctic wetland soil, Svalbard, Norway (786°N). Int J Syst Microbiol 56:541–547

Wells LE (2008) Cold active viruses. In: Margesin R, Schinner F, Marx JC, Gerday C (eds) Psychrophiles: from biodiversity to biotechnology. Springer, Berlin, Heidelberg, pp 157–176

Wharton RA Jr, McKay CP, Simmons GM Jr, Parker CB (1985) Cryoconite holes on glaciers. Bioscience 35:499–503

Whyte LG, Hawari J, Zhou E, Bourbonnere L, Inniss WE, Greer CW (1998) Biodegradation of variable-chain-length alkanes at low temperatures by a psychrotrophic *Rhodococcus*. Appl Environ Microbiol 64:2578–2584

Wierzchos J, de los Ríos A, Sancho LG, Ascaso C (2004) Viability of endolithic microorganisms in rocks from the McMurdo Dry Valleys of Antarctica established by confocal and fluorescence microscopy. J Microsc Oxford 216:57–61

Woo JH, Hwang YO, Kang SG, Lee HS, Cho JC, Kim SJ (2007) Cloning and characterization of three epoxide hydrolases from a marine bacterium, *Erythrobacter litoralis* HTCC2594. Appl Microbiol Biotechnol 76:365–375

Yakimov MM, Giuliano L, Gentile G, Crisafi E, Chernikova TN, Abraham W-R, Lünsdorf H, Timmis K.N, Golyshin PN (2003) *Oleispira antarctica* gen. nov., sp. nov., a novel hydrocarbonoclastic marine bacterium isolated from Antarctic coastal sea water. Int J Syst Evol Microbiol 53:779–785

Yan BQ, Chen XL, Hou XY, He H, Zhou BC, Zhang YZ (2009) Molecular analysis of the gene encoding a cold-adapted halophilic subtilase from deep-sea psychrotolerant bacterium *Pseudoalteromonas* sp. SM9913: cloning, expression, characterization and function analysis of the C-terminal PPC domains. Extremophiles 13:725–733

Yang SH, Lee JH, Ryu JS, Kato C, Kim SJ (2007) *Shewanella donghaensis* sp. nov., a psychrotrophic, piezosensitive bacterium producing high levels of polyunsaturated fatty acid, isolated from deep-sea sediments. Int J Syst Evol Microbiol 57:208–212

Yu Y, Li H, Zeng Y, Chen B (2009) Extracellular enzymes of cold-adapted bacteria from Arctic sea ice, Canada Basin. Polar Biol 32:1539–1547

Yumoto I, Nakamura A, Iwata H, Kojima K, Kusumoto K, Nodasaka Y, Matsuyama H (2002) *Dietzia psychralkaliphila* sp. nov., a novel, facultatively psychrophilic alkaliphile that grows on hydrocarbons. Int J Syst Evol Microbiol 52:85–90

Yumoto I, Hirota K, Sogabe Y, Nodasaka Y, Yokota Y, Hoshino T (2003) *Psychrobacter okhotskensis* sp. nov., a lipase-producing facultative psychrophile isolated from the coast of the Okhotsk Sea. Int J Syst Evol Microbiol 53:1985–1989

Zachariassen KE, Lundheim R (1999) Application of antifreeze proteins. In: Margesin R, Schinner F (eds) Biotechnological applications of cold adapted organisms. Springer, Berlin, Heidelberg, pp 319–332

Zakaria MM, Ashiuchi M, Yamamoto S, Yagi T (1998) Optimization for beta-mannanase production of a psychrophilic bacterium, *Flavobacterium* sp. Biosci Biotechnol Biochem 62:655–660

Zakhia F, Jungblut AD, Taton A, Vincent WF, Wilmotte A (2008) Cyanobacteria in cold ecosystems. In: Margesin R, Schinner F, Marx JC, Gerday C (eds) Psychrophiles: from biodiversity to biotechnology. Springer, Berlin, Heidelberg, pp 121–135

Halophilic microorganisms from man-made and natural hypersaline environments: physiology, ecology, and biotechnological potential

Madalin Enache[1], Gabriela Popescu[1], Takashi Itoh[2] and Masahiro Kamekura[3]

[1]Institute of Biology of the Romanian Academy, Bucharest, Romania
[2]Japan Collection of Microorganisms, RIKEN BioResource Center, Saitama, Japan
[3]Halophiles Research Institute, Noda, Japan

1 Introduction

What are hypersaline environments? Geologists or geochemists define saline lakes *sensu lato* as bodies of water with salinity more than 3 g/l (0.3%), while those *sensu stricto* (hypersaline) are bodies of water that exceed the modest 35 g/l (3.5%) salt of oceans (Williams 1998). Many microbiologists use the term hypersaline to denote the well-known salt lakes, such as the Dead Sea and the Great Salt Lake or crystallizer ponds of solar salterns, environments almost saturated with salt.

A lot of hypersaline environments are found in nature throughout the world, natural or man-made (Javor 1989a). Rock salt from salt deposits has been a source of sodium chloride for human beings from prehistoric period (Multhauf 1978). The subterranean salt deposit of the Mediterranean Sea is the result of Messinian salinity crisis 5.96–5.33 My ago (Duggen et al. 2003). Strong brines occur in both marine-derived (thalassic) and nonmarine (athalassic) systems. In arid coastal zones, large scale sabkhas, salt flats, or strong brines of tiny scales are observed (Javor 1989b). These hypersaline environments are too harsh for normal life to exist, but a variety of microbes, both Bacteria and Archaea, survive.

2 Halophilic microorganisms

When microbiologists are confronted with the question of how to define halophilic microorganisms, an answer is the classical definition by Kushner (1985), who categorized them into slight, moderate and extreme halophiles, depending on the NaCl concentration that supported optimal growth. The term "halotolerant" is generally used for organisms that are able to grow in media without added NaCl, but are able to grow at high salt concentration, while "halophilic" is for those that require addition of NaCl or other salts to media for their growth. The definition seems clear, but the fact is that many "halotolerant" microorganisms are able to grow at higher salt concentrations than some "halophilic" microorganisms.

Another answer is an operational definition by Oren (2002): "microorganisms that are able to grow well above 100 g/l salt." Since the time Oren compiled the data, numerous papers on the "halophilic microorganisms" have been published proposing new genera and species, both Bacteria and Archaea. In this chapter, we arbitrarily define halophilic microorganisms as those that require NaCl higher than 30 g/l for growth and are able to grow well above 200 g/l, irrespective of the salt concentration of the origin of isolation. Although most origins are hypersaline, above 100 g/l salt as defined by Oren, sometimes data of salt concentration of bodies of water are not described in the original papers. When the sources are soil samples (Ventosa et al. 2008) or materials collected on seashores or leaves of plants, it is difficult to measure the exact salt concentration of the very small niches in those samples where halophilic microorganisms are thriving.

2.1 Haloarchaea

Haloarchaea, members of the family *Halobacteriaceae*, are a group of extremely halophilic microorganisms, forming a part of the domain Archaea. The early investigations on the general bacteriology of red pigmented halophilic bacteria as the cause of the reddening of salted hides (red heat) and fish (red eye) began in the 1920s. The scientists came to the conclusion that marine or solar salts and rock salts are contaminated by the "halophilic bacteria" (Anderson 1954; Juez 1988). The eighth edition of Bergey's Manual of Determinative Bacteriology placed rod-shaped extreme halophiles in the genus *Halobacterium* with two species, *Halobacterium salinarium* and *Halobacterium halobium*, and coccoid extreme halophiles in the genus *Halococcus*, with only the species *Halococcus morrhuae* (Gibbons 1974). Soon, *Hbt. halobium* was suggested to be a member of *Archaebacteria* by Magrum et al. (1978). Tindall et al. (1980) were the first to isolate alkaliphilic haloarchaea that grow only in media of pH higher than 7.5, and they introduced the novel genera *Natronobacterium* and *Natronococcus* (Tindall et al. 1984). Two more genera, *Haloarcula* and *Haloferax*, were proposed to accommodate some new isolates and several species of the genus *Halobacterium* (Torreblanca et al. 1986). Since then, numerous strains have been isolated from hypersaline environments distributed all over the world. Neutrophilic strains were differentiated at the generic

Table 1A. Halophilic archaea and their distribution around the world[a]

Country	Saline environment	Archaea
Algeria	Ezzemoul sabkha	*Halorubrum ezzemoulense*
Antarctica	Deep Lake	*Halorubrum lacusprofundi*
Argentine	Salt flats (not specified)	*Haloarcula argentinesis* *Halomicrobium mukohataei*
Australia	Hamelin Pool, Shark Bay, Western Australia	*Halococcus hamelinensis* *Haloferax elongans* *Haloferax mucosum*
	Cheetham Salt Works, Geelong, Victoria	*Halonotius pteroides* *Haloquadratum walsbyi* *Halorubrum coriense* *Natronomonas moolapensis*
Austria	Salt mine, Altaussee	*Halobacterium noricense*
	Rock salt, Bad Ischl salt mine	*Halococcus dombrowskii* *Halococcus salifodinae*
Canada	Salted cowhide	*Halobacterium salinarum*[b]
Chile	Lake Tebenquiche, Atacama Saltern	*Halomicrobium katesii* *Halorubrum tebenquichense*
China	Saline soil, Daqing, Heilongjiang Province	*Haloterrigena daqingensis*
	Fuqing solar saltern, Fujian Province	*Halorubrum litoreum*
	Rudong marine solar saltern, Jiangsu Province	*Haladaptatus litoreus* *Halogeometricum rufum* *Halogranum rubrum* *Halopelagius inordinatus* *Haloplanus vescus* *Halosarcina limi*
	Solar saltern, Zhoushan archipelago, Zhejiang Province	*Haloferax larsenii*
	Sea salt, Qingdao, Shandong Province	*Halococcus qingdaonensis*
China (Inner Mongolia Autonomous Region)	Baerhu Soda Lake	*Natronolimnobius baerhuensis* *Natronolimnobius innermongolicus*
	Lake Bagaejinnor	*Halorubrum kocurii*
	Lake Chagannor, 17(C, pH 10.5	*Halorubrum luteum* *Natronorubrum sediminis*
	Chahannao soda lake	*Halobiforma nitratireducens* *Natrialba chahannaoensis*
	Lake Ejinor	*Halorubrum ejinorense* *Halorubrum orientale*

(*continued*)

Table 1A. (continued)

Country	Saline environment	Archaea
		Halovivax asiaticus
		Natrinema ejinorense
	Jilantai salt lake	Halobacterium jilantaiense
	Lake Shangmatala	Halopiger xanaduensis
	Lake Xilin Hot	Halostagnicola larsenii
		Haloterrigena salina
		Halovivax ruber
	Unnamed soda lake, Hulunbeir prefecture	Natrialba hulunbeirensis
China (Tibet Autonomous Region)	Bange salt-alkaline lake, pH 10	Natronorubrum bangense
		Natronorubrum tibetense
	Lake Zabuye, pH 9.4	Halalkalicoccus tibetensis
		Halorubrum tibetense
China (Xinjiang Uygur Autonomous Region)	Aibi (or Ebinur) salt lake	Haloarcula amylolytica
		Halorubrum lipolyticum
		Haloterrigena limicola
		Haloterrigena longa
		Haloterrigena saccharevitans
		Natrinema versiforme
		Natronorubrum aibiense
	Aiding salt lake	Halorubrum aidingense
		Natronorubrum sulfidifaciens
	Xiao-Er-Kule Lake	Halorubrum xinjiangense
	Saline lake (sampling site not specified)	Halorubrum alkaliphilum
	Ayakekum salt lake, Altun Mountain, pH 7.8	Halobiforma lacisalsi
		Halorubrum arcis
		Natrinema altunense
Egypt	Brine pool, Sinai	Haloarcula quadrata
	Saline soil, Aswan	Halobiforma haloterrestris
		Halopiger aswanensis
		Natrialba aegyptiaca
	Solar saltern, Alexandria	Haloferax alexandrinus
	Wadi Natrun	Natronomonas pharaonis
Israel/Jordan	The Dead Sea	Haloarcula marismortui
		Halobaculum gomorrense
		Halococcus morrhuae[c]
		Haloferax volcanii
		Haloplanus natans
		Halorubrum sodomense

(continued)

Table 1A. (continued)

Country	Saline environment	Archaea
Italy	"Red heat" in salted hides	Natrinema pellirubrum[d]
	Solar salt, Trapani, Sicily	Halorubrum trapanicum[d]
Japan	Solar salt, Niigata	Natronoarchaeum mannanilyticum
	Salt field, Ishikawa	Haloarcula japonica
	Sea sand (sampling site not specified)	Natrialba asiatica
Kenya	Lake Magadi	Halorubrum vacuolatum
		Natrialba magadii
		Natronobacterium gregoryi
		Natronococcus amylolyticus
		Natronococcus occultus
Mexico	Solar saltern, Baja California	Halorubrum chaoviator
Phillipines	Solar salt (sampling site not specified)	Halarchaeum acidiphilum
Puerto Rico	Solar saltern, Cabo Rojo	Halogeometricum borinquense
		Haloterrigena thermotolerans
Red Sea	Shaban Deep, bride–sediment interface (depth of 1,447 m, pH 6.0)	Halorhabdus tiamatea
Romania	Telega Lake, Prahova	Haloferax prahovense
South Korea	Jeotgal (salty condiment)	Haladaptatus cibarius
		Halalkalicoccus jeotgali
		Halorubrum cibi
		Haloterrigena jeotgali
		Natronococcus jeotgali
Spain	Fuente de Piedra salt lake, Malaga	Haloterrigena hispanica
	San Fernando solar saltern, Cadiz	Halococcus saccharolyticus
	Santa Pola solar saltern, Alicante	Haloarcula hispanica
		Haloferax gibbonsi
		Haloferax lucentense
		Haloferax mediterranei
Taiwan	Solar salt (sampling site not specified)	Natrialba taiwanensis
Thailand	Fermented salty foods (Ka-pi, Nam-pla, Pla-ra)	Halobacterium piscisalsi
		Halococcus thailandensis
		Natrinema gari
Turkmenistan	Saline soil	Halorubrum distributum
		Halorubrum terrestre
		Haloterrigena turkmenica
USA	The Great Salt Lake, Utah	Halorhabdus utahensis
	Death Valley, California	Haloarcula vallismortis

(continued)

Table 1A. (continued)

Country	Saline environment	Archaea
	Saltern, San Francisco Bay, California	*Haloferax denitrificans*
		Halorubrum saccharovorum
	Cargill Solar Salt Plant, Newark, California	*Halorubrum californiense*
	Rock salt crystals, Carlsbad, New Mexico	*Halosimplex carlsbadense*
	Zodletone Spring, Oklahoma	*Haladaptatus paucihalophilus*
		Haloferax sulfurifontis
		Halosarcina pallida

[a]*Natrinema pallidum* NCIMB 777 was isolated from salted cod fish, but the site of isolation is not clear
[b]Lochhead (1934)
[c]Kocur and Hodgkiss (1973)
[d]On-line catalog of NCIMB

level based on physiological characteristics and on the presence or absence of specific membrane lipids, phosphatidylglycerosulfate (PGS) and glycolipids (Kates 1995; Grant et al. 2001). Alkaliphilic strains, on the other hand, were devoid of detectable amount of glycolipids. The introduction of the PCR technique (Saiki et al. 1988) made analysis of 16S rRNA gene sequences easy in microbial taxonomy (Weisburg et al. 1991). The first phylogenetic tree of haloarchaea was reconstructed in 1993 using 19 sequences available at that time, demonstrating that species of the genera *Halobacterium, Halococcus, Haloarcula, Haloferax,* and *Natronobacterium* formed coherent clusters (Kamekura and Seno 1993).

At present, haloarchaeal strains are classified in 119 species of 32 genera (as of February 2010; Table 1A) (Enache et al. 2007b; Burns et al. 2010; Minegishi et al. 2010; Shimane et al. 2010). All strains of halophilic Archaea require NaCl higher than at least 4.7% (0.8 M) and are able to grow up to at least 23% (4.0 M) NaCl. Although the majority are neutrophiles, alkaliphilic species have been isolated from soda lakes or Wadi Natrun in Kenya, China, India, Egypt, etc. (Horikoshi 1999; Rees et al. 2004). They are accommodated in the genera *Natronobacterium, Natronococcus, Natronomonas* (Kamekura et al. 1997), *Halalkalicoccus* (Xue et al. 2005) and *Natronolimnobius* (Itoh et al. 2005). The genera *Halobiforma, Halorubrum, Natrialba* and *Natronorubrum* consist of both neutrophilic and alkaliphilic species.

Halophilic Archaea are believed to survive for long times in fluid inclusions of halite (Grant et al. 1998; Kunte et al. 2002; Stan-Lotter et al. 2003; Park et al. 2009). Recently, Fendrihan et al. (2009b) suggested the use of Raman spectroscopy as a potential method for the detection of extremely halophilic Archaea embedded in halite.

Table 1B. Halophilic bacteria and their distribution around the world

Country	Saline environment	Bacteria	Range of NaCl (%)	Optimum NaCl (%)
Algeria	Ezzemoul sabkha	Halomonas sabkhae	5–25	7.5
		Salicola salis	10–25	15–20
Canada	Contaminant on agar plate	Actinopolyspora halophile	10–33	15–20
Chile	Solar saltern, Cahuil, Pichilemu	Halomonas nitroreducens	3–20	5–7.5
China	Xiaochaidamu salt lake, Qinghai province	Gracilibacillus halophilus	7–30	15
China (Inner Mongolia Autonomous Region)	Lake Chagannor, 17°C, pH 10.5	Bacillus chagannorensis	3–20 (salts)	7 (salts)
		Salsuginibacillus kocurii	3–20 (salts)	10 (salts)
	Lake Shangmatala	Aquisalibacillus elongatus	3–20	10
	Lake Xilin Hot	Virgibacillus salinus	3–20	10
	Xiarinaoer soda lake	Salsuginibacillus halophilus	9–30	19
China (Xinjiang Uygur Autonomous Region)	Aiding salt lake	Alkalibacillus salilacus	5–20	10–12
		Bacillus aidingensis	8–33	12
		Lentibacillus halodurans	5–30	8–12
		Prauserella sedimina	5–20	10
		Salinibacillus aidingensis	5–20	10
	Qijiaojing Lake	Haloechinothrix alba	9–23	15
	Saline lake (sampling site not specified)	Bacillus salarius	3–20	10–12
		Lentibacillus lacisalsi	5–25	12–15
		Saccharopolyspora halophilia	3–20	10–15
	Saline soil (sampling site not specified)	Alkalibacillus halophilus	5–30	10–20
		Nocardiopsis salina	3–20	10
		Prauserella halophila	5–25	10–15
		Saccharomonospora paurometabolica	5–20	10
		Streptomonospora alba	5–25	10–15
		Streptomonospora amylolytica	5–20	10
		Streptomonospora flavalba	5–25	10
		Streptomonospora halophila	5–20	10
Congo	Oil-well head sample	Halanaerobium congolense	4–24	10
Egypt	Wadi Natrun	Natranaerobius thermophilus	18–29	19–23
		Natranaerobius trueperi	19–31	22
		Natronovirga wadinatrunensis	19–31	23
		Thiohalospira alkaliphila	3–23	12
France	Salin-de-Giraud saltern, Camargue	Halanaerobacter salinarius	5–30	14–15
		Halorhodospira neutriphila	6–30	9–12
		Thiohalocapsa halophila	3–20	7

(continued)

Table 1B. (continued)

Country	Saline environment	Bacteria	Range of NaCl (%)	Optimum NaCl (%)
Greece	Saltworks, Mesolongi	Bacillus halochares	6–23	15
Iran	Lake Aran-Bidgol	Lentibacillus persicus	3–25	7.5–10
	Howz Soltan Lake	Bacillus persepolensis	5–20	10
Iraq	Saline soil	Actinopolyspora iraqiensis	5–20	10–15
Israel/Jordan	The Dead Sea	Rhodovibrio sodomensis	6–20	12
		Salisaeta longa	5–20	10
		Selenihalanaerobacter shriftii	10–24	21
		Virgibacillus marismortui	5–25	10
Japan[a]	Solar salt	Nesterenkonia halobia	3–25	–
	Salted foods	Chromohalobacter japonicus	5–25	7.5–12.5
		Halanaerobium fermentans	7–25	10
Kenya	Lake Magadi	Natroniella acetigena	10–26	12–15
Kuwait	Salt marsh soil	Saccharomonospora halophila	10–30	–
Mexico	Solar saltern, Baja California	Halospirulina tapeticola	3–20	10
	Brine water, Gulf of Mexico	Halanaerobium acetethylicum	5–22	10
Mongolia	Barun-Davst-Nur	Halovibrio denitrificans	12–30	12–15
Nauru (South Pacific)	Sea shore wood	Salimicrobium halophilum	3–30	–
Peru	Maras salterns, Andes	Salicola marasensis	10–30	15
Portugal	Terminal pond of a saltern	Rhodovibrio salinarum	3–24	9–15
Puerto Rico	Black mangrove, solar saltern of Cabo Rojo	Halobacillus mangrove	5–20	10
Russia	Kulunda Steppe, Altai	Halospina denitrificans	12–30	15–18
		Methylohalomonas lacus	3–23	12
		Thiohalorhabdus denitrificans	9–23	18
		Thiohalospira halophila	12–30	15–18
		Thiomicrospira halophila	3–20	9
Senegal	Retba Lake	Halanaerobium lacusrosei	7.5 to saturated	18–20
South Korea	Byunsan solar saltern, Yellow Sea	Lentibacillus salinarum	3–24	10–12
	Solar saltern, Yellow Sea	Alkalibacillus flavidus	4–26	10
	Seawater, Yellow Sea	Salinisphaera dokdonensis	4–21	10
	Kunsan solar saltern	Nocardiopsis kunsanensis	3–20	10
	Jeotgal (salty condiment)	Lentibacillus jeotgali	3–20	10–15

(continued)

Table 1B. (continued)

Country	Saline environment	Bacteria	Range of NaCl (%)	Optimum NaCl (%)
Spain	Soil from Fuente de Piedra, saline wetland, Malaga	Halomonas fontilapidosi	3–20	5–7.5
	Cabo de Gata solar saltern, Almeria	Halomonas almeriensis	5–25	7.5
		Kushneria indalinina	3–25	7.5–10
		Salinicola halophilus	3–25	7.5–10
	Mallorca solar saltern, Balearic Islands	Salinibacter rubber	15–33 (salts)	20–30 (salts)
	Santa Pola solar saltern, Alicante	Halomonas cerina	7.5–20	7.5–10
		Virgibacillus salexingens	7–20 (salts)	10 (salts)
Thailand	Fermented salty foods (Ka-pi, Nam-pla, Pla-ra)	Lentibacillus halophilus	12–30	20–26
		Lentibacillus juripiscarius	3–30	10
		Lentibacillus kapialis	5–30	15
Tunisia	Chott El Guettar	Halothermothrix orenii	4–20	10
	Chott El-Djerid	Halanaerobaculum tunisiense	14–30	20–22
Ukraina	Lake Sivash, Crimea	Halanaerobium saccharolyticum	3–30	10
		Halocella cellulosilytica	5–20	15
		Orenia sivashensis	5–25	7–10
USA	The Great Salt Lake, Utah	Halomonas variabilis	7–29	9
	Death Valley, California	Actinopolyspora mortivallis	5–30	10–15
	Saltern, San Francisco Bay, California	Halanaerobacter chitinivorans	3–30	12–18
	Evaporated sea water, Oregon	Rhodothalassium salexigens	5–20	12
	Saline oil field brine, Oklahoma	Arhodomonas aquaeolei	6–20	15
		Halanaerobium salsuginis	6–24	9
	Searles Lake, California	Halarsenatibacter silvermanii	20 to saturated	saturated

[a]The following species have been isolated from ordinary, nonsaline soil samples taken in Japan: *Alkalibacillus silvisoli* 5–25% (10–15%), *Geomicrobium halophilum* 5–25% (10–15%), and *Halalkalibacillus halophilus* 5.0–25% (10–15%)

2.2 Halophilic Bacteria

Table 1B is a summary of saline and hypersaline environments and the halophilic Bacteria isolated from these sites. The range of salt concentrations, mostly NaCl, or a mixture of salt in some cases, that permitted growth and optimum concentration are also indicated. The species listed in Table 1B belong to the following classes: Cyanobacteria of the phylum Cyanobacteria, Alphaproteobacteria and Gammaproteobacteria of the phylum Proteobacteria, Clostridia and "Bacilli" of the phylum

Firmicutes, Actinobacteria of the phylum Actinobacteria, and Flavobacteria of the phylum Bacteroidetes. References for each species are not given in the list because of the huge number which would have to be cited. Readers are recommended to consult the very useful web site "List of Prokaryotic names with Standing in Nomenclature," maintained by J. P. Euzéby at http://www.bacterio.cict.fr/for relevant papers.

2.3 Romanian hypersaline environments

Hypersaline environments are widely distributed also in Romania, either in solid form or in liquid form: salt lakes and salt mines located in Prahova county, the Techirghiol lake nearby to Black Sea coast, the Balta Albă lake in Buzău county, etc. (Fig. 1). Some of these environments have been well described some time ago (Broşteanu 1901) and today also constitute an attractive research area either for

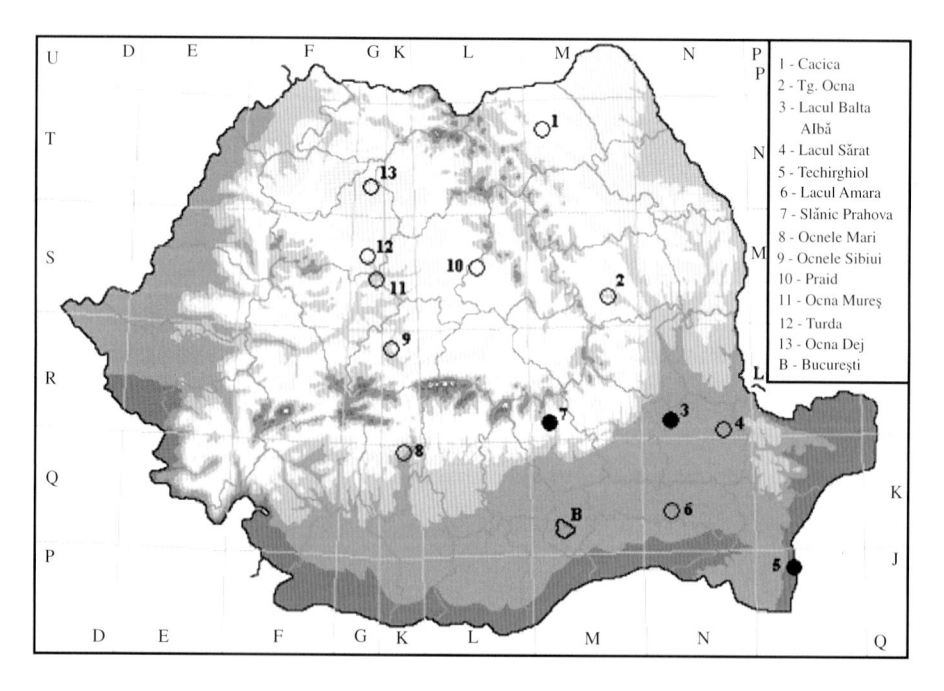

Fig. 1. Geographical positions of major hypersaline environments in Romania. The filled points represent the examined areas described in the text. **1**, salt mine at Cacica; **2**, salt mine at Targu; **3**, salt lake (Lacul Balta Alba); **4**, hypersaline salt lake (Lacul Sarat); **5**, salty therapeutical mud lake at Techirghiol (near the coast of the Black Sea); **6**, haloalkaline lake (Lacul Amara); **7**, lakes and salt mine of Slănic (Slănic Prahova); **8**, salt mine at Ocnele Mari (Ocnele Mari); **9**, lakes and salt mine at Sibiu (Ocnele Sibiu); **10**, salt mine at Praid (Praid); **11**, salt mine at Mures (Ocna Mures); **12**, salt mine at Turda (Turda); **13**, salt mine at Dej (Ocna Dej); B, Bucharest, capital of Romania

geologists or for biologists (Sencu 1968; Faghi et al. 1999; Teodosiu et al. 1999; Har et al. 2006).

Salt exploitation in the Romanian Carpathian area has been conducted by various methods from antiquity until today (Drăgănescu and Drăgănescu 2001), due to the presence of lots of salt massifs with characteristics that supported their continued use such as surface proximity, superior purity of NaCl, or large reserves. In some places, the massifs are located at the surface as small salt mountains, for example at Slănic, Praid, Sărata Monteoru (Fig. 1). In the Slănic Prahova area for example, salt exploitation started in 1685 by using bell type exploitation technology. After the eighteenth century, some areas of exploitation were abandoned, resulting in various man-made salt lakes with different depths (varying from 3 until to 40 m, see Table 2) and widths, along to the left or right side of the Slănic valley (Drăgănescu 1990). On the other hand, in Telega, where similar technologies were used, salt extraction started before 1685 and relinquished exploitation areas

Table 2. Chemical features of examined lakes and colony forming units (c.f.u.), as determined on agar plates containing 12.5% NaCl and 16% $MgCl_2 \cdot 6H_2O$, from surface water samples

Lake	Maximum depth (m)	pH	Density	Chloride content (g/l)	c.f.u./ml
Red Bath	3	7.9	1.06	74.9	2500
Green Bath	40	9.0	1.10	138.5	1050
Shepherd Bath	7.25	8.7	1.07	97.4	2400
Bride Cave	32	8.3	1.20	254.6	750
Techirghiol	9	7.0	–	60	–
Telega (Palada)	36	8.3	1.15	161	1100
White Bath	–	8.6	1.10	44	890

Table 3. Characteristics of salt lakes from Telega (Gâştescu 1965)

Lake	Height above the sea (m)	Surface (m^2)	Maximum width (m)	Maximum depth (m)
Doftana	413	9200	84	26
Central Bath	414	1344	38	45
Sweet Lake	424	1480	35	21
Stavrică	415	1740	52	107.5
Mocanu	415	630	22	14
Palada (Telega)	416	1416	30	36

Fig. 2. Salt massifs in the Slănic Prahova (**a**, **b**) and Praid (**c**) areas. The red color may be possible due to the presence of microorganisms or their remnants inside of the salt massifs in Slănic (**b**)

resulted in various man-made salt lakes, as shown in Table 3 (Gâştescu 1971; Enache et al. 2008a).

The salt deposit in Slănic Prahova formed in the Neogene period (24 My ago) and is located at around 100 km north of Bucharest in a sub-Carpathian hillock area (Figs. 1 and 2). This salt deposit has various lengths and thicknesses as described previously (Drăgănescu 1990; Enache et al. 2008a). Investigations for the presence of halophilic microorganisms were conducted in several salt lakes which formed in the relinquished salt mine exploitations in this salt deposit (see Sect. 2.4). These lakes are known today as Bride Cave, Red Bath, Green Bath, and Shepherd Bath.

The salt deposit of Telega (Doftana-Telega), which formed also in the Neogene period, is located at a similar distance and position from Bucharest such as the Slănic deposit, and is a mixture of crystals with colors varying from white to gray and swarthy, having a surface of 2.1 km^2. Various salt lakes which resulted in the mouth of an abandoned salt mine in this deposit are detailed in Table 3.

Another hypersaline environment with large reserves located at Ocnele Mari (Vâlcea county) has been exploited by dissolving salt, and the brine was channeled directly into a petrochemical plant by pipelines for the production of various chemicals (NaOH, Na_2CO_3, HCl, Cl_2, NaOCl, etc.). Consequently, a large subterranean void resulted which nowadays is under control to preserve collapsing. A huge man-made salt lake, estimated as larger than 50 ha, is expected to be present. This man-made hypersaline environment will constitute a further subject for halophilic research.

Another examined salt lake is Techirghiol lake, located near the Black Sea coast. This lake is characterized by variable concentrations of salt, ranging from negligible to 60 g/l. The lake is relatively well known for its biological communities and saline regimes, and has been well investigated over time, because of either the important therapeutical properties of the mud present in the lake or the dynamics of moderately halophilic microbial communities.

Another hypersaline environment, recently investigated by our team, is represented by the Balta Albă salt lake, located near the town of Râmnicu Sărat. The area is characterized by the presence of various salty (not extremely) soils. Various hypotheses have been elaborated on the origin of a salt lake in this area, arguing for either an ancient origin or the consequence of intense evaporation processes which took place in this area. Some geological data supported the latter hypothesis (Gâştescu 1965). The salinity of this lake was 40 g/l in our surface sample water.

2.4 Halophiles from salty environments in Romania

The chloride concentration of water samples taken from these salt lakes ranged from 44 to 254.6 g/l (Table 2) and pH values between 7.0 and 9.0. A number of these environments have been found to contain halophilic microorganisms able to grow in the presence of high concentrations of NaCl and $MgCl_2 \cdot 6H_2O$. The numbers of colonies ranged from 750 c.f.u./ml in water samples from Bride Cave until to around 2,500 c.f.u./ml in Red Bath (Table 1). In other lakes, the colony numbers were relatively low, when compared to those reported for various salt lakes with similar salt concentrations (Enache et al. 2008b). Investigations have been conducted (Enache et al. 1999b, 2000a, b) and a new species *Haloferax prahovense* was proposed recently (Enache et al. 2007a).

The diversity of halophiles, mainly in crystallizer ponds of solar salterns, remains one of the important research targets, although the constant NaCl concentrations of these artificial hypersaline environments over the years could be considered as a selective pressure for halophilic microorganisms. Due to various climatic and other natural conditions, different salt concentrations could be found in the salt lakes.

Although various molecular techniques were applied to understand the ecological principles of salt lakes, the microorganisms which could be isolated and cultivated in the laboratory appear to play the most important role in the ecological economy of these lakes.

The number of colony forming units decreased with increasing chloride and sodium concentrations in the examined salt lakes, but the colonies assigned to be Archaea had no apparent correlation with the concentrations of these two ions (Enache et al. 2008b). The predominant presence of *Haloferax* species in these lakes (Enache et al. 2008a, b) suggests that members of this genus play an important role in the ecology of salt lakes, even though the largest numbers of species, which have been identified in hypersaline environments, are of the genus *Halorubrum*. On the other hand, the microbiota of subterranean rock salt from the Slănic area are characterized by the presence of some *Halorubrum* and *Haloarcula* species, and some isolates appear to be closely related to *Hbt. noricense*, a strain isolated from an ancient evaporite formed in the Permo-Triassic period (Gruber et al. 2004). The *Haloferax* members, observable in hypersaline lakes located at the surface of the Slănic salt deposit, were identified also in subterranean rock salt, but with decreasing numbers (unpublished data).

3 Biotechnological potential

The physiological and biochemical features specific for halophilic Archaea and Bacteria as well as their capacity to produce biopolymers, enzymes, osmoprotectors of industrial interest, with properties superior to those synthesized by nonhalophilic species, make them a promising group of unlimited biotechnological potential (Rodriguez-Valera 1992). In the following sections, we would like to describe the potential for biotechnological application of halophilic microorganisms, including those isolated from Romanian hypersaline environments. Readers are also advised to consult previous general reviews (Rodriguez-Valera 1992; Ventosa et al. 1998; Horikoshi 1999; Mellado and Ventosa 2003; Ventosa 2004; Borgne et al. 2008).

3.1 Bioremediation

Biodegradation of organic pollutants by halophilic Bacteria and Archaea has been reviewed by Borgne et al. (2008). A few moderately halophilic bacteria were found to show the capacity, to various degrees, of decontaminating the pesticide dichlorvos ($C_4H_7O_4Cl_2P$) from saline environments (Oncescu et al. 2007). These strains, which were isolated from Shepherd Bath, Telega (Palada), and Techirghiol lakes, were tentatively assigned as *Halomonas* species according to their preliminary biochemical and physiological characterization. The kinetics of dichlorvos degradation appeared to follow the mechanism of "compatible solutes", used by large numbers of organisms to cope with osmotic stress generated by the presence of high concentrations of salt. Considerable efforts are still necessary in order to estimate the true potential of these halophilic microorganisms to be applied in environmental processes and in the remediation of contaminated hypersaline ecosystems. This effort should also be focused on basic research to understand the overall degradation mechanism, to identify the enzymes involved in the degradation process and the metabolic regulation.

3.2 Nanobiotechnology

A novel approach involving halophilic microorganisms appears to be their potential for nanobiotechnology. The interaction between some silica and titanium nanotubes with moderately halophilic microorganisms, for example *Virgibacillus halodenitrificans*, has recently been reported (Merciu et al. 2009). Variation in methods for the preparation of nanotubes (hydrothermal method or sol–gel method in the presence of templates) and different chemical treatments (thermal treatment and acid washing) after synthesis appear to be correlated with a putative antibacterial effect of these nano-materials toward various halophilic strains. Although a bacteriocin (halocin)-like attack mechanism is likely, further investigation will be necessary concerning the interaction between nano-materials and halophiles.

3.3 S-layers

Crystalline cell surface layers are commonly observed cell envelope structures of several Bacteria and Archaea, and they have numerous applications in biotechnology and nanotechnology. The extremely halophilic Archaea lack a peptidoglycan component in their cell wall and contain simple S-layers external to the cell membrane (Trachtenberg et al. 2000; Eichler 2003). S-layers consist of identical protein or glycoprotein subunits and completely cover the cell surface during all stages of growth and division. Most S-layers are 5–15 nm in thickness and possess pores of identical size and morphology in the 2–8 nm range (Sleytr and Sara 1997; Schuster et al. 2005).

Isolated S-layer subunits of various microorganisms have the intrinsic ability to self-assemble into highly defined monomolecular arrays either in suspension, at air/water interfaces or liquid/surface interfaces, including lipid films, liposomes, and solid supports such as silicon wafers (Schuster and Sleytr 2000; Schuster et al. 2005). The relative simplicity, regularity, and symmetry within the monolayer plane of the S-layer make it an attractive subject for nanobiotechnological studies with targets for medical applications (Sleytr et al. 1999; Trachtenberg et al. 2000).

Our investigations on S-layers were carried out using a halophilic archaeon, *Haloferax* sp. strain GR 2 (deposited as JCM 13922), isolated from the Bride Cave Lake in Prahova county. Preliminary investigations related to the binding of S-layer to some porous silicon substrates were performed. The biochemical characterizations by protein content and chemical treatment had demonstrated the presence of S-layer in the isolated strain. Transmission electron microscopic examination of the isolated S-layer showed the existence of the monomolecular crystalline lattice with a highly ordered arrangement in the dense form, while in relaxed form after treatment with 4 M urea (Dumitru et al. 2007). The S-layer proteins attached to both hydrophilic and hydrophobic surfaces of all plates of porous silicon, which were investigated, but it seemed that the hydrophobic surfaces were more favorable. Thus, the treatment of silicon plates with hexamethyldisilazane, which promotes the hydrophobicity and organic character of the porous silicon surface, increased the amount of attached S-layer protein (Sleytr et al. 1999; Dumitru et al. 2007; Kleps et al. 2009).

3.4 Extracellular enzymes

A great deal of information on eukaryotic and bacterial halophilic enzymes is currently available, for examples, on amylases, lipases, nucleases, nucleotidases, and proteases (Ventosa et al. 2005). For the production of Thai fish sauce (*nam-pla*), a condiment similar to *Garum* or *Liquamen* of the ancient Roman society, prepared from fish in concentrated brine, proteases of haloarchaea which are present in solar salt, plays an important role in the degradation of fish protein into amino acids (Thongthai et al. 1992). Further studies on haloarchaeal enzymes are expected to contribute to the elucidation of the properties of these extracellular enzymes.

Amylases were produced by some strains isolated from Bride Cave and Techirghiol lakes (Enache and Faghi 1999, 2009). The enzyme of *Hfx.* sp. GR1 purified by ethanol precipitation and differential chromatography showed maximum activity at pH 6.5 and 50°C in the presence of 3.5 M NaCl, and lost its activity below 1.5 M NaCl (Enache et al. 2001). Amylases produced by strains of *Haloferax* and *Halorubrum* isolated from Techirghiol lake, a low salt environment, showed higher activity with increasing concentration of $MgCl_2$ in the presence of the relatively low 2.1 M NaCl, but activity decreased with increasing Mg concentrations at the higher concentration of 3.4 M NaCl (Enache et al. 2009).

Extracellular lipase activity was detected in some strains isolated from lakes Shepherd Bath, Green Bath, Red Bath, and Bride Cave. Among them, the enzyme produced by *Hfx.* sp. GR1 was influenced by NaCl concentrations in the growth media and had a maximum activity at 3 M NaCl. The activity was lost at NaCl concentrations below 2.5 M (Enache et al. 2004a).

The molecular mechanisms of the adaptation of enzyme proteins to high salt have been described in detail (Vellieux et al. 2007, Yamamura et al. 2009).

3.5 Halocins

The halocins, proteinaceous antibiotics, which are haloarchaeal equivalents of bacteriocins, were first discovered by Rodriguez-Valera et al. (1982). The wide variety of activity spectra detected for halocins (H1, H4, H6, S8, C8, etc.) may imply that a great number of different halocins are produced and probably show various mechanisms of action (Torreblanca et al. 1994; O'Connor and Shand 2002). Halocins were also detected in some strains isolated from Romanian salt lakes, and they showed a variety of action spectra (Enache et al. 1999a). When compared as halocin producers and targets, some strains showed identical patterns, supporting the tight clustering of strains in the phylogenetic tree reconstructed from 16S rRNA gene sequences (Enache et al. 2004b, 2008b).

Halocin 6 (H6) is a protein of 32 kDa produced by *Hfx. gibbonsii* SH7 and blocks the Na^+/H^+ antiporter in sensitive strains. Quite interestingly, H6 has been shown to inhibit in vitro the Na^+/H^+ exchanger of mammalian cells and to exert in vivo a cardio-protective effects against ischemia and reperfusion injury (Lequerica et al. 2006).

3.6 Exopolysaccharides

Halophilic microorganisms are able to synthesize extracellular polysaccharide (EPS), which are biopolymers combining, in an excellent manner, the rheological properties (high viscosity and pseudoplasticity) with a remarkable resistance to extreme salinity, temperature, and pH values, conditions which are encountered in several industrial processes and which make usage of biopolymers produced by nonhalophilic microorganisms impossible (Rodriguez-Valera 1992; Ventosa et al. 1998). The physical and chemical properties of the EPS produced by halophilic microorganisms enable their utilization in the food, textile, and dye industries, also for the production of pharmaceuticals and cosmetic products, in the oil extracting industry as well as for processes for the removal of toxic compounds.

The biosynthetic activity for EPS was detected in some *Haloferax* strains isolated from Telega (Palada) lake such as *Hfx. prahovense* and *Hfx.* sp. TL5. The optimal conditions for EPS production by *Hfx. prahovense* were the same as those resulting in highest cell growth. The maximum EPS yield (0.475 g%) was obtained in medium with 3% glucose as single carbon source at 2 M NaCl, under stirring at 200 rpm, at 37°C, after 7 days of incubation. The strain produced EPS also in media with galactose, lactose, maltose, sucrose, or fructose as carbon source. Higher salt concentrations (5 M) and higher temperature (45°C) had an inhibitory effect on both growth and EPS synthesis. Synthesis of EPS started during the early exponential growth phase, increased concomitantly with a rise in the number of viable cells, and then decreased after 7 days of cultivation. The monomer composition of the EPS from *Hfx. prahovense* was similar to the composition of EPS synthesized by halophilic Archaea of the genus *Haloferax* (Anton et al. 1988). The polymer of *Hfx. prahovense* was a heteropolysaccharide containing mainly glucose, fructose, galactose, and mannose as was observed by TLC. Differential scanning calorimetry revealed that the polymer was stable up to 207°C; the chemical composition observed by TLC was confirmed by FTIR investigations. FTIR also showed the presence of uronic acids and sulfate in the polymer (Popescu et al. unpublished results). A similar highly thermostable EPS was isolated and characterized from cultures of some moderately halophilic bacteria isolated from Shepherd Bath (Cojoc et al. 2009).

3.7 Resistance to heavy metals

Several halophilic Archaea (Dumitru et al. 2002) and Bacteria (Enache et al. 2000c) isolated in Romania were shown to be resistant to heavy metals. The data suggested that moderately halophilic Bacteria exhibited a higher tolerance to metallic ions as compared to halophilic Archaea. The investigated haloarchaeal strains were susceptible to Zn and Hg, but moderately resistant to Cr and Ni, being classified as tolerant according to the criteria proposed by Nieto (1991).

The metal tolerance level of our isolate *Haloferax* sp. TL5 (assigned as a strain of *Hfx. prahovense*, see Enache et al. 2008a) was compared with that of *Hfx. mediterranei*. Strain *Hfx.* sp. TL5 showed a similar behavior to the collection strain *Hfx. mediterranei*; both strains tolerated 5.0 mM Cr and 2.5 mM Ni and Pb. Strain *Hfx.* sp. TL5 had a higher susceptibility for Zn ions, compared with *Hfx. mediterranei*.

We also measured the capacity of these strains to reduce the concentration of several heavy metal ions from media. Both strains showed the capacity to reduce the concentration of Pb, Cr, Zn, and Ni ions from media with high salinity (Popescu and Dumitru 2009). The two strains produced higher cell densities when grown in media with metal ions and 2–2.5% glucose than in media without glucose. This suggests that EPS synthesized in the presence of glucose may protect the cells against the toxicity of heavy metals. The two *Haloferax* strains showed the same capacity to reduce the concentration of Pb ion; for example, the initial concentration of 331 mg Pb/l was reduced to 5 mg/l after 10 days of cultivation. *Hfx.* sp. TL5 has a higher biosorption capacity of Cr and Ni ions from medium with or without glucose than *Hfx. mediterranei*.

Hfx. mediterranei presented a higher removal activity of Zn ion from media with or without glucose than *Hfx.* sp. TL5 (Popescu and Dumitru 2009). The results revealed that the synthesis of EPS enhanced the reduction activity of Cr, Zn, and Ni by the haloarchaeal strains which were investigated. The anionic nature of EPS synthesized by *Haloferax* strains, based on their high sulfate and uronic-acid contents (Rodriguez-Valera 1992; Mellado and Ventosa 2003), is similar to that of EPS synthesized by other halophilic microorganisms and may be responsible for the capacity of these strains to bind and remove heavy metals from solutions with high NaCl concentrations. This property would make these biopolymers a viable alternative to the more aggressive physical and chemical methods, and they could be used as bioadsorbents in polluted hypersaline environments.

3.8 Therapeutical value

The salt lakes in Romania have been used also for various economical, recreational, and therapeutical purposes. The therapeutical use of the mud from salt lakes started in 1840 (Bulgăreanu 1993). Although attributed to the accumulation of sapropelic material, the mechanisms of mud formation and their microbiota are poorly understood. A few attempts in our laboratory to elucidate the mechanisms, using the mud from Techirghiol lake, were fruitless. An industry has been developed for the exploitation of mud mainly from Balta Albă and Techirghiol lakes.

In Middle Europe and Eastern Europe, the advantageous effect of some natural salt caves on lung diseases has been known since the nineteenth century; possibly, salt-miners knew it far earlier based on the observation that injured animals went to caves for recovering. The beneficial effect of salt (speleotherapy) was reported by the Polish doctor F. Bochkowsky in 1843. Speleotherapy is based on the in-patient treatment in salt caves possessing a specific microclimate. The effects of salt mine treatment on health in the village of Solotvino in the Carpathian Mountains have been investigated by Russian scientists (Simyonka 1989). Several salt caves are used for speleotherapy in Middle Europe and Eastern Europe as follows: Salzgrotten, Saliseum/Vienna (Austria); Slanic, Turda, and Praid (Romania); Wieliczka (Poland); Nakhlichevan (Azerbaidjan); Chon-Tous (Kirghizstan); Cave Berezniki in Perm (Russia); Solotvino (Ukraine). Halotherapy is a form of speleotherapy, a science aimed to create somewhat similar conditions in a microclimatic environment as in salt caves. In the 1980s, Russia was the pioneer in creating the first salt chambers. These are specially prepared rooms with walls and basements covered with halite, a crystal form of salt (Chervinskaya et al. 1995; Hedman et al. 2006; Nica et al. 2007).

4 Concluding remarks

Research in the field of halophilic microbiology attracted huge interest during the last decade, yielding more than 18 new genera (from 2000 to the present) in the archaeal family *Halobacteriaceae* and numerous bacterial genera (Table 1). Many of them have their origin in man-made "hypersaline" environments developed for various purposes, e.g. commercial solar salt and salted food. These sources have a connection to marine environments, which apparently are not "hypersaline." We would like to remind the readers of the fact that a relatively small number of the validly published halophilic species of Archaea and Bacteria come from truly hypersaline lakes (such as the Dead Sea or the Great Salt Lake). Another fact is that no rigid evidence of the existence of viable haloarchaeal cells in seawater has been presented, except perhaps for halococci (Rodriguez-Valera et al. 1979). Taking into account these facts, future research will be necessary to give an answer to the question of how the halophilic microorganisms originated during the early stages in the evolution of life and how they diversified and were distributed throughout the world, in the past and present (Oren 2004). Although the molecular clock of organisms isolated from inclusion bodies of ancient salt crystals is difficult to be correlated with the geological time of evaporation, the question of longevity of halophilic Archaea and Bacteria continues to be a tremendous fascination to all microbiologists (Grant et al. 1998; Fendrihan et al. 2009a, b; Park et al. 2009).

References

Anderson H (1954) The reddening of salted hides and fish. Appl Microbiol 2:64–69
Anton J, Meseguer I, Rodriguez-Valera F (1988) Production of an extracellular polysaccharide by *Haloferax mediterranei*. Appl Environ Microbiol 54:2381–2386
Borgne SL, Paniagua D, Vazquez-Duhalt R (2008) Biodegradation of organic pollutants by halophilic bacteria and archaea. J Mol Microbiol Biotechnol 15:74–92
Broşteanu C (1901) Our salt mine. Historic, juridical and economical study towards saline exploitations and salt monopoly on Romans and Romanian (in Romanian). In: Lazareanu GA (ed) Bucharest
Bulgăreanu VAC (1993) The protection and management of saline lakes of therapeutic value in Romania. Int J Salt Lake Res 2:165–171
Burns DG, Janssen PH, Itoh T, Kamekura M, Echigo A, Dyall-Smith ML (2010) *Halonotius pteroides* gen. nov., sp. nov., an extremely halophilic archaeon recovered from a saltern crystallizer in southern Australia. Int J Syst Evol Microbiol 60:1196–1199
Chervinskaya AV, Zilber NA (1995) Halotherapy for treatment of respiratory diseases. J Aerosol Med 8:221–232
Cojoc R, Merciu S, Oancea P, Pincu E, Dumitru L, Enache M (2009) Highly thermostable exopolysaccharide produced by the moderately halophilic bacterium isolated from a man-made young salt lake in Romania. Pol J Microbiol 58:289–294

Drăgănescu L (1990) Dates from historic salt exploitation at Slanic-Prahova (in Romanian). Rev Muzeelor 27:68–71

Drăgănescu L, Drăgănescu S (2001) The history of the evolution of salt working methods in Romania, from antiquity to the present. In: 17th Intl Mining Congress and Exhibition of Turkey–IMCET, pp 627–633

Duggen S, Hoernle K, van den Bogaard P, Rüpke L, Morgan JP (2003) Deep roots of the Messinian salinity crisis. Nature 422:602–606

Dumitru L, Teodosiu G, Enache E (2002) The tolerance of extremely halophilic archaea to heavy metals. Proc Inst Biol 4:261–267

Dumitru L, Teodosiu-Popescu G, Enache M, Cojoc R, Kleps I, Ignat T (2007) S-layer of *Haloferax* sp. GR 2 (JCM 13922): isolation, characterization and binding to silicon nanostructurated substrates. In: Kleps I, Ion AC, Dascalu D (eds) Progress in nanoscience and nanotechnologies, Seria vol 11, Micro and nanoengineering. Ed. Acad Române, Bucharest, pp 146–152

Eichler J (2003) Facing extremes: archaeal surface-layer (glyco) proteins. Microbiology 149:3347–3351

Enache M, Faghi AM (1999) Detection of extracellular halophilic amylase activity from the extreme halophilic genus *Haloferax*. Proc Inst Biol 2:143–146

Enache M, Dumitru L, Faghi AM (1999a) Occurrence of halocins in mixed archaebacteria culture. Proc Inst Biol 2:151–154

Enache M, Teodosiu G, Dumitru L, Faghi AM, Zarnea G (1999b) Diversity of halobacteria in accordance with their membrane lipids composition. Proc Inst Biol 2:147–149

Enache M, Teodosiu G, Faghi AM, Dumitru L (2000a) Identification of halophilic Archaebacteria isolated from some Romanian salts lakes on the basis of lipids composition. Rev Roum Biol Biol Veg 45:93–99

Enache M, Faghi AM, Teodosiu G, Dumitru L, Zarnea G (2000b) Effect of temperature and NaCl concentration of growth media on neutral lipid in *Haloferax volcanii* and *Haloferax* sp. GR1 strain. Proc Inst Biol 3:251–256

Enache E, Faghi AM, Teodosiu G, Dumitru L, Zarnea G (2000c) The halophilic microorganisms response to heavy metals. Rev Roum Biol Biol Veg 45:111–116

Enache M, Faghi AM, Dumitru L, Zarnea G (2001) Halophilic α-amylase from the extremely halophilic archaea *Haloferax* sp. strain GR1. Proc Rom Acad Series B 2:107–109

Enache M, Teodosiu G, Dumitru L, Zarnea G (2004a) The effect of NaCl concentrations on the growth and lipase activity at *Haloferax* sp. GR1. Proc Inst Biol 6:233–236

Enache M, Faghi AM, Dumitru L, Teodosiu G, Zarnea G (2004b) Halocin HF 1 a bacteriocin produced by *Haloferax* sp. GR 1. Proc Rom Acad Series B 1:27–32

Enache M, Itoh T, Kamekura M, Teodosiu G, Dumitru L (2007a) *Haloferax prahovense* sp. nov., an extremely halophilic archaeon isolated from a Romanian salt lake. Int J Syst Evol Microbiol 57:393–397

Enache M, Itoh T, Fukushima, Usami R, Dumitru L, Kamekura M (2007b) Phylogenetic relationship within the family *Halobacteriaceae* inferred from $rpoB'$ gene and protein sequences. Int J Syst Evol Microbiol 57:2289–2295

Enache M, Itoh T, Kamekura M, Popescu G, Dumitru L (2008a) Halophilic archaea of *Haloferax* genus isolated from anthropocentric Telega (Palada) salt lake. Proc Rom Acad Series B 10:11 16

Enache M, Itoh T, Kamekura M, Popescu G, Dumitru L (2008b) Halophilic archaea isolated from man-made young (200 years) salt lakes in Slănic, Prahova, Romania. Cent Eur J Biol 3:388–395

Enache M, Popescu G, Dumitru L, Kamekura M (2009) The effect of Na^+/Mg^{2+} ratio on the amylase activity of haloarchaea isolated from Techirghiol lake, Romania, a low salt environment. Proc Rom Acad Series B 11:3–7

Faghi AM, Teodosiu G, Dumitru L (1999) The dynamic of the halophilic microorganisms growth at different values of pH and temperature. I. Moderately halophilic bacteria. Proc Inst Biol 2:155–161

Fendrihan S, Bérces A, Lammer H, Musso M, Rontó G, Polacsek TK, Holzinger A, Kolb C, Stan-Lotter H (2009a) Investigating the effects of simulated Martian ultraviolet radiation on *Halococcus dombrowskii* and other extremely halophilic archaebacteria. Astrobiol 9:104–112

Fendrihan S, Musso M, Stan-Lotter H (2009b) Raman spectroscopy as a potential method for the detection of extremely halophilic archaea embedded in halite in terrestrial and possibly extra-terrestrial samples. J Raman Spectr 40:1996–2003

Gâștescu P (1965) On the origin of salt lakes from Romanian Plain (in Romanian). Natura–Seria Geografie–Geologie 2:42–45

Gâștescu P (1971) The lake from Romania (in Romanian). Ed Acad Rep Soc România, Bucharest

Gibbons NE (1974) Family V. Halobacteriaceae fam. nov. In: Buchanan RE, Gibbons NE (eds) Bergey's manual of determinative bacteriology, 8th edn. Williams & Wilkins, Baltimore, pp 269–273

Grant WD, Gemmell RT, McGenity TJ (1998) Halobacteria: the evidence for longevity. Extremophiles 2:279–287

Grant WD, Kamekura M, McGenity TJ, Ventosa A (2001) The order *Halobacteriales*. In: Boone DR, Castenholz RW (eds) Bergey's manual of systematic bacteriology, vol 1, 2nd edn. Springer, New York, pp 294–334

Gruber C, Legat A, Pfaffenhuemer M, Radax C, Weidler G, Busse HJ, Stan-Lotter H (2004) *Halobacterium noricense* sp. nov., an archaeal isolate from a bore core of an alpine Permian salt deposit, classification of *Halobacterium* sp. NRC-1 as a strain of *H. salinarum* and emended description of *H. salinarum*. Extremophiles, 8:431–439

Har N, Barbu O, Codrea V, Petrescu I (2006) New data on the mineralogy of the salt deposit from Slănic Prahova (Romania). Studia UBB, Geologia 51:29–33

Hedman J, Hagg T, Sandell J, Haahtela T (2006) The effect of salt chamber treatment on bronchial hyperresponsiveness in asthmatics. Allergy 61:605–610

Horikoshi K (1999) Alkaliphiles: some applications of their products for biotechnology. Microbiol Mol Biol Rev 63:735–750

Itoh T, Yamaguchi T, Zhou P, Takashina T (2005) *Natronolimnobius baerhuensis* gen. nov., sp. nov., and *Natronolimnobius innermongolicus* sp. nov., novel haloalkaliphilic archaea isolated from soda lakes in Inner Mongolia, China. Extremophiles 9:111–116

Javor B (1989a) Chapter 1. Geology and chemistry. In: Hypersaline environments. Microbiology and biogeochemistry. Springer-Verlag, Berlin, Heidelberg, New York, pp 5–25

Javor B (1989b) Chapter 15. Gavish Sabkha and other hypersaline marine sabkhas, pools and lagoons. In: Hypersaline environments. Microbiology and biogeochemistry. Springer-Verlag, Berlin, Heidelberg, New York, pp 222–235

Juez G (1988) Taxonomy of extremely halophilic archaebacteria. In: Rodriguez-Valera F (ed) Halophilic Bacteria, vol II. CRC Press, Boca Raton, FL, pp 3–24

Kamekura M, Seno Y (1993) Partial sequence of the gene for a serine protease from a halophilic archaeum *Haloferax mediterranei* R4, and nucleotide sequences of 16S rRNA encoding genes from several halophilic archaea. Experientia 49:503–513

Kamekura M, Dyall-Smith ML, Upasani V, Ventosa A, Kates M (1997) Diversity of alkaliphilic halobacteria: proposal for transfer of *Natronobacterium vacuolatum*, *Natronobacterium magadii*, and *Natronobacterium pharaonis* to *Halorubrum*, *Natrialba* and *Natronomonas* gen. nov., respectively, as *Halorubrum vacuolatum* comb. nov., *Natrialba magadii* comb.

nov., and *Natronomonas pharaonis* comb. nov., respectively. Int J Syst Bacteriol 47:853–857

Kates M (1995) Adventures with membrane lipids. Biochem Soc Transac 23:697–709

Kleps I, Ignat T, Miu M, Simion M, Teodosiu-Popescu G, Enache M, Dumitru L (2009) Protein–mesoporous silicon matrix obtained by S-layer technology. Phys Status Solidi C 6:1605–1609

Kocur M, Hodgkiss W (1973) Taxonomic status of the genus *Halococcus* Schoop. Int J Syst Bacteriol 23:151–156

Kunte HJ, Trueper HG, Stan-Lotter H (2002) Halophilic microorganism. In: Horneck G, Baumstark-Khan C (eds) Astrobiology. The quest for the conditions of life. Springer Verlag, Berlin, Heidelberg, pp 185–200

Kushner DJ (1985) The Halobacteriaceae. In: Woese CR, Wolfe RS (eds) The Bacteria, vol VIII. Academic Press, New York, pp 171–214

Lequerica JL, O'Connor JE, Such L, Alberola A, Meseguer I, Dolz M, Torreblanca M, Moya A, Colom F, Soria B (2006) A halocin acting on Na^+/H^+ exchanger of Haloarchaea as a new type of inhibitor in NHE of mammals. J Physiol Biochem 62:253–262

Lochhead AG (1934) Bacteriological studies on the red discoloration of salted hides. Can J Res 10:275–286

Magrum LJ, Luehrsen KR, Woese CR (1978) Are extreme halophiles actually "bacteria"? J Mol Evol 11:1–8

Mellado E, Ventosa A (2003) Biotechnological potential of moderately and extremely halophilic microorganisms. In: Barredo JL (ed) Microorganisms for health care, food and enzyme production. Research Signpost, Kerala, pp 233–256

Merciu S, Vacaroiu C, Filimon R, Popescu G, Preda S, Anastasescu C, Zaharescu M, Enache M (2009) Nanotubes biologically active in media with high salt concentration. Biotechnol Biotechnol Eq 23:827–831

Minegishi H, Echigo A, Nagaoka S, Kamekura M, Usami R (2010) *Halarchaeum acidiphilum* gen. nov., sp. nov., a moderately acidophilic haloarchaeon isolated from commercial solar salt. Int J Syst Evol Microbiol (in press). DOI: 10.1099/ijs.0.013722-0

Multhauf RP (1978) Neptune's gift. A history of common salt. The Johns Hopkins University Press, Baltimore and London

Nica SA, Meilă AM, Macovei L (2007) Speleotherapy (in Romanian). Rom J Rheumatol 4:269–273

Nieto JJ (1991) The response of halophilic bacteria to heavy metals. In: Rodriguez-Valera F (ed) General and applied aspects of halophilic microorganisms. Plenum Press, New York and London, pp 173–179

O'Connor EM, Shand RF (2002) Halocins and sulfolobicins: the emerging story of archaeal protein and peptide antibiotics. J Ind Micrbiol Biotechnol 28:23–31

Oncescu T, Oancea P, Enache M, Popescu G, Dumitru L, Kamekura M (2007) Halophilic bacteria are able to decontaminate dichlorvos, a pesticide, from saline environments. Cent Eur J Biol 2: 563–573

Oren A (2002) Preface. In: Halophilic microorganisms and their environments. Kluwer Academic Publishers, Dordrecht, pp xv–xviii

Oren A (2004) Prokaryote diversity and taxonomy: current status and future challenges. Philos Trans R Soc Lond B 359:623–638

Park JS, Vreeland RH, Cho BC, Lowenstein TK, Timofeeff MN, Rosenzweig WD (2009) Haloarchaeal diversity in 23, 121 and 419 MYA salts. Geobiology 7:1–9

Popescu G, Dumitru L (2009) Biosorption of some heavy metals from media with high salt concentrations by halophilic Archaea. Biotechnol Biotechnol Eq 23:791–795

Rees CH, Grant DW, Jones EB, Heaphy S (2004) Diversity of Kenyan soda lake alkaliphiles assessed by molecular methods. Extremophiles 8:63–71

Rodriguez-Valera F (1992) Biotechnological potential of halobacteria. Biochem Soc Symp 58:135–147

Rodriguez-Valera F, Ruiz-Berraquero F, Ramos-Cormenzana A (1979) Isolation of extreme halophiles from seawater. Appl Environ Microbiol 38:164–165

Rodriguez-Valera F, Juez G, Kushner DJ (1982) Halocins: salt-dependent bacteriocins produced by extremely halophilic rods. Can J Microbiol 28:151–154

Saiki RK, Gelfand DH, Stoffel S, Scharf SJ, Higuchi R, Horn GT, Mullis KB, Erlich HA (1988) Primer-directed enzymatic amplification of DNA with a thermostable DNA polymerase. Science 239:487–491

Schuster B, Sleytr UB (2000) S-layer-supported lipid membranes. Rev Mol Biotechnol 74:233–254

Schuster B, Györvary E, Pum D, Sleytr UB (2005) Nanotechnology with S-layer protein. In: Vo-Dinh T (ed) Methods in molecular biology. vol 300: Protein nanotechnology, protocols, instrumentation and applications. Humana Press Inc., Totowa, NJ, pp 101–123

Sencu V (1968) Salt Mount from Slănic Prahova (in Romanian). Ocrot Nat 12:167–179

Shimane Y, Hatada Y, Minegishi H, Mizuki T, Echigo A, Miyazaki M, Ohta Y, Usami R, Grant WD, Horikoshi K (2010) *Natronoarchaeum mannanilyticum* gen. nov., sp. nov., an aerobic, extremely halophilic member of the Archaea isolated from commercial salt made in Niigata, Japan. Int J Syst Evol Microbiol (in press). DOI: 10.1099/ijs.0.016600-0

Simyonka YM (1989) Some particular features of infections and inflammatory processes, and immune status in patients with infection-dependent bronchial asthma during speleotherapy in salt-mine microclimate (in Russian). In: Bronchial asthma. Leningrad, pp 136–140

Sleytr UB, Sara M (1997) Bacterial and archaeal S-layer proteins: structure–function relationships and their biotechnological applications. Trends Biotechnol 15:20–26

Sleytr UB, Messner P, Pum D, Sara M (1999) Crystalline bacterial cell surface layers (S-layers): from supramolecular cell structure to biomimetics and nanotechnology. Angew Chem 38:1034–1054

Stan-Lotter H, Radax C, Gruber C, Legat A, Pfaffenhuemer M, Wieland H, Leuko S, Weidler G, Kömle N, Kargle G (2003) Astrobiology with haloarchaea from Permo-Triassic rock salt. Int J Astrobiol 1:271–284

Teodosiu G, Dumitru L, Faghi AM (1999) The dynamic of the halophilic microorganisms growth at different values of pH and temperature. II. Extremely halophilic archaea. Proc Inst Biol 2: 163–168

Thongthai C, McGenity TJ, Suntinanalert P, Grant WD (1992) Isolation and characterization of an extremely halophilic archaebacterium from traditionally fermented Thai fish sauce (nam-pla). Lett Appl Microbiol 14:111–114

Tindall BJ, Mills AA, Grant WD (1980) An alkalophilic red halophilic bacterium with low magnesium requirement from a Kenyan soda lake. J Gen Microbiol 116:257–260

Tindall BJ, Ross HNM, Grant WD (1984) *Natronobacterium* gen. nov. and *Natronococcus* gen. nov., two new genera of haloalkaliphilic archaebacteria. Syst Appl Microbiol 5:41–57

Torreblanca M, Rodriguez-Valera F, Juez G, Ventosa A, Kamekura M, Kates M (1986) Classification of non-alkaliphilic halobacteria based on numerical taxonomy and polar lipid composition, and description of *Haloarcula* gen. nov. and *Haloferax* gen. nov. Syst Appl Microbiol 8:89–99

Torreblanca M, Meseguer I, Ventosa A (1994) Production of halocin is a practically universal feature of archaea halophilic rods. Lett Appl Microbiol 19:201–205

Trachtenberg S, Pinnick B, Kessel M (2000) The cell surface glycoprotein layer of the extreme halophile *Halobacterium salinarum* and its relation with *Haloferax volcanii*: cryo-electron tomography of freeze-substituted cells and projection studies of negatively stained envelopes. J Struct Biol 130:10–26

Vellieux F, Madern D, Zaccai G, Ebel C (2007) Molecular adaptation to high salt. In: Gerday C, Glansdorff N (eds) Physiology and biochemistry of extremophiles. ASM Press, Washington, DC, pp 240–253

Ventosa A (2004) Halophilic microorganisms. Springer-Verlag, Berlin, Heidelberg

Ventosa A, Nieto JJ, Oren A (1998) Biology of moderately halophilic aerobic bacteria. Microbiol Mol Biol Rev 62:504–544

Ventosa A, Sanchez-Porro C, Martin S, Mellado E (2005) Halophilic archaea and bacteria as a source of extracellular hydrolytic enzymes. In: Gunde-Cimerman N, Oren A, Plemenitas A (eds) Adaptation to life at high salt concentrations in archaea, bacteria, and eukarya. Springer, Dordrecht, pp 337–354

Ventosa A, Mellado E, Sanchez-Porro C, Marquez MC (2008) Halophilic and halotolerant microorganism from soils. In: Dion P, Nautiyal CS (eds) Microbiology of extreme soils. Springer-Verlag, Berlin, Heidelberg, pp 87–115

Weisburg WG, Barns SM, Pelletier DA, Lane DJ (1991) 16S ribosomal DNA amplification for phylogenetic study. J Bacteriol 173:697–703

Williams WD (1998) Guidelines of lake management, vol 6: Management of inland saline waters. In: International lake environment committee foundation and the United Nations environment programme, Kusatsu, Japan

Xue Y, Fan H, Ventosa A, Grant WD, Jones BE, Cowan DA, Ma Y (2005) *Halalkalicoccus tibetensis* gen. nov., sp. nov., representing a novel genus of haloalkaliphilic archaea. Int J Syst Evol Microbiol 55:2501–2505

Yamamura A, Ichimura T, Kamekura M, Mizuki T, Usami R, Makino T, Ohtsuka J, Miyazono K, Okai M, Nagata K, Tanokura M (2009) Molecular mechanism of distinct salt-dependent enzyme activity of two halophilic nucleoside diphosphate kinases. Biophys J 96:4692–4700

Applications of extremophiles in astrobiology: habitability and life detection strategies

Felipe Gómez and Víctor Parro

Centro de Astrobiología (CSIC-INTA), Carretera de Ajalvir, Km 4, Torrejón de Ardoz, Madrid, Spain

1 Introduction

One of the most interesting questions that science will have to confront in the new century is the existence of life elsewhere in the Universe. The possibility of the existence of life on other planetary bodies apart from Earth such as Mars or Jupiter's moon Europa will have to be addressed during future space missions. Now is the time for preparing the techniques for next generation tools for automated life detection. Next Mars missions will be focused mainly on the potential habitability of the surface of Mars and the unambiguous identification of signs of life, but future missions to Europa are also planned. The return of Mars samples to Earth for state-of-the-art laboratory analysis is beginning to be considered in the past years by international collaborative consortia.

In assessing the possibility of life to exist in other planets or moons, we have to define clearly what we are looking for. We have to address the perception of life and its limits on Earth as the only example we know. Therefore, extreme environments are important natural laboratories where to gain knowledge about the limits of life and the real possibility of automated identification of life in these conspicuous and challenging surroundings.

The current exploration of Mars is producing a considerable amount of information which requires comparison with terrestrial analogs in order to interpret and evaluate compatibility with possible extinct and/or extant life on the planet. The first astrobiological mission specially designed to detect life on Mars, the Viking mission, thought life there was unlikely, considering the amount of UV radiation which is bathing the surface of the planet, the resulting oxidative conditions, and the lack of adequate atmospheric protection.

The necessity for exploration of the surface of Europa originates from the idea of the existence of a water ocean in its interior. The surface of Europa presents evidence of an active geology showing many tectonic features that seem to be connected with some liquid interior reservoir. Life needs several requirements for its establishment, but the only sine qua non element is water, taking into account our experience with extreme ecosystems on Earth.

The discovery of extremophiles on Earth widened the window of possibilities for life to develop in the Universe, and as a consequence on Mars. The compilation of data produced by the ongoing missions (Mars Global Surveyor, Mars Odyssey, Mars Express, and Mars Exploration Rovers Spirit and Opportunity) offers a completely different view: signs of an early wet Mars and rather recent volcanic activity. The discovery of important accumulations of sulfates and the existence of iron minerals such as jarosite, goethite, and hematite in rocks of sedimentary origin have allowed specific terrestrial models related to this type of mineralogy to come into focus. Río Tinto (Southwestern Spain, Iberian Pyritic Belt) is an extremely acidic environment, a product of the chemolithotrophic activity of microorganisms that thrive in the massive pyrite-rich deposits of the Iberian Pyritic Belt. The high concentrations of ferric iron and sulfates, products of the metabolism of pyrite, generate a collection of minerals, mainly gypsum, jarosite, goethite, and hematites, all of which have been detected in different regions of Mars (Fernández-Remolar et al. 2004).

But where to look for life on other planetary bodies? The surfaces of planets or icy moons are adverse for life. Harsh conditions for life to deal with are similar to those harsh conditions found on primordial Earth during the time when the origin of life occurred. In the latter case, life originated under conditions of high irradiation, meteorite bombardment, and high temperature. Some particularly protective environments or elements such as iron solutions could shield organic molecules and the first bacterial life forms (Gómez et al. 2007). Work with terrestrial analogs could help us to increase our understanding.

2 Extremophiles and astrobiology

2.1 Extremophiles

Some lessons can be learnt from extremophiles in order to interpret the real possibilities of life elsewhere in the Universe. If life is finally found in the Martian surface or the subsurface of Europa, the identified form of life probably would be what we know on Earth as an extremophile or a microorganism that is able to live in harsh environmental physico-chemical conditions from the human being point of view, such as pH, pressure, radiation conditions, temperature, water activity, salinity, and oxidative stress. Of particular interest from the astrobiological point of view are microorganisms living in protected niches such as permafrost, locations under ice shield layers on polar areas, hot springs and hot vent geothermal fields, and acidic environments. Of special scientific interest are psychrophiles, hyperthermophiles, acidophiles, halophiles, and microorganisms tolerating high osmotic pressures and also alkaliphiles and microorganisms with adaptations to high radiation doses are of interest for astrobiology (Rothschild and Mancinelli 2001). The discovery of extremophiles gave also important inputs to biotechnological industries. Practical applications from microorganisms that have to deal with difficult environmental conditions can be applied in commercial uses. Enzymes which are able to work under high temperatures or conditions of high pressure, or biopolymers with industrial applications, can be obtained from extreme environments.

Liquid water is a requisite for life. Extreme conditions are those situations which make the existence of water and the existence of a water activity level sufficient for life challenging. Tremendous discussions about the nature of an extremophile and the philosophical constraints about the human point of view for the definition of extremophiles have been previously reported, but the availability of water is an important element for such definitions. High temperatures above the boiling point of water imply water restriction, but oxidative stress, cold and high salinity are also characteristics which involve limitations on water availability, all of them being included in the description of extreme environments.

Furthermore, extremophiles are organisms that are able to survive in an environment with one "extreme" characteristic, but polyextremophiles also exist, since some microorganisms survive in environments with several "extremes," e.g., temperature and pH boundaries, in anaerobic conditions. For example, alkalithermophilic bacteria can be widely distributed or located in particular and isolated environments. Included in this group is the Gram-positive *Bacillus–Clostridium* subbranch, which may be considered from a philosophical aspect, since short doubling times are not common in extreme environments. However, *Thermobrachium celere* is able to live at pH above 9.0 and temperatures higher than 55°C, with a doubling time as short as 10 min (Wiegel 1998).

Among the previously mentioned extreme environments of special interest with respect to biomarkers and life identification, we would like to point out permafrost, cryptoendolithic microenvironments – especially those whose origins are precipitated minerals from acidic environments – hot vent hydrothermal environments and low water activity environments such as the Atacama Desert (Fig. 1).

The environmental conditions reported for the Martian surface – oxidative stress, high UV radiation levels, etc. (Klein 1978) suggest that the realistic possibility for the development of life there is very small. Identification of water–ice in the subsurface on Mars has been reported using the Thermal Emission Spectrometer onboard of the Mars Odyssey (Kieffer and Titus 2001) and from the High Energy Neutron Detector (Litvak et al. 2006). These recently published data have important astrobiological connotations, because in addition to being a potential source for water, these locations are shielding habitats

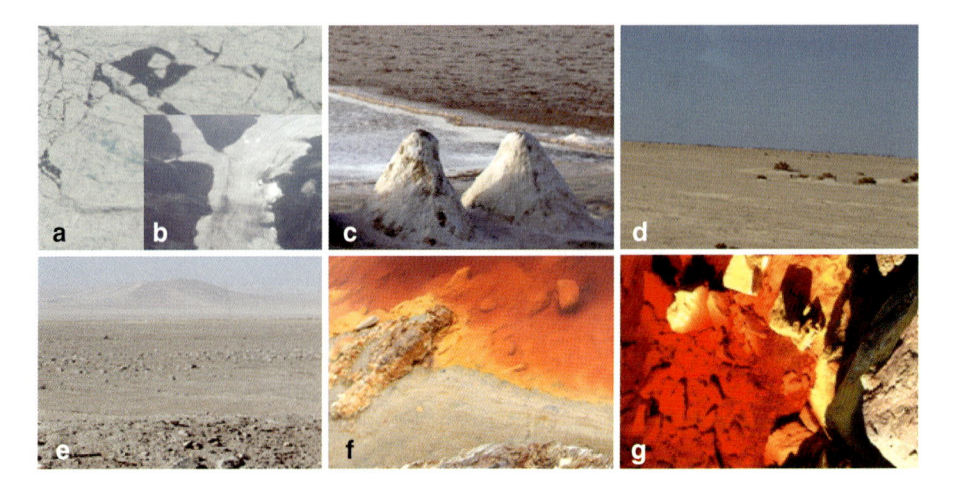

Fig. 1. Extreme environments on Earth. They can be characterized by extreme cold as the icy chaotic terrains around the North Pole (**a**) or glacial moraines (**b**), extremely salty environments such as Chott El Jerid, Tunisia (**c**), extreme dryness such as the sandy desert in Tunisia (**d**), low water activity in the Atacama Desert (**e**), extremely low pH as in the Rio Tinto in Spain (**f**) or in the La Union Mines (**g**), also in Spain

against the harsh conditions existing on the planet, such as UV radiation (Gomez et al. 2007). Similarities between terrestrial and Martian permafrost (Frolov 2003) and other structures that could play a protective role for subsurface ecosystems (Gilichinsky et al. 2007) have been discussed. Recently, Smith et al. (2009) reported the presence of spheroids on a strut of the Phoenix Lander's leg. Phoenix landed on May 25, 2008 with the objective of studying the habitability potential of the ice-rich soil in the Martian Arctic. These spheroids were interpreted as liquid saline water droplets in Phoenix Lander's legs (Smith et al. 2009). Studies in ground facilities have demonstrated the stability of liquid saline water on the Phoenix landing site on the surface of Mars (Zorzano et al. 2009). Permafrost environments are colonized basically by anaerobic bacteria including methanogens and are preserving well organic matter, which is of special interest for biomarkers and automated tools for the detection of life (see below).

Cryptoendolithic microenvironments are simple niches located in pores and empty spaces inside, but near the surface, of rocks. These protected microenvironments normally harbor photosynthetic life, which takes advantage of the solar radiation that penetrates into the rock. This particular case highlights the tremendous importance of the correct localization where to look for signs of life on other planetary bodies, when no human operators are involved. Sedimentary deposits from acidic environments are of special interest due to the well-reported difficulties in the preservation of organic matter, due to the oxidative stress (Fig. 1).

Hot vent hydrothermal environments are especially of interest with respect to early Mars and the moon Europa. The chemosynthetic ecosystems surrounding submarine hydrothermal vents are basal niches, where life is based on the oxidation of inorganic compounds in a protected and isolated environment. The possible existence of a silicate rock core nucleus in contact with a water layer as a subsurface ocean below the ice (Khurana et al. 2002) forming the inner structure of Jupiter's moon promises a high astrobiological potential and a hot spot for looking for life away from planet Earth. Future astrobiology missions to Europa will require not only new tools in the search for life, but also new ice drilling technology with the objective of reaching the subsurface environment.

Desert areas with low water activity such as the Atacama Desert are habitats which represent one of the limits of life on Earth (Navarro-González et al. 2003). The difficulty of detecting organic molecules in the Atacama soil suggested that it is a perfect Martian soil simulant for organics and also useful for application of extant life detection strategies, protocols and technologies.

2.2 Habitability

Habitability was formerly defined as the capacity of a place to harbor life in optimal conditions. The implications of the "habitability" definition for the search for life in other planetary bodies have drastically changed after the boom of research on extremophiles and extreme environments in the past decade. The vision of the Viking experiments was completely different from what might be a similar experiment on Mars in the future. The main difference originates from the point of view with regard to extremophiles. The Viking experiments were focused on putative microbial growth using heterotrophic media for the enrichment of cultures. This idea came from the available knowledge of that time. Nowadays, the design of the experiments would be radically different. The possible existence of life outside of Earth encompasses the existence of extremophiles. By definition, a possible form of life present on any other planetary body would be an extremophile itself, due to the harsh environmental conditions (high radiation levels, pressure, temperature, and water availability, among other characteristics) which it would have to deal with. Then, the definition of habitability should be completely different from that of the 70s to this moment.

Following the ideas of human habitability (Mahdavi 1998), we could extrapolate the same ideas to microbial populations living in harsh environments. The availability of an "optimal" space where processes for gaining energy for metabolism can be developed is absolutely necessary for the final definition of habitability. Controversy has arisen about the question if microorganisms could achieve the optimal "design" of this space by the generation of habitability conditions or homeostatic mechanisms (Fernández-Remolar et al. 2008a). Homeostatic mechanisms and the investment by microorganisms could play an essential role in the adaptation of harsh environmental conditions to "optimal" conditions. Some results in adaptation to acidic extreme environments and the generation of subsurface ecosystems are of astrobiological relevance. Subsurface ecosystems are ideally protected environments (Gilichinsky et al. 1992; Fernández-Remolar et al. 2008b) which shield not only microbial populations from external radiation and harsh conditions, but also some particular physico-chemical elements (minerals, precipitates, iron solutions, etc.) could be used as shielding materials (Gilichinsky et al. 1993; Gómez et al. 2007). Extremophiles are essential for a suitable definition of "habitability," when focusing on the correct places for the search for life in future astrobiological missions. Another important factor with respect to habitability is an energy source. Lovelock (1965) has pointed out the coexistence of both oxidized and reduced gases in the atmosphere as an indicator of the presence of living organisms. The necessity of redox couples from where life could "extract" enough redox potential for the generation of proton gradients is of crucial importance for the sustainability of life on Earth. The metabolic diversity reported on Earth demonstrates the wide diversity in redox couples that can be used as energy source. Again, new lessons can be learnt from extremophiles regarding their capability for gaining energy from redox couples with low differences of redox potentials.

Finally, a sine qua non requisite of life is the existence of a solvent where redox and organic chemistry reactions, on which life on Earth is based, could take place. A low level of water activity limits the manifestation of life; water activities below 0.9 are very restrictive levels which only extremophiles can afford.

Life detection strategies should be based on the previously described parameters that could affect habitability. To look for metabolic activity through the occurrence of redox chemistry has been previously tested on the Viking missions, but new designs for future missions will also be based on this approach. The question about the existence of extant life out of the Earth could only be answered by missions testing metabolic activity. Final positive results concerning the presence of active microbes would be assured with these approaches.

3 Extremophiles as sources of biomarkers for the search for extraterrestrial life

The only life we know is that on Earth and since it is difficult to produce an exact definition of life, we describe it instead. This is an additional challenge when we try to identify life elsewhere: how can we be sure we are detecting life? Probably the final answer will come after an extensive compilation of an inventory of proofs, matching the set of features that we understand as life. Extremophilic microorganisms can help in the search of such proofs. By knowing their metabolisms and how they live and survive in their specific environments on Earth, we can argue and assume that similar ways of life can be found in similar environments on other planetary bodies. We also assume that similar ways of life means similar morphologies, metabolic activities, and molecular components, in essence, similar or identical biomarkers. From a terrestrial point of view, Mars or other extraterrestrial bodies (e.g., Jupiter's moon Europa) can be considered as extreme environments where several or all environmental parameters seem to be very far from "normal" values. Some terrestrial environments share physico-chemical characteristics with those on the extant or ancient Mars: hydrothermal regions, acidic iron- and sulfur-rich areas, permafrost, extremely dry deserts such as Atacama (in Chile) or hyperarid deserts in China, dry polar environments (dry valleys in Antarctica, or Devon Island in the Arctic), and subsurface hydrothermal environments (Deep South African gold mines). The notion of a hypothetical Martian biota occurring in subsurface environments is strengthened by evidence suggesting recent hydrothermal and volcanic activities on the red planet (Márquez et al. 2004; Hauber et al. 2005). The discovery of sulfate- and iron-containing minerals, which can be formed only in the presence of water, on Mars (Klingelhöfer et al. 2004) suggests that iron and sulfur metabolisms may have played a key role as energy source and during the formation of structural components (metalloproteins, Fe–S proteins) for a hypothetical Martian life (Boston et al. 1992). The Río Tinto area (in southwestern Spain) has been also considered a good Martian analog, at least concerning iron and sulfur mineralogy (Fernández-Remolar et al. 2004, 2005). Chemolithoautotrophic prokaryotes such as *Leptospirillum ferrooxidans* and *Acidithiobacillus ferrooxidans*, which are abundant in the Tinto River ecosystem, have very simple nutritional requirements (Balashova et al. 1974); they do not need light for life and one can argue that similar microorganisms could be inhabiting similar Martian environments, either on the present or on the ancient and wetter Mars. Both bacteria are able to fix CO_2 and N_2 (Mackintosh 1978; Norris et al. 1995; Parro and Moreno-Paz 2003) and they obtain energy from iron (Fe^{2+}) oxidation. *A. ferrooxidans* is a very versatile bacterium that can grow chemolithoautotrophically not only on Fe^{2+}/O_2 or H_2/O_2 under aerobic conditions, but also on H_2/Fe^{3+}, H_2/S_0, or S_0/Fe^{3+} under anaerobic conditions, with Fe^{3+} and S_0 being the electron acceptors (Ohmura et al. 2002). Iron and sulfur cycles are taking place in the Río Tinto ecosystem (González-Toril et al. 2003).

We hypothesize that microorganisms living in similar environments should share similar molecular mechanisms to deal with such conditions and, consequently, will provide molecules that could be good biomarker targets to search for life elsewhere. Strategies for life detection should take this into consideration when designing missions and instrumentation. The better we know terrestrial analog environments and extremophilic microorganisms living there, the better we can focus the search for extraterrestrial biomarkers. For example, by combining geochemical, genomic, and environmental transcriptomic studies we identified putative target biomarkers to be searched for in similar planetary environments (Parro et al. 2007). From that work, we can also conclude that the prokaryotes in Rio Tinto are in fact poikilotolerant (*poikilos*, "diverse" or "variable") microorganisms in the sense that they can live under several extremes: low pH (<1.8), high metal concentrations ($>20\,\mathrm{g\,l^{-1}}$ Fe), high oxidative stress, high osmotic pressure ($>80\,\mathrm{g\,l^{-1}}$ of sulfate in some places), and low nutrient availability (i.e., oligotrophy), see Fig. 2. This is highly relevant for a hypothetical Martian biota which eventually would have to deal with low temperatures and low water activity, probably in a salty solution to avoid freezing.

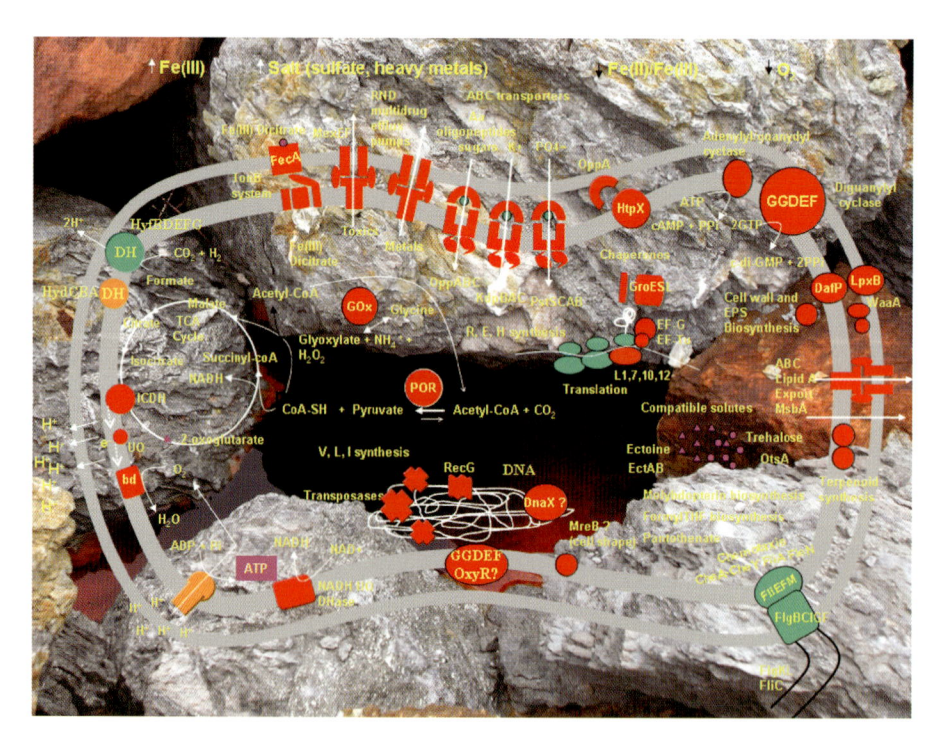

Fig. 2. Environmental transcriptome analysis of extremophiles. This figure represents the main environmental transcriptomic fingerprint of the acidophilic iron oxidizing bacterium *Leptospirillum ferrooxidans* in its natural medium (Río Tinto, Southwestern Spain). The environment in the picture is characterized by low pH (1.8), high iron and sulfur concentrations (20 g l^{-1} and 80 g l^{-1}, respectively), which constitute very extreme environmental conditions (from Parro et al. 2007). Red figures represent those proteins or metabolites whose genes are upregulated in this site when compared to other sites with milder environmental conditions

3.1 Preservation of molecular biomarkers

Terrestrial life is basically made of four types of molecules and their polymers or combinations: carbohydrates, lipids, nucleo-bases, and amino acids. We define molecular biomarker as any molecule (monomer or polymer) of unequivocal biological origin. Very often the main problem is not the detection of molecules, but the determination of their biotic or abiotic origin. This is the case for most amino acids, some sugars, and other compounds, which are found also in meteorites (Irvine 1998; Pizzarello 2007). In these cases the sole presence of this type of molecule is not sufficient to be considered as molecular biomarkers. Whichever the case, after the death of an organism, the degree and rate of degradation of living matter can be influenced by factors such as (i) enzymatic hydrolysis and in situ microbial re-assimilation; (ii) high irradiation (UV and other types); (iii) high rates of oxidation; (iv) attacks by metal ions; (v) Maillard reactions (condensation of the carbonyl group of reducing sugars and primary amino group of amino acids); (vi) high temperatures or optimal pHs for degrading agents. By contrast, conditions avoiding or slowing such degradation contribute to maintain the stability and integrity of biomolecules, such as (i) low temperatures, which diminish the metabolic rates of microorganisms and the catalytic activities of enzymes (Gilichinski et al. 1992); (ii) inclusion in hypersaline solutions, salt crystals and polymerized resins such as amber, thus retarding catalytic degradation and favoring desiccation; (iii) rapid burial which produces an anoxic environment (in general, anoxic environments have lower metabolic rates than aerobic ones), or confines and protects against UV radiation and oxidation; (iv) precipitation or binding of molecules to the surface of colloids, mineral particles, or organic macromolecular aggregates; (v) rapid dehydration which severely restricts catalytic activities; (vi) mild pH values (extreme pH severely affects the preservation of molecules and favors, for example, the depurinization of DNA at low pH or the degradation of RNA at alkaline pH); (vii) absence of reactive and degradative metal ions (the structure of biomolecules is severely affected by certain metal ions or radicals). All these types of environmental factors are, in general, more critical for the overall molecular preservation than the age of the sample (Tuross and Stathoplos 1993).

3.1.1 Molecular biomarkers for extant life

Depending on the type of molecular biomarkers, they can indicate the existence of present life at the time of sampling or just be molecular fossils of extinct forms of life. Two main intrinsic features of life as we know it are complexity and compartmentalization. The molecular complexity is represented by biological polymers (polysaccharides, polypeptides, polynucleotides, phospholipids, glycolipids, etc.), while compartmentalization results from structures such as membranes, intermembrane spaces, nucleoids, whole cells, and even biofilms. Compartmentalization must have been critical for the origin and evolution of life: membranes provide the appropriate bilayer for hydrophobic interactions as well as an excellent isolation from the environment; extracellular matrices of exopolysaccharides (EPSs) protect cells from the harsh conditions of the surrounding environment and also help to trap nutrients, salts, water, or ions necessary for life; intracellular compartments allow the accumulation of energy sources or special solutes and salts for internal bodies such as magnetosomes.

The number of microorganisms in nature is widely variable, ranging from no viable cells in some places of extremely dry environments such as the Atacama Desert, to 10^{10} in some rich soils and sediments (Table 1; Vorobyova et al. 1997). In order to generate enough amounts of molecular biomarkers in a given sample, a minimal concentration of microorganisms must have inhabited this environment; otherwise dilution effects due to the mineral substrate make their detection very difficult or impossible. Microorganisms in nature show an irregular distribution and they prefer to live in microbial communities, forming biofilms and attaching to solid supports. An extracellular matrix composed of a mixture of EPS, proteins, DNA, and

Table 1. Number of microbial cells in extreme environments

Environment	Number of bacteria	
	Total counts (cells/g dw[a])	Viable counts (cfu/g dw[a])
Atacama Desert[b]	–	$0–10^6$
Permafrost sediments		
Arctic (>300 m)	$10^7–10^9$	$10^2–10^8$
Antarctic (20 m)	$10^7–10^8$	$10^1–10^4$
Permafrost buried soils	10^9	$10^4–10^6$
Bottom sediments of lakes		
Upper layers	$10^8–10^9$	$10^4–10^5$
Layers at 3–8 m	10^8	$10^2–10^3$
Upper layer of sediments	10^7	$10^2–10^6$
Environment	Number of bacteria	
	Total counts (cells/ml)	Viable counts (cells/ml)
Central part of the ocean		
Water	$10^2–10^6$	0–10
Antarctic glaciers (up to 1800 m)	–	1 cell/l

[a] dw dry weight and cfu colony-forming units. Data modified from Vorobyova et al. (1997)
[b] From Navarro-González et al. (2003)

water (Zhang et al. 1998) is the support where the cells are held together and even communicate with each other. The synthesis of EPS is dependent on several environmental factors such as water availability, desiccation stress (Roberson et al. 1993), the availability of nutrients (carbon, nitrogen, potassium, and phosphate), or slow growth (Sutherland 2001). After the death of a biofilm, many of its compounds are degraded, although part of it might be trapped between mineral or crystal interfaces and can be well preserved in those places under poor diagenetic

Table 2. Molecules considered as biomarkers for extant and extinct life

Type of molecules	Possible origin
Extant life	
Biochemicals. Biological polymers and metabolites: EPS, lipopolysaccharides, peptidoglycan, teichoic acids, teichuronic acids, proteins, oligo- and polysaccharides, nucleic acids, porphyrins, flavins, vitamins, antibiotics, and compatible solutes	From the cell wall components, energy and storage compounds. Indicative of actual or recent metabolism. For example, compatible solutes are small molecular weight metabolites (ectoine, trehalose, betaine, and some amino acids) accumulated intracellularly as response to high salt or dryness
Metabolic products: methane and acetate	From methanogens and acetogens
Macro and microscopical forms and structures. Whole cells	From replication structures such as buds, chains of cells, septa, fruiting bodies, spores. Some biominerals. Macrostructures such as biofilms and stromatolite-like structures
Cellular compartments	Structural components: cell walls, membranes, and organelles
Extinct life	
Aliphatic and cyclic hydrocarbons: *n*-alkanes, algaenans, cycloalkanes, branched alkanes, pristane, phytane, acyclic isoprenoids, phytol, pentamethylicosane, highly branched isoprenoids, alkylthiophenes, alkylcyclohexanes, and cyclopentanes	Aliphatic: From degradation products of aliphatic macromolecules of unicellular green algae. Indicative of marine and lacustrine habitats. Cyclic: From sulfuration of sedimentary lipids in hypersaline environments, or from cyclohexyl fatty acids of thermophilic bacteria
Aromatic carotenoids: okenane, chlorobactane, etc. Arylisoprenoid derivatives	Diagenetic and catagenetic products of carotenoids from pigmented (green, brown) chlorobiaceae
Hopanoids and other pentacyclic triterpenes	Form the membrane of bacteria and blue green algae. Found in recent and old sediments
PAHs and humic acids	From prebiotic or biological origin (soil and marine environments). Very complex PAH derived from triglycerides
Steroids (sterane, diasterane, and norcholestanes)	Saturated version of eukaryotic sterols. Diasterane is produced by rearrangement of sterane during diagenesis
Porphyrins and maleimides: bacteriochlorophylls, heme group, and maleimides	Photosynthetic bacteria, cytochromes, siderophores. Maleimides are the diagenetic product from tetrapyrrole ring of chlorophylls
Amino acids and nucleotides: All 20 proteinogenic L-amino acids, modified amino acids, dNTPs, other modified nucleotides, purine, and pyrimidines	Thymine dimers are produced by UV radiation. Some amino acids, purines, and pyrimidines could be produced by abiotic chemistry

processes. Consequently, EPS or any of its monomers or oligomers is an excellent biomarker for present life (Table 2). Also, environmental DNA can be protected from hydrolysis by DNases through the interaction of nucleic acids with clay minerals and soil colloidal particles (Cai et al. 2006). Shang and Tiessen (1998) reported that particle size, crystallinity of iron oxides and microaggregation of oxides, organic matter, and other minerals play an important role in stabilizing the organic matter of semiarid tropical soils. Nitrogen-containing compounds are found in aged soils, recent and even fossilized sediments, and the survival of peptide-like structures is a ubiquitous phenomenon (Knicker 2004). Peptide material and other proteinaceous material can survive microbial degradation when forming part of the nitrogen-containing refractory organic matter (Knicker and Hatcher 1997; Knicker et al. 2001).

3.1.2 Molecular biomarkers for extinct life

Intact or structurally modified molecular biomarkers may be preserved to different degrees with passage of time. Lipids and carbohydrates or their diagenetic products can persist for millions of years (Stankiewicz et al. 1997; Brocks et al. 2003). Additionally, these products can be used as taxonomic criteria to identify the kind of microorganism and the environment where they originated (Table 2). Most carbohydrate-containing compounds are water soluble and highly susceptible to degradation, by either enzymatic hydrolysis, oxidation, or Maillard reactions. Nevertheless, carbohydrates can form insoluble complex structures which are highly resistant to degradation, such as lignins (from plants), chitins (amino sugars from exoskeletons of insects and arthropods, or from the cell wall of fungi), algaenans (part of the cell walls of algae) or bacterial spore coats, respectively. In addition, EPS from biofilms might form complex structures which are highly resistant biopolymers.

Nucleic acids (DNA and RNA) are the most unstable of the biological macromolecules and are easily degraded, depending upon the level of water, oxygen, and most importantly, the temperature of the local environment. Willerslev et al. (2004) showed how DNA and other organic molecules survive longer in colder environments, since the rate of decay was slowed down by an order of magnitude for every $10°C$ drop in temperature. Consequently, although the microorganisms were not viable, nucleic acids should theoretically survive over long periods of time (several hundred thousands and even a million of years) in ice and permafrost. The extrapolation of these findings to Mars seems straightforward: Martian permafrost might be a good environment for the preservation of organic and biological matter at the subsurface (Gilichinsky et al. 2007). Additionally, DNA and other macromolecules have a strong affinity to minerals, such that they can be removed from solution and increase their stability by binding to mineral surfaces (Sykes et al. 1995).

Proteins and their amino acid constituents were identified in fossils from the Paleozoic (De Jong et al. 1974), and immunoreactivity was demonstrated against preserved biopolymeric structures from the Mesozoic (Collins et al. 1991) and mammoth fossils (Schweitzer et al. 2002). Schweitzer et al. (2005) reported the presence of well-preserved soft tissues and cells from demineralized bone matrix of *Tyrannosaurus rex* fossil samples. They detected collagen I protein (Schweitzer et al. 2007) and they obtained the amino acid sequence of peptides extracted from 68-million-year-old *T. rex* (Asara et al. 2007) and 80-million-year-old Campanian hadrosaur fossils (Schweitzer et al. 2009). Collins et al. (2003) reported the survival of peptide bonds and epitopes (the sites of an antigen to which the antibody binds) in fossil shells from a semicontinuous New Zealand brachiopod sequence extending for 3 Ma.

Lipids and derivatives are transformed with time to highly stable long chain or cyclic hydrocarbons by losing side chains and functional groups. Lipid biomarkers provide informative biosignatures in well-preserved ancient sedimentary organic matter and can be used for tracing the evolution of biological processes over geological time (Summons 1988; Sherman et al. 2007). Some lipids such as fatty acids, metabolites, and cell membrane components, can be rapidly degraded after cell death. However, other lipids are highly stable for billions of years, such as the hopanol derivatives (Brocks et al. 1999), and can be used as unambiguous fossil molecular biomarkers.

3.2 Stability of organic and living matter in planetary bodies: the case of Mars

The negative results of the attempts for the detection of organics with the gas chromatography – mass spectrometer (GC–MS) on the Viking 1976 missions to Mars were surprising in light of the relatively high amounts of reduced carbon estimated to reach the Martian surface each year by meteorites (up to 2.4×10^8 g; Flynn 1996). Many of these compounds are volatile and they should have been detected by the Viking's GC–MS (Sephton et al. 1998). This failure of detection of organic matter on the Martian surface was interpreted as a consequence of the action of a powerful oxidant capable to degrade all organics to carbon dioxide. Ultraviolet radiation cleaves water to H^+ and HO^- radicals which can react directly with organic molecules, or generate peroxides (H_2O_2) or other oxidizing species by combining with other elements in the Martian soil. By using an instrument designed for studying the oxidation state of Martian soils, Quinn et al. (2005) detected a highly oxidative and acidic activity in one of the most extreme environments on Earth, which is also good terrestrial analog for Mars, the dry core of the Atacama Desert. They concluded that water abundance plays a key role in controlling the reaction kinetics of the oxidizing acids, and their results were consistent with the presence of strong acids in the accumulated dust.

The organic compounds known to be present on Mars either via meteorites or generated on Mars from nonbiological or hypothetical biological synthesis are most likely converted to carboxylic acid derivatives by means of the highly oxidizing conditions on the Martian surface (Benner et al. 2000). The expected metastable products should be acetates, oxalate, benzenecarboxylic acid, or phthalic acid. Due to the lack of volatility, the salts of organic carboxylic acids are not easily detected by GC–MS. Therefore, even if they had been present on the Martian surface, they would not have been directly detectable by the Viking GC–MS experiment. Other research has shown that the stability of organic macromolecules such as tholins (abiotic origin) and humic acids (biological origin) is different in the presence of H_2O_2 (McDonald et al. 1998), and some of them could have been stable on the Martian surface during the whole history of Mars, most probably in the polar regions.

In the search for life on Mars, other biomolecules such as amino acids, nucleobases, and fatty acids, may constitute excellent biomarkers, but they are much less resistant to oxidative degradation than tholin-like or humic acid-like compounds. Additionally, photodegradation may destroy a significant fraction of organics initially present in the upper layers of the regolith (Oró and Holzer 1979). Consequently, a search for molecular biomarkers on Mars should include environments protected from surface oxidants and UV radiation, such as subsurface samples or the interior of rocks. Kminek and Bada (2006) showed that amino acids can be protected from radiolytic decomposition as long as they are adequately shielded from space radiation. They considered an average amino acid radiolysis constant of $0.113\,MGy^{-1}$ ($MG = 10^6$ Gray), and they estimated that it is necessary to drill to a depth of 1.5–2 m for detecting amino acids signature of life forms that became extinct about 3 billions years ago. The detection of any remnants in the uppermost half meter of the Martian subsurface would explain an extinction event younger = between present and 100–500 millions years ago. The relative proportion between D and L *iso*-forms of amino acids and/or sugars may indicate their prebiotic or biological origin. A microfabricated capillary electrophoresis device for the determination of amino acid chirality in extraterrestrial exploration was incorporated into the so-called UREY instrument, initially selected by the European Space Agency (ESA) as part of the Pasteur payload for the ExoMars mission (Aubrey et al. 2008).

4 Life detection strategies

We can learn many lessons from extremophiles that could be very useful for the design and implementation of new strategies to detect life or its remains on other planetary bodies. They show us where we could look for similar microorganisms (subsurface, inside a crystal, in a saltern, an acidic pond, etc.), what we could detect (a metabolic activity, a molecular biomarker, a morphology, an isotopic signature, etc.), and how we should address that search (remote sensing, in situ analysis, noncontact or contact techniques, etc.). Any possible ecosystem apart from Earth would have to be, by definition, an extreme environment, and any protocol for life detection has to be tested and judged with the perspective of extremophiles in mind.

4.1 Immunological systems for the detection of signs of life

4.1.1 Antibody microarray immunoassays

Antibodies are glycoproteins capable to specifically recognize and bind other molecules (antigens) through a noncovalent highly specific interaction. Antibodies are large molecules (150 kDa), and they can be modified at their constant regions by other substances, such as fluorophores, other proteins, or even other antibodies, without affecting the binding affinity to the antigen. It is possible to produce antibodies against a wide variety of compounds ranging from small molecules such as amino acids, pesticides, sugars, and lipids, to large polymers and whole cells. Antibodies can be highly specific, such that they can discriminate between enantiomers of the same molecule (Shabat et al. 1995) or different metals coordinated to the same chelator (Corneillie et al. 2006).

DNA and protein microarrays allow the stable immobilization of up to thousands of affinity receptors to a solid support, such as glass, nitrocellulose, nylon, and gold. Microarray immunoassays can be performed in several formats depending on the nature and the size of the target molecule (the antigen or analyte; Kusnezow et al. 2003; Fernández-Calvo et al. 2006). When the target molecules contain only one epitope or antigenic determinant (this is the case for xenobiotics, pesticides, steroid hormones, small metabolites, explosives, etc.), a competitive assay has to be performed (Knecht et al. 2004; Fernández-Calvo et al. 2006). On the other hand, a sandwich assay is only possible for those molecules or complexes containing at least two epitopes. The test sample is incubated directly with an antibody microarray to allow antigens to bind to the immobilized antibodies. In a second incubation the same antibody (fluorescent or enzyme labeled) can bind to free epitopes of the retained antigens, generating a kind of sandwich (Parro et al. 2005; Rivas et al. 2008; Parro 2009). The positive reactions can be detected following fluorochrome excitation and image capturing. In a direct immunoassay, antigens present in a sample are previously labeled, incubated with the antibody microarray and, after a washing step and excitation with the laser, an image is captured and analyzed. All these types of immunoassays – direct, sandwich, competitive, and displacement –

can be performed in an antibody microarray format (Knecht et al. 2004; Parro et al. 2005; Fernández-Calvo et al. 2006; Rivas et al. 2008). We developed a multiarray competitive immunoassay (MACIA) for the simultaneous detection of compounds of a wide range of molecular sizes, from single aromatic ring derivatives or polyaromatic hydrocarbons (PAHs) to small peptides, proteins, or whole spores and cells (Fernández-Calvo et al. 2006; Parro et al. 2008a).

4.1.2 Antibody microarray-based instrumentation for in situ life detection

An antibody microarray-based instrument has been proposed since several years for planetary exploration (Steele et al. 2001; Parro et al. 2005, 2008a). Up to now, all techniques and instrumentation developed for the detection of organic molecules are based on the analysis of volatile compounds, mainly due to little or no requirement for sample preparation. However, complex polymeric compounds and other refractory material (the most abundant organic matter even in meteorites is highly refractory) are nonvolatile and are even destroyed or irreversibly modified by the high temperatures (over 200°C) required for volatilization. The 1976 Viking's GC–MS (Biemann 1974), the Cassini/Huygens GC–MS (Israël et al. 2005), the SAM (Sample Analysis at Mars) of NASA's MSL mission, or the Urey instrument (Aubrey et al. 2008) proposed for ESA's ExoMars mission are or were all based on the analysis of volatile compounds. Sandwich antibody microarray immunoassays can detect nonvolatile compounds either in free form or as part of complex biological polymers (Parro et al. 2007, 2008b; Rivas et al. 2008).

We suggest two approaches for the selection of biomarkers and antibody development: (i) a direct strategy, in which well-known molecular structures are used to produce antibodies, and (ii) a shotgun strategy, where astrobiologically relevant environmental samples are subjected to biochemical extractions and fractionations and then used as immunogens to produce antibodies (Fig. 3). The rationale of this latter strategy is to produce highly specific and reactive antibodies against environmental molecules which are present at the time of sampling and which are frequently modified compared with those produced or isolated in the laboratory. These modifications (acylations, methylations, etc.) may not be detected by the antibodies raised against apparently the same molecules isolated in the laboratory from other sources.

Consequently, we have produced polyclonal antibodies against extracts from different extreme environments and from extremophilic microorganisms (Rivas et al. 2008). For example, we took samples for the extraction and production of antibodies from the extremely acidic environment of Rio Tinto: water, sediments, mineral deposits (sulfate precipitates, jarosite, hematite), and rocks from a pyrite-rich subsurface drill core (between 100 and 160 m deep), obtained during the field campaign of the NASA-funded MARTE project (Mars Astrobiology Research and

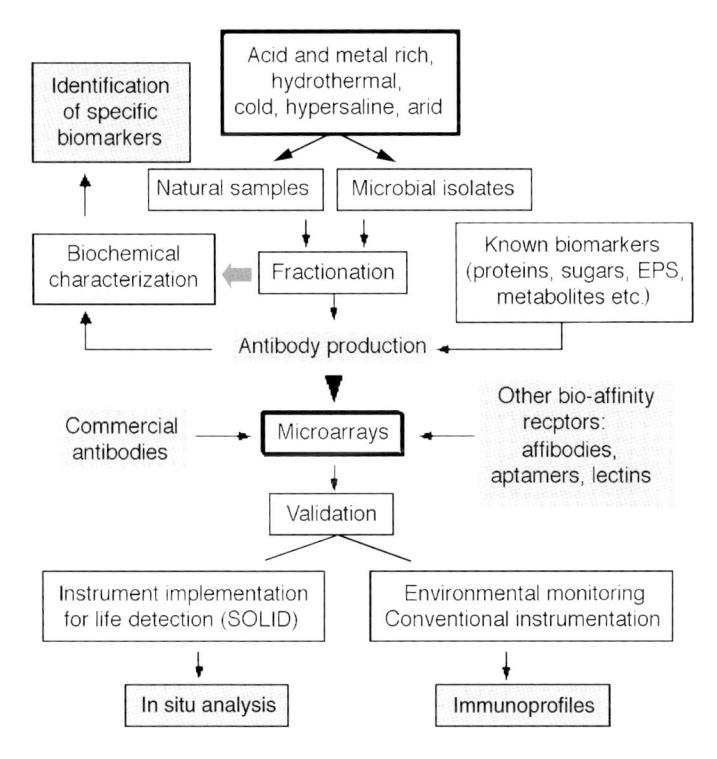

Fig. 3. Strategy for antibody development and antibody microarray-based instrumentation for in situ life detection in astrobiology. Extremophiles and extreme environments are excellent sources of target biomarkers for the production of antibodies (modified from Rivas et al. 2008)

Technology Experiment; http://marte.arc.nasa.gov/; Stoker et al. 2008). At the same time, we were performing biochemical fractionations from pure bacterial cultures, in order to produce antibodies against different kinds of macromolecules (EPS, anionic polymers, and cell wall components). We have also produced antibodies against pure bacterial cultures (from type collections), which were isolated from cold (Arctic and Antarctic), hydrothermal or hypersaline environments, and tested them against environmental samples from around the world (Fig. 4). We have also produced antibodies against compounds which have been detected both in meteorites and in the interstellar space (naphthalene, phenylphenol, etc.; Fernández-Calvo et al. 2006).

Convinced of the potential of this technology, we developed, explored the concept and constructed the instrument called SOLID (for "Signs Of LIfe Detector"; Parro et al. 2005). A field version (SOLID2; http://cab.inta.es/solid/) was successfully tested in different campaigns. It was also one of the analytical instruments participating in a robotic Mars drilling simulation on September 2005

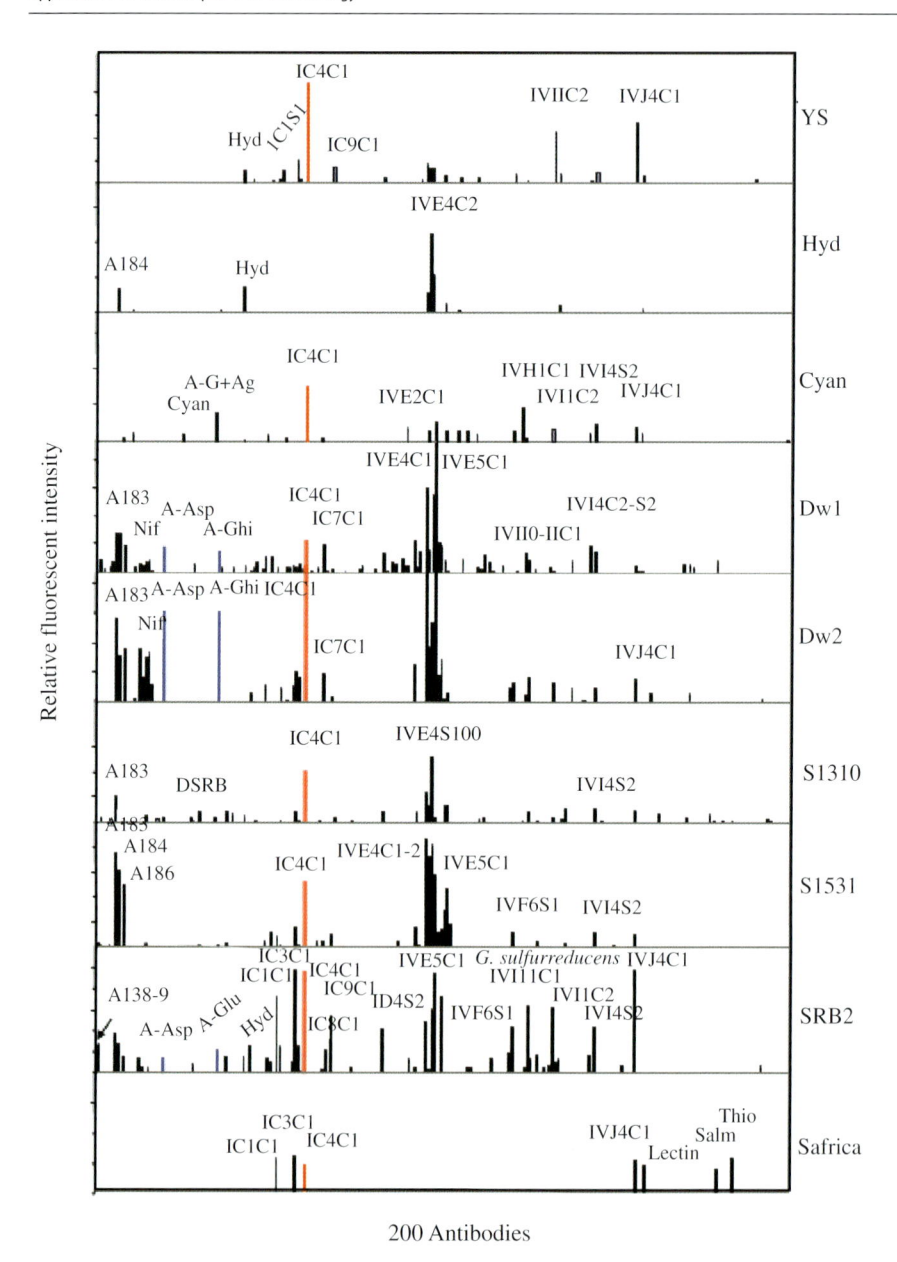

Fig. 4. Antibody microarrays allow immunoprofiling of samples from extreme environments around the world. Universal biomarkers can be inferred as well as specific fingerprints (modified from Rivas et al. 2008). The samples analyzed by antibody microarrays were: *YS*, an extract from Yellowstone National Park (USA); *Hyd*, a sample enriched with *Hydrogenobacter* spp. from Yellowstone National Park; *Cyan*, a sample enriched with *Cyanidium* spp. from Yellowstone National Park; *Dw1* and *Dw2*, two samples from the Dry Valleys (Antarctica) enriched in cyanobacteria; *S1310*, *S1531* and *SRB2*, extracts from hydrothermal samples from Iceland; *Safrica*, samples from biofilms in deep mines of South Africa

(Parro et al. 2008b) during the development of the MARTE project (Stoker et al. 2008). A new version has recently been constructed, the SOLID3 (Parro et al., 2011) following the specifications for ESA's Life Marker Chip (LMC) experiment, which was initially considered for the ExoMars mission. The SOLID3 instrument can perform both sandwich and competitive fluorescence-based immunoassays. We have currently a collection of more than 300 antibodies (the LDCHIP300, for Life Detector CHIP), and both the LDCHIP300 and SOLID3 have already been tested in field campaigns in the Atacama Desert (Chile) and on Deception Island (Antarctica) (to be published elsewhere).

4.2 Detection systems for extant life based on metabolic activity

Future astrobiological missions will have to confront the question about the identification of extant life. The first real astrobiological mission in addressing that issue was the Viking mission in the 70s. Tremendous controversies were reported from the results obtained by the mission, especially those which related to the tests developed for the Viking Landers probing active metabolism of microbes (Levin and Straat 1981).

4.2.1 Viking mission

The Viking mission was composed of two spacecrafts, Viking 1 and Viking 2, both of which consisted of an orbiter and a lander. Three primary scientific objectives were dealt with during the mission: obtaining high resolution images of the Martian surface (for the first time from a Mars mission), characterization of the atmosphere and structure of the surface of the red planet, detection of signs of life. Viking 1 was launched on August 20, 1975, and arrived at Mars on June 19, 1976. Viking 2 was launched on September 9, 1975, arriving August 7, 1976. Viking 1 lander touched down Chryse Planitia at 22.48°N, 49.97°W. Viking 2 lander touched down Utopia Planitia at 47.97°N, 225.74°W.

The procedures used for the detection of metabolic activity on this mission can be summarized as follows: an assimilation test for CO and CO_2, release of CO_2 from labeled substrates and measurements of the detection of gas exchange, with another biological experiment which tried to detect microbial growth using light scattering (Klein et al. 1972).

The first experiment devoted to the detection of metabolic activity was called carbon assimilation or Pyrolytic Release (PR) (Horowitz et al. 1976). The sample was placed in an incubation cell under an atmosphere containing labeled carbon dioxide and carbon monoxide. Water was added as vapor. After several days of incubation under light (wavelength cut-off below 320 nm) the sample was pyrolyzed at 600°C and liberated carbon dioxide was measured. Organics were analyzed in order to study labeled Mars atmospheric CO and CO_2 into organics. There was a weakly positive result in the sample Chryse 1 from Viking 1 lander. It was not confirmed after a second trial with similar sample. The reaction was demonstrated to be thermolabile and inhibited by moisture.

The gas exchange experiment consisted of a Martian soil sample in one of the incubation chambers whose head space of Martian gases was removed and substituted by helium. Water with organic and inorganic compounds was added and applied over the soil sample. Periodically, the evolution of the head space gases was analyzed by a GC for identification and concentration measurements of elements as oxygen, nitrogen, or compounds such as methane.

The labeled release experiment consisted of the addition of nutrient solution of 7 compounds labeled with ^{14}C (Levin and Straat 1976). After incubation the evolution of labeled gas was monitored in the detector chamber. This experiment was very controversial because positively labeled gas was clearly detected (Levin and Straat 1976), but repeated experiments failed to produce labeled gas. Several years later similar results were reported using the same conditions in the laboratory and oxidative compounds, which led to a nonbiological interpretation of the Viking results (Levin and Straat 1981). The Viking organic analysis instrument (GC–MS) failed to detect organics in Martian surface samples.

The biological Viking experiments results were finally interpreted as responses to the oxidative nature of the Martian soil, based on UV irradiation which generated compounds such as superoxide radical ions in the Martian surface regolith.

4.2.2 Looking for metabolic activity in the future

Apart from the Viking missions no other mission has employed direct detection strategies for active life. It is necessary to develop automated tools for the detection of extant microbial life in future missions. Next generation tools for the detection of metabolic activity need, in our experience, to be based on extremophiles as the most "intricate" organisms since they can survive really harsh environmental conditions and are able to colonize almost inaccessible niches as part of their adaptation mechanisms. Previously ideas focused on several approaches for life detection, conceptually ranging from minimal to moderately geocentric. Levin and Perez (1967) proposed five live detection strategies from different metabolic perspectives: (1) strategies which involve the use of radioactive substrates and their metabolism with the evolution of labeled gases (an approach which was later used in the Viking missions), (2) approaches for the detection of photosynthesis in a heterotrophic–autotrophic system, (3) the detection of photosynthesis in strict autotrophs, (4) strategies for the detection of energetic molecules as adenosinetriphosphate, and finally (5) detection of phosphorus and its metabolic uptake.

We propose new experimental designs and development of an automated system for enrichment and identification of life forms. The system will consist of a two-module system for microbial life detection. The first module would be focused on microbial enrichment. The growth of microbes will be followed by redox or conductivity parameters. Actually, one of the objectives of this research will be the optimization of the growth of biomass by these two physical parameters. The second module is being designed as a microbial population's identification module. Molecular ecology techniques are being applied to microbe's identification and used in an automated way ("in situ" hybridization with DNA probe techniques).

4.3 Other systems for the detection of signs of life

A variety of technologies and instruments have been suggested for the detection of some signatures of life in future planetary exploration missions. Knowledge of the biology of extremophiles can help to focus on certain expected biosignatures and to select an appropriate technology for detecting them. For example, spectroscopic techniques such as Raman, infrared, or fluorescence spectroscopy, allow the analysis of multiple samples and are excellent methods for the detection of biological matter associated with minerals or rocky substrates, requiring none or very little sample preparation. Raman spectroscopy is based on the study of vibrational, rotational, and other low-frequency modes in a given system following the excitation with monochromatic light (usually a laser in the visible, near infrared, or near ultraviolet range). A Raman spectrum is collected after illuminating a sample with a laser beam (which can be as small as microns of spot diameter), collecting the light from the illuminated spot with a lens and sending it through a monochromator. Raman spectroscopy is sensitive to molecular structures and composition, therefore different compounds cause different spectra. The interaction with different molecular environments produces changes in the spectral band signatures. Raman spectroscopy can detect organic and inorganic components in heterogenous systems, which allows detection of biological (Edwards et al. 2003) and geological markers (Sharma et al. 2002). It has been used to characterize mineralogical samples as well as to detect biological molecular signatures in several extreme environments: extremely acidic, such as Río Tinto (Edwards et al. 2007), extremely arid and cold, such as Antarctica (Ellery and Wynn-Williams 2003) or arid with high temperature shifts such as the Atacama Desert (Jorge-Villar and Edwards 2006). Molecular biomarkers from extremophilic microorganisms have been detected by Raman spectroscopy, such as pigments (carotenoids and chlorophylls), lipidic compounds, peptide-like compounds (scytonemin and mycosporine, which are produced by certain cyanobacteria as protection against UV radiation; Garcia-Pichel et al. 1992). A miniaturized Raman spectrometer has been suggested several years ago as a suitable apparatus for in situ planetary analysis (see Tarcea et al. 2008 for a review), and it was selected as one of the key instruments for the European Space Agency's (ESA's) ExoMars mission (Sobron et al. 2008).

Powerful techniques such as capillary electrophoresis have also been proposed for the detection of organic compounds containing amino groups, such as amino acids, amines, amino sugars, or nucleobases. Based on this analytical technique, an instrument called Urey was proposed and initially selected as part of the Pasteur payload of the ESA's ExoMars rover mission. Besides the detection of NH_2-containing compounds, another objective of Urey is to determine whether these compounds are of biotic or abiotic origin by using their compositional and chiral characteristics (Aubrey et al. 2008). In addition, the instrument is provided with a system for the detection of oxidants (MOI, Mars Oxidant Instrument), whose action can directly affect the organic compounds and the interpretation of the results.

The SAM is one of the instruments that form part of NASA's Mars Science Laboratory (MSL), to be launched in 2011. The SAM instrument is the primary MSL analytical tool for the organic analysis, consisting of a set of instruments that includes a quadrupole mass spectrometer (QMS), a GC, and a tunable laser spectrometer (TLS). SAM will perform in situ measurements of organic molecules in gases released from solid samples or directly sampled from the atmosphere. Methane is one of the gases that SAM can detect. Methane is a potential biomarker because it can be a product of microbial metabolism. A large portion of the methane on Earth comes from biological sources and is the byproduct of extremophilic methanogens. Methane has also been detected in the Martian atmosphere by ground-based observations (Mumma et al. 2009) and from ESA's Mars Express orbiter (Formisano et al. 2004). The Martian methane can be an excellent biomarker for both extant life and extinct life; however, there are also plausible abiotic sources such as serpentinization at several km depths, volcanic emissions, or exogenous delivery from cometary sources. The finding of methanogens in terrestrial analogs such as the Rio Tinto subsurface will help us to evaluate and better understand the origin of Martian methane.

The terrestrial subsurface is widely known as a habitat for both heterotrophic life and chemosynthetic microbial life (Stevens and McKinley 1996). Subsurface microorganisms are often tightly attached to cavities and voids of rocks, where they form small and patched biofilms that eventually can build so-called subsurface filament fabrics (SFFs), that is, mineral incrustations following the filamentous and three-dimensional network of the biofilms (Hofmann 2008). The SFFs as well as stromatolites can be relatively easy recognized by microscopical techniques such as the close-up imagers (50 μm/pixel resolution) and microscopes (3 μm/pixel) proposed for different Mars lander missions (MER, MSL, Phoenix, ExoMars).

Here again, knowledge of the terrestrial subsurface extremophilic life will give us important clues for recognizing similar forms, from either extant life or extinct life, in the paleosubsurface of other planetary bodies. Different geological morphologies can be considered as biosignatures (Altermann 2008; Westall 2008). By studying the interaction between growth patterns, morphology, and facies association, a biological origin of stromatolite laminations preserved in Archaean cherts and carbonates has been suggested. Sedimentary facies associations have been identified in several morphologies of the Precambrian lithified microbial mats (Altermann 2008). Sedimentation and mineralization processes are mediated very often by biological processes, leading to the formation of so-called biogenerated rock structures, such as stromatolites, onkolites, oolites, or cementing structures of sandstones preserving biological structures. The extracellular polymeric substances (EPSs) produced by complex microbial communities (biofilms) can act as traps for mineral grains and thus affect and drive the final mineralization pattern. This is the case for the biosedimentary structures called onkolites, which are calcifying upon total decay of cyanobacterial biofilms (Krumbein 2008). Due to the fact that both Earth and Mars shared similar geological environments in the past, the study of biosignatures in rocks from early Earth can give important information for the search for remains of life on Mars. Additionally, extensive knowledge of the microscopic morphologies present in meteorites together with the assessment of all the processes that can affect or alter their structure and composition, will contribute to a better understanding of life on Earth and elsewhere. In addition to the macro- and microfossilized structures, the biochemistry of life also left after diagenetic processes a collection of molecular fossils (Summons et al. 2008).

It is well known that life prefers light isotopes to heavier ones. For example, microbial sulfur metabolisms produce mass-dependent sulfur isotope fractionation. In fact, variations in sulfur isotope ratios have been used as biosignatures both for early rocks and for Martian meteorites (Ono 2008) and stable isotope ratios can be used as biomarkers on Mars (Van Zuilen 2008). It is critical, then, to know and to understand the biological processes and environments on Earth leading to certain isotopic fingerprints.

4.4 Signs of life by remote sensing systems

The detection of methane on Mars either by the spectrometers on board of the Mars Express Orbiter (Formisano et al. 2004) or the detection of methane on Mars from Earth by telescopes (Mumma et al. 2009) constitutes a good example of the search for chemical biosignatures using remote sensing. Formisano et al. (2004) reported the detection of 10 ± 5 parts per billion by volume methane in the Martian atmosphere by means of the Planetary Fourier Spectrometer onboard ESA's Mars Express orbiter. Mumma et al. (2009) also reported the detection of methane on Mars, directly from Earth, by using high-dispersion infrared spectrometers at three ground-based telescopes. About 90% of the methane of the Earth's atmosphere is produced by living systems and the remainder by geochemical processes. There is ongoing research as well as consideration about the plausible sources of Martian methane (Onstott et al. 2006; Lefèvre and Forget 2009). By now we cannot exclude any suggestion – it could be either biogenic or nonbiogenic, including past or present subsurface microorganisms, it could be due to hydrothermal activity or to cometary impacts. Meanwhile, new reports about methanogenic extremophiles on Earth expand the possibilities for methane producing biota on the subsurface of Mars (Morozova et al. 2007; Potter et al. 2009).

A better understanding of the limits of life on Earth can also help to define the concept of habitability on a galactic scale, the so-called galactic habitable zone (Prantzos 2008). The basic idea is that the temporal and spatial positions of the Milky Way may strongly determine several physical processes that favor the development of complex life, or its destruction. For example, the risk of a supernova explosion close enough to be a threat for life is larger in the inner than in the outer Galaxy, and the probability for such an event was higher in the past than in the present. Another important factor is the amount of elements heavier than hydrogen and helium (metallicity) in the interstellar medium, which varies across the disk of our Galaxy and may be important for the existence of Earth-like planets. In fact, it has been observed that the hot stars of the giant extrasolar planets discovered so far are generally more metal-rich than the stars without planets in our cosmic surrounding (Fischer and Valenti 2005). However, all these studies are in their infancy and there are many aspects to be investigated to understand the role of metallicity in the formation and survival of Earth-like planets.

As the detection techniques for extrasolar planets improve, an increase in the number of discovered Earth-like planets is expected. A better understanding of Earthshine by remote sensing will contribute to evaluate the possibility of life on other planets. For example, we know that the association of simple molecules such as O_2, O_3, CO_2, and CH_4 in the atmosphere may be considered a global biomarker. Similarly, the spectral signature of the photosynthetic pigments (from either oxygenic or anoxygenic photosynthesis) can be detected at a global scale and it is also considered a biosignature for the search for extrasolar life (Kiang et al. 2007). However, the technology to see Earth-like planets as resolved disks and to study the spectral profile of their atmospheres is not yet ready, although possible designs have already been considered. For example, a 150 km wide "hypertelescope" in space would provide 40 resolution elements (pixels) across an Earth at a distance of 10 light years in yellow light (Labeyrie 1999). The arrangement of 150 telescopes, each with a mirror of 3 m in diameter, would collect enough photons in 30 min to produce an image from 300 to thousands of resolution elements. With this spatial resolution it will be possible to identify clouds, continents, oceans, or even large areas of vegetation.

Acknowledgments

This work was supported by the Centro de Astrobiología (INTA-CSIC) and the Spanish Ministerio de Educación y Ciencia, grants nos. ESP2006-08128, ESP2006-06640, AYA2008-4013, and by FP7 EC GA no. 228319 EuroPlaNeT RI and no. 211700 CAREX Coordination Action.

References

Altermann W (2008) Accretion, trapping and binding of sediment in Archean stromatolites – morphological expression of the antiquity of life. Space Sci Rev 135:55–79

Asara JM, Schweitzer MH, Freimark LM, Phillips M, Cantley LC (2007) Protein sequences from mastodon and *Tyrannosaurus rex* revealed by mass spectrometry. Science 316:280–285

Aubrey AD, Chalmers JH, Bada JL, Grunthaner FJ, Amashukeli X, Willis P, Skelley AM, Mathies RA, Quinn RC, Zent AP, Ehrenfreund P, Amundson R, Glavin DP, Botta O, Barron L, Blaney DL, Clark BC, Coleman M, Hofmann BA, Josset JL, Rettberg P, Ride S, Robert F, Sephton MA, Yen A (2008) The Urey instrument: an advanced *in situ* organic and oxidant detector for Mars exploration. Astrobiology 8:583–595

Balashova UV, Vedina I, Markosyan GE, Zavarzin GA (1974) *Leptospirillum ferrooxidans* and aspects of its autotrophic growth. Mikrobiologiya 43:581–585

Benner SA, Devine KG, Matveeva LN, Powell DH (2000) The missing organic molecules on Mars. Proc Natl Acad Sci USA 97:2425–2430

Biemann K (1974) Test results on the Viking gas chromatograph–mass spectrometer experiment. Orig Life 5:417–430

Boston PJ, Ivanov MV, McKay CP (1992) On the possibility of chemosynthetic ecosystems in subsurface habitats on Mars. Icarus 95:300–308

Brocks JJ, Logan GA, Buick R, Summons RE (1999) Archean molecular fossils and the early rise of eukaryotes. Science 285:1033–1036

Brocks JJ, Summons RE, Buick R, Logan GA (2003) Origin and significance of aromatic hydrocarbons in giant iron ore deposits of the late Archean Hamersley Basin, Western Australia. Org Geochem 34:1161–1175

Cai P, Huang QY, Zhang XW (2006) Interactions of DNA with clay minerals and soil colloidal particles and protection against degradation by DNase. Environ Sci Technol 40:2971–2976

Collins MJ, Muyzer G, Westbroek P, Curry GB, Sandberg PA, Xu SJ, Quinn R, Mackinnon D (1991) Preservation of fossil biopolymeric structures: conclusive immunological evidence. Geochim Cosmochim Acta 55:2253–2257

Collins MJ, Walton D, Curry GB, Riley MS, Von Wallmenich TN, Savage NM, Muyzer G, Westbroek P (2003) Long-term trends in the survival of immunological epitopes entombed in fossil brachiopod skeletons. Org Geochem 34:89–96

Corneillie TM, Whetstone PA, Meares CF (2006) Irreversibly binding anti-metal chelate antibodies: artificial receptors for pretargeting. J Inorg Biochem 100:882–890

De Jong EW, Westbroek P, Westbroek JW, Bruning JW (1974) Preservation of antigenic properties of macromolecules over 70 Myr. Nature 252:63–64

Edwards HG, Newton EM, Dickensheets DL, Wynn-Williams DD (2003) Raman spectroscopic detection of biomolecular markers from Antarctic materials: evaluation for putative Martian habitats. Spectrochim Acta A Mol Biomol Spectrosc 59:2277–2290

Edwards HGM, Vandenabeele P, Jorge-Villar SE, Carter EA, Perez FR, Hargreaves MD (2007) The Rio Tinto Mars analogue site: an extremophilic Raman spectroscopic study. Spectrochim Acta A Mol Biomol Spectrosc 68:1133–1137

Ellery A, Wynn-Williams D (2003) Why Raman spectroscopy on Mars? – a case of the right tool for the right job. Astrobiology 3:565–579

Fernández-Calvo P, Rivas LA, Näke C, García-Villadangos M, Gómez-Elvira J, Parro V (2006) A multi-array competitive immunoassay for the detection of broad-range molecular size organic compounds relevant for astrobiology. Planet Space Sci 54:1612–1621

Fernández-Remolar DC, Gómez-Elvira J, Gómez F, Sebastián E, Martín J, Manfredi JA, Torres J, González Kesler C, Amils R (2004) The Tinto River, an extreme acidic environment under control of iron, as an analog of the Terra Meridiani hematite site of Mars. Planet Space Sci 52:239–248

Fernández-Remolar DC, Morris RV, Gruener JE, Amils R, Knoll AH (2005) The Río Tinto Basin, Spain: mineralogy, sedimentary geobiology, and implications for interpretation of outcrop rocks at Meridiani Planum, Mars. Earth Planet Sci Lett 240:149–167

Fernández-Remolar DC, Gómez F, Prieto-Ballesteros O, Schelble, RT, Rodríguez N, Amils R (2008a) Some ecological mechanisms to generate habitability in planetary subsurface areas by chemolithotrophic communities: the Río Tinto subsurface ecosystem as a model system. Astrobiology 8:157–173

Fernández-Remolar DC, Prieto-Ballesteros O, Rodríguez N, Gómez F, Amils R, Gómez-Elvira J, Stocker C (2008b) Underground habitats in the Río Tinto basin: a model for subsurface life habitats on Mars. Astrobiology 8:1023–1047

Fischer D, Valenti J (2005) The planet–metallicity correlation. Astrophys J 622:1102–1117

Flynn GJ (1996) The delivery of organic matter from asteroids and comets to the early surface of Mars. Earth Moon Planets 72:469–474

Formisano V, Atreya S, Encrenaz T, Ignatiev N, Giuranna M (2004) Detection of methane in the atmosphere of Mars. Science 306:1758–1761

Frolov A (2003) A review of the nature and geophysical studies of the thick permafrost in Siberia: relevance to exploration on Mars. J Geoph Res 108(e4):8039–8046. DOI: 10.1029/2002je001881

Garcia-Pichel F, Sherry ND, Castenholz RW (1992) Evidence for an ultraviolet sunscreen role of the extracellular pigment scytonemin in the terrestrial cyanobacterium *Chlorogloeopsis* sp. Photochem Photobiol 56:17–23

Gilichinsky DA, Vorobyova EA, Erokhina LG, Fyordorov-Davydov DG, Chaikovskaya NR, Fyordorov-Dayvdov DG (1992) Long-term preservation of microbial ecosystems in permafrost. Adv Space Res 12:255–263

Gilichinsky DA, Soina VS, Petrova MA (1993) Cryoprotective properties of water in the Earth cryolithosphere and its role in exobiology. Orig Life Evol Biosph 23:65–75

Gilichinsky DA, Wilson GS, Friedmann EI, McKay CP, Sletten RS, Rivkina RM, Vishnivetskaya TA, Erokhina LG, Ivanushkina NE, Kochkina GA, Shcherbakova VA, Soina VS, Spirina EV, Vorobyova EA, Fyodorov-davydov DG, Hallet B, Ozerskaya SM, Sorokovikov VA,

Laurinavichyus KS, Shatilovich AY, Chanton JP, Ostroumov VE, Tiedje JM (2007) Microbial populations in antarctic permafrost: biodiversity, state, age, and implication for astrobiology. Astrobiology 7:275–311

Gómez F, Aguilera A, Amils R (2007) Soluble ferric iron as an effective protective agent against UV radiation: implications for early life. Icarus 191:352–359

González-Toril E, Llobet-Brossa E, Casamayor EO, Amann R, Amils R (2003) Microbial ecology of an extreme acidic environment, the Tinto River. Appl Environ Microbiol 69:4853–4865

Hauber E, van Gasselt S, Ivanov B, Werner S, Head JW, Neukum G, Jaumann R, Greeley R, Mitchell KL, Muller P, HRSC Co-Investigator Team (2005) Discovery of a flank caldera and very young glacial activity at Hecates Tholus, Mars. Nature 434:356–361

Hofmann BA (2008) Morphological biosignatures from subsurface environments: recognition on planetary missions. Space Sci Rev 135:245–254

Horowitz NH, Hobby GL, Hubbard JS (1976) The Viking carbon assimilation experiment: interim report. Science 194:1321–1322

Irvine WM (1998) Extraterrestrial organic matter: a review. Orig Life Evol Biosph 28:365–383

Israël G, Szopa C, Raulin F, Cabane M, Niemann HB, Atreya SK, Bauer SJ, Brun JF, Chassefiere E, Coll P, Conde E, Coscia D, Hauchecorne A, Millian P, Nguyen MJ, Owen T, Riedler W, Samuelson RE, Siguier JM, Steller M, Sternberg R, Vidal-Madjar C (2005) Complex organic matter in Titan's atmospheric aerosols from in situ pyrolysis and analysis. Nature 438:796–799

Jorge-Villar SE, Edwards HG (2006) Raman spectroscopy in astrobiology. Anal Bioanal Chem 384:100–113

Khurana KK, Kivelson MG, Russell CT (2002) Searching for liquid water in Europa by using surface observatories. Astrobiology 2:93–103

Kiang NY, Segura A, Tinetti G, Govindjee, Blankenship RE, Cohen M, Siefert J, Crisp D, Meadows VS (2007) Spectral signatures of photosynthesis. II. Coevolution with other stars and the atmosphere on extrasolar worlds. Astrobiology 7:252–274

Kieffer HH, Titus TN (2001) TES mapping of Mars' north seasonal cap. Icarus 154:162–180

Klein HP (1978) The Viking biological experiments on Mars. Icarus 34:666–674

Klein HP, Lederberg J, Rich A (1972) Biological experiments: the Viking Mars Lander. Icarus 16:139–146

Klingelhöfer G, Morris RV, Bernhardt B, Schröder C, Rodionov DS, de Souza PA Jr, Yen A, Gellert R, Evlanov EN, Zubkov B, Foh J, Bonnes U, Kankeleit E, Gütlich P, Ming DW, Renz F, Wdowiak T, Squyres SW, Arvidson RE (2004) Jarosite and hematite at Meridiani Planum from Opportunity's Mössbauer spectrometer. Science 306:1740–1745

Kminek G, Bada JL (2006) The effect of ionizing radiation on the preservation of amino acids on Mars. Earth Planet Sci Lett 245:1–5

Knecht BG, Strasser A, Dietrich R, Märtlbauer E, Niessner R, Weller MG (2004) Automated microarray system for the simultaneous detection of antibiotics in milk. Anal Chem 76:646–654

Knicker H (2004) Stabilization of N-compounds in soil and organic-matter-rich sediments – what is the difference? Mar Chem 92:167–195

Knicker H, Hatcher PG (1997) Survival of protein in an organic-rich sediment: possible protection by encapsulation in organic matter. Naturwissenschaften 84:231–234

Knicker H, del Río JC, Hatcher PG, Minard RD (2001) Identification of protein remnants in insoluble geopolymers using TMAH thermochemolysis/GC–MS. Org Geochem 32:397–409

Krumbein WE (2008) Biogenerated rock structures. Space Sci Rev 135:81–94

Kusnezow W, Jacob A, Walijew A, Diehl F, Hoheisel JD (2003) Antibody microarrays: an evaluation of production parameters. Proteomics 3:254–264

Labeyrie A (1999) Snapshots of alien worlds – the future of interferometry. Science 285:1864–1865

Lefèvre F, Forget F (2009) Observed variations of methane on Mars unexplained by known atmospheric chemistry and physics. Nature 460:720–723

Levin GV, Perez GR (1967) The search for extraterrestrial life. In: Advances in the astronautical sciences series V 22. American Astronautical Society, Publications Office, Tarzana, CA

Levin GV, Straat P (1976) Viking labeled release biology experiment: interim report. Science 194:1322–1329

Levin GV, Straat P (1981) A search for a nonbiological explanation of the Viking labeled release life detection experiment. Icarus 45:494–516

Litvak ML, Mitrofanov AS, Kozyrev AB, Sanin VI, Tretyakov WV, Boynton NJ, Kelly D, Hamara C, Shinohara RS, Saunders R (2006) Comparison between polar regions of Mars from HEND/Odyssey data. Icarus 180:23–37

Lovelock JE (1965) A physical basis for life detection experiments. Nature 207:568

Mackintosh ME (1978) Nitrogen fixation by *Thiobacillus ferrooxidans*. J Gen Microbiol 105:215–218

Mahdavi A (1998) Steps to a general theory of habitability. Hum Ecol Rev 5:23–30

Márquez A, Fernández C, Anguita F, Farelo A, Anguita J, de la Casa MA (2004) New evidence for a volcanically, tectonically, and climatically active Mars. Icarus 172:573–581

McDonald GD, De Vanssay E, Buckley JR (1998) Oxidation of organic macromolecules by hydrogen peroxide: implications for stability of biomarkers on Mars. Icarus 132:170–175

Morozova D, Möhlmann D, Wagner D (2007) Survival of methanogenic archaea from Siberian permafrost under simulated Martian thermal conditions. Orig Life Evol Biosph 37:189–200

Mumma MJ, Villanueva GL, Novak RE, Hewagama T, Bonev BP, Disanti MA, Mandell AM, Smith MD (2009) Strong release of methane on Mars in northern summer. Science 323:1041–1045

Navarro-González R, Rainey FA, Molina P, Bagaley DR, Hollen BJ, de la Rosa J, Small AM, Quinn RC, Grunthaner FJ, Cáceres L, Gomez-Silva B, McKay CP (2003) Mars-like soils in the Atacama desert, Chile, and the dry limit of microbial life. Science 7:1018–1021

Norris PR, Murrell JC, Hinson D (1995) The potential for diazotrophy in iron- and sulfur-oxidizing acidophilic bacteria. Arch Microbiol 164:294–300

Ohmura N, Sasaki K, Matsumoto N, Saiki H (2002) Anaerobic respiration using Fe^{3+}, S^0, and H_2 in the chemolithoautotrophic bacterium *Acidithiobacillus ferrooxidans*. J Bacteriol 184:2081–2087

Ono S (2008) Multiple-sulphur isotope biosignatures. Space Sci Rev 135:203–220

Onstott TC, McGown D, Kessler J, Lollar BS, Lehmann KK, Clifford SM (2006) Martian CH_4: sources, flux, and detection. Astrobiology 6:377–395

Oró J, Holzer G (1979) The photolytic degradation and oxidation of organic compounds under simulated Martian conditions. J Mol Evol 14:153–160

Parro V (2009) Antibody microarrays for environmental monitoring. In: Timmis K (ed) Handbook of hydrocarbon and lipid microbiology, vol 3, Chapter 22. Springer-Verlag, Berlin, Heidelberg.

Parro V, de Diego-Castilla G, Rodríguez-Manfredi JA, Rivas LA, Blanco-López Y, Sebastián E, Romeral J, Compostizo C, Herrero PL, García-Marín A, Moreno-Paz M, García-Villadangos M, Cruz-Gil P, Peinado V, Martín-Soler J, Pérez-Mercader J, Gómez-Elvira J (2011) SOLID3: a multiplex antibody microarray-based optical sensor instrument for in situ life detection in planetary exploration. Astrobiology 11(1):15–28. Epub 2011 Feb 6.

Parro V, Moreno-Paz M (2003) Gene function analysis in environmental isolates: the nif regulon of the strict iron oxidizing bacterium *Leptospirillum ferrooxidans*. Proc Natl Acad Sci USA 100:7883–7888

Parro V, Rodríguez-Manfredi JA, Briones C, Compostizo C, Herrero PL, Vez E, Sebastián E, Moreno-Paz M, García-Villadangos M, Fernández-Calvo P, González-Toril E, Pérez-Mercader J, Fernández-Remolar D, Gómez-Elvira J (2005) Instrument development to search for biomarkers on mars: terrestrial acidophile, iron-powered chemolithoautotrophic communities as model systems. Planet Space Sci 53:729–737

Parro V, Moreno-Paz M, González-Toril E (2007) Analysis of environmental transcriptomes by DNA microarrays. Environ Microbiol 9:453–464

Parro V, Rivas LA, Gómez-Elvira J (2008a) Protein microarrays-based strategies for life detection in astrobiology. Space Sci Rev 135:293–311

Parro V, Fernández-Calvo P, Rodríguez Manfredi JA, Moreno-Paz M, Rivas LA, García-Villadangos M, Bonaccorsi R, González-Pastor JE, Prieto-Ballesteros O, Schuerger AC, Davidson M, Gómez-Elvira J, Stoker CR (2008b) SOLID2: an antibody array-based life-detector instrument in a Mars Drilling Simulation Experiment (MARTE). Astrobiology 8:987–999

Pizzarello S (2007) The chemistry that preceded life's origin: a study guide from meteorites. Chem Biodivers 4:680–693

Potter EG, Bebout BM, Kelley CA (2009) Isotopic composition of methane and inferred methanogenic substrates along a salinity gradient in a hypersaline microbial mat system. Astrobiology 9:383–390

Prantzos N (2008) On the "galactic habitable zone". Space Sci Rev 135:313–322

Quinn RC, Zent AP, Grunthaner FJ, Ehrenfreund P, Taylor CL, Garry JRC (2005) Detection and characterization of oxidizing acids in the Atacama Desert using the Mars Oxidation Instrument. Planet Space Sci 53:1376–1388

Rivas LA, García-Villadangos M, Moreno-Paz M, Cruz-Gil P, Gómez-Elvira J, Parro V (2008) A 200-antibody microarray biochip for environmental monitoring: searching for universal microbial biomarkers through immunoprofiling. Anal Chem 80:7970–7979

Roberson EB, Chenu C, Firestone MK (1993) Microstructural changes in bacterial exopolysaccharides during desiccation. Soil Biol Biochem 25:1299–1301

Rothschild LJ, Mancinelli RL (2001) Life in extreme environments. Nature 409:1092–1101

Schweitzer M, Hill CL, Asara JM, Lane WS, Pincus SH (2002) Identification of immunoreactive material in mammoth fossils. J Mol Evol 55:696–705

Schweitzer MH, Wittmeyer J, Avci R, Pincus S (2005) Experimental support for an immunological approach to the search for life on other planets. Astrobiology 5:30–47

Schweitzer MH, Suo Z, Avci R, Asara JM, Allen MA, Arce FT, Horner JR (2007) Analyses of soft tissue from *Tyrannosaurus rex* suggest the presence of protein. Science 316:277–280

Schweitzer MH, Zheng W, Organ CL, Avci R, Suo Z, Freimark LM, Lebleu VS, Duncan MB, Vander Heiden MG, Neveu JM, Lane WS, Cottrell JS, Horner JR, Cantley LC, Kalluri R, Asara JM (2009) Biomolecular characterization and protein sequences of the Campanian hadrosaur *B. canadensis*. Science 324:626–631

Sephton MA, Pillinger CT, Gilmour I (1998) $\delta 13C$ of free and macromolecular aromatic structures in the Murchison meteorite. Geochim Cosmochim Acta 62:1821–1828

Shabat D, Itzhaky H, Reymond JL, Keinan E (1995) Antibody catalysis of a reaction otherwise strongly disfavoured in water. Nature 374:143–146

Shang C, Tiessen H (1998) Organic matter stabilization in two semiarid tropical soils: size, density, and magnetic separations. Soil Sci Soc Am J 62:1247–1257

Sharma SK, Angel SM, Ghosh M, Hubble HW, Lucey PG (2002) Remote pulsed laser Raman spectroscopy system for mineral analysis on planetary surfaces to 66 meters. Appl Spectrosc 56:699–705

Sherman LS, Waldbauer JR, Summons RE (2007) Improved methods for isolating and validating indigenous biomarkers in Precambrian rocks. Org Geochem 38:1987–2000

Smith PH, Tamppari LK, Arvidson RE, Bass D, Blaney D, Boynton WV, Carswell A, Catling DC, Clark BC, Duck T, DeJong E, Fisher D, Goetz W, Gunnlaugsson HP, Hecht MH, Hipkin V, Hoffman J, Hviid SF, Keller HU, Kounaves SP, Lange CF, Lemmon MT, Madsen MB, Markiewicz WJ, Marshall J, McKay CP, Mellon MT, Ming DW, Morris RV, Pike WT, Renno N, Staufer U, Stoker C, Taylor P, Whiteway JA, Zent AP (2009) H_2O at the Phoenix landing site. Science 325:58–61

Sobron P, Sobron F, Sanz A, Rull F (2008) Raman signal processing software for automated identification of mineral phases and biosignatures on Mars. Appl Spectrosc 62:364–370

Stankiewicz BA, Briggs DEG, Evershed RP, Duncan IJ (1997) Chemical preservation of insect cuticle from the Pleistocene asphalt deposits of California, USA. Geochim Cosmochim Acta 61: 2247–2252

Steele A, Toporski J, McKay DS, Schweitzer M, Pincus S, Pérez-Mercader J, Parro V (2001) Microarray assays for solar system exploration. ESA SP 496:91–97

Stevens T, McKinley J (1996) Hydrogen-based microbial ecosystems in the Earth. Science 272: 896–897

Stoker CR, Cannon HN, Dunagan SE, Lemke LG, Glass BJ, Miller D, Gomez-Elvira J, Davis K, Zavaleta J, Winterholler A, Roman M, Rodriguez-Manfredi JA, Bonaccorsi R, Bell MS, Brown A, Battler M, Chen B, Cooper G, Davidson M, Fernández-Remolar D, Gonzales-Pastor E, Heldmann JL, Martínez-Frías J, Parro V, Prieto-Ballesteros O, Sutter B, Schuerger AC, Schutt J, Rull F (2008) The 2005 MARTE robotic drilling experiment in Río Tinto, Spain: objectives, approach, and results of a simulated mission to search for life in the Martian subsurface. Astrobiology 8:921–945

Summons RE (1988) In: Broadhead TW (ed) Molecular evolution and the fossil record. The Paleontological Society, Univ Tennessee, Knoxville, USA, pp 98–113

Summons RE, Albrecht P, McDonald G, Moldowan JM (2008) Molecular biosignatures. Space Sci Rev 135:133–159

Sutherland I (2001) Biofilm exopolysaccharides: a strong and sticky framework. Microbiology 147:3–9

Sykes GA, Collins MJ, Walton DI (1995) The significance of a geochemically isolated intracrystalline organic fraction within biominerals. Org Geochem 23:1059–1065

Tarcea N, Frosch T, Rösch P, Hilchenbach M, Stuffler T, Hofer S, Thiele H, Hochleitner R, Popp J (2008) Raman spectroscopy – a powerful tool for in situ planetary science. Space Sci Rev 135: 281–292

Tuross N, Stathoplos L (1993) Ancient proteins in fossil bones. Methods Enzymol 224: 121–129

Van Zuilen (2008) Stable isotope ratios as a biomarker on Mars. Space Sci Rev 135:221–232

Vorobyova E, Soina V, Gorlenko M, Minkovskaya N, Zalinova N, Mamukelashvili A, Gilichinsky D, Rivkina E, Vishnivetskaya T (1997) The deep cold biosphere: facts and hypothesis. FEMS Microbiol Rev 20:277–290

Westall F (2008) Morphological biosignatures in early terrestrial and extraterrestrial materials. Space Sci Rev 135:95–114

Wiegel J (1998) Anaerobic alkalithermophiles, a novel group of extremophiles. Extremophiles 2: 257–267

Willerslev E, Hansen AJ, Poinar HN (2004) Isolation of nucleic acids and cultures from fossil ice and permafrost. Trends Ecol Evol 19:141–147

Zhang XQ, Bishop PL, Kupferle MJ (1998) Measurement of polysaccharides and proteins in biofilm extracellular polymers. Water Sci Technol 37:345–348

Zorzano MP, Mateo-Martí E, Prieto-Ballesteros O, Osuna S, Renno N (2009) The stability of liquid saline water on present day Mars. Geophys Res Lett 36:L20201. DOI: 10.1029/2009GL040315

Extremophiles in spacecraft assembly clean rooms

Christine Moissl-Eichinger

Institute for Microbiology and Archaea Center, University of Regensburg, Regensburg, Germany

1 Introduction

1.1 Planetary protection

Over the past decades, the search for life beyond Earth has become one of the greatest motivations for mankind to travel to space. Mars, in particular, being one of the most fascinating planets in our solar system in reach of modern spacecraft, could deliver answers to many open questions. Nevertheless, travel to and landing on Mars is challenging for life detection missions. Since Mars could possibly provide biotopes for its own or terrestrial life, a contamination via microbial hitchhikers from Earth could have severe consequences. The scientific field of planetary protection is concerned about a possible contamination of extraterrestrial environments by terrestrial biomolecules and life forms. Additionally, a reverse contamination of Earth by extraterrestrial material is also a fundamental concern: "States parties shall pursue studies of outer space, including the Moon and other celestial bodies, and conduct explorations of them so as to avoid their harmful contamination and also adverse changes in the environment of the Earth resulting from the introduction of extraterrestrial matter and, when necessary, adopt appropriate measures for this purpose" (UN Outer Space Treaty; Anonymous 1967). This scope of ESA's and NASA's planetary protection policies emphasizes that forward contamination by terrestrial life and even by biomolecules needs to be avoided, in order to preserve extraterrestrial bodies and to prevent confounding of future life detection experiments on other planets.

According to the recommendations of COSPAR (Committee of Space Research), space missions are divided into five categories considering the scientific interest and also the probability of possible contamination of other planets and, in case of return missions, also Earth (Table 1) (Anonymous 2002, amended 2005).

Landing missions on Mars are generally assigned to category IV with subcategories a, b, and c. Although Mars is very cold and most likely too dry for life,

Table 1. Planetary protection mission categories

Category	Mission type	Possible targets
I	Missions to a target body without direct interest for understanding the process of chemical evolution or the origin of life. Since no protection of these bodies is warranted, no planetary protection requirements are necessary.	Venus, undifferentiated asteroids
II	Missions to target bodies with significant interest relative to the process of chemical evolution and the origin of life, but in which there is only a remote chance of contamination. Planetary protection requirements include mainly simple documentation and passive contamination control (clean room assembly).	Jupiter, Saturn, Uranus, comets
III	Flyby and orbiter missions, targeting a body of significant interest for chemical evolution and/or origin of life, with high risk of contamination that could jeopardize future search for life missions. COSPAR requirements are a documentation include also a possible bioburden reduction if necessary. Furthermore, an inventory of the microbial community present is required if an impact is very probable.	Mars, Europa, Enceladus
IV	Mostly probe and lander missions, targeting bodies of high interest concerning chemical evolutions and/or origin of life, with a significant chance of contamination. Category IV lander missions are separated into three subcategories (a, b, and c) with different requirements based on the location of the landing site and the objectives of that mission. IVb and c have the strictest bioburden limits and require detailed documentations, bioassays for biobuden measurements, (partial) sterilization of hardware, agressive cleaning, protection from recontamination and aseptic assembly.	Mars, Europa, Enceladus
V	All Earth return missions, distinguishing unrestricted and restricted Earth return, depending on the probability of the presence of indigenous life forms on the visited solar body. Restricted Earth return missions require strict containment of samples.	Unrestricted Earth return: Moon; Restricted Earth return: Mars, Europa

some "special regions" could support life from Earth or indigenous Martian life, should it exist (Rummel 2009). These special regions are locations that might allow the formation and maintenance of liquid water on or under the surface of Mars. Missions that are intended to land in such a region are to be placed under strictest microbial control and limitations. In particular, the search for life could be affected by the contamination of landing spacecraft and their sensitive biosensors. False positives could possibly mask present signatures of Martian life and therefore inhibit the successful search for extraterrestrial forms of life. Additionally, although not very likely, organisms from Earth could possibly proliferate and contaminate Martian biotopes, competing with potential indigenous life.

Therefore, complete sterility of a spacecraft is a desirable goal. Nowadays, sensitive instruments and detectors onboard do not allow to heat-sterilize the entire spacecraft as was done with the Viking landers in the 1970s ($111.7°C \pm 1.7°C$, 23–30 h; Puleo et al. 1977). Instead, all assembly procedures are performed in microbiologically controlled clean rooms for integration of precleaned and (as far as possible) presterilized spacecraft hardware.

Spacecraft assembly clean rooms are quite similar to pharmaceutical or hospital clean rooms. In the pharmaceutical industry, clean rooms are required for aseptic production, and the monitoring of microbial and also particle counts are part of good manufacturing practices (Nagarkar et al. 2001).

1.2 Clean rooms

In order not to affect or even confound future life detection missions on celestial bodies, which are of interest for their chemical and biological evolution, all spacecraft are constructed in so-called clean rooms and are subject to severe cleaning processes and microbiological controls before launch (Crawford 2005). Clean rooms are certified according to ISO14644-1. Therefore, the clean room class ISO 5 corresponds to the former clean room class 100 (US FED STD 209E), allowing a maximum of 3.5 particles with a maximum 0.5 μm diameter per liter of air.

During assembly, test, and launch operations (ATLOs) of e.g. Mars landers, appropriate cleanliness and sterility levels must be guaranteed: The proper maintenance of the clean room includes a repeated cleaning with antimicrobial agents, particulates are filtered from the air using HEPA (high efficiency particulate air filter) filtering, and even staff working in the clean room must take appropriate actions to minimize any particulate and microbial contamination. Clean room personnel must follow specific access procedures (air locks and tacky mats) to minimize the influx of particulate matter. Staff has to wear special suits, use sterile tools, observe possible biocontamination risks and even undergo frequent health checks.

1.3 Contamination control and examinations

To date, space missions in preparation have to follow an implementation plan describing all actions necessary to reduce and measure bioburden. This plan includes also the requirement of (daily) sampling of the spacecraft and hardware using swabs and wipes. The bacterial spore count is then assessed by culturing a heat-shocked sample according to a standard protocol, and aims to reflect the most resistant component of the aerobic, heterotrophic and mesophilic microbial community present (Anonymous 1999). The current NASA standard is based on the methods originally developed for the Viking missions in the 1970s. In brief, the surface of a spacecraft is either swabbed (with cotton swabs) or wiped (for larger surfaces). The sampling tools are extracted in water by a combination of vortexing/shaking and sonication. After a heat shock (80°C, 15 min) the suspension is pour plated using trypticase soy agar (TSA) and incubated at 32°C. After a final count (72 h), the resulting plate count is used as a basis for the calculation of the overall microbial cleanliness of the spacecraft surface. Standard cotton swabs as used for the current

NASA procedure reveal some problems with handling and residues (Probst et al. 2010b); therefore, NASA and ESA are developing novel protocols for the detection of biocontamination.

ESA's new standard methodology is based on the usage of the nylon-flocked swab. Additionally, for better cultivation results of low-nutrient adapted clean room microorganisms, the cultivation medium is R2A instead of TSA, as given in the new ESA standard protocol (Anonymous 2008; Probst et al. 2010b). The recommended sampling size for swabs is only 25 cm^2, whereas polyester wipes are used for the sampling of larger surfaces. If surfaces of space hardware are contaminated above the accepted levels, biocleaning is necessary: alcohols (isopropyl alcohol), disinfectants, and UV exposure are some methods applied to reduce existing contaminants. Furthermore, bioshields can be used to enclose certain (clean) hardware, or the entire spacecraft to avoid contamination (Debus 2006).

In case of Mars, limits on bioburden are based on requirements first imposed on the Viking missions. Therefore, the acceptable microbial contamination is set to about 3×10^5 bacterial spores per Mars landing spacecraft or 300 spores per m^2 on exposed surfaces (Viking presterilization biological burden levels; Anonymous 2002; Pillinger et al. 2006). For instance, for Beagle 2, an ESA IVa mission, the overall surface bioburden was estimated to be 2.3×10^4 spores, the total bioburden was 1.01×10^5 spores and bioburden density approximately 20.6 spores per m^2 and therefore within the acceptable range (Pillinger et al. 2006).

1.4 Clean room microbiology

Examination protocols for assembly of spacecraft in clean rooms focus on the detection and enumeration of culturable mesophilic and heterotrophic organisms. Nevertheless, clean rooms are unique environments for microbes: due to low-nutrient levels (oligotrophic), desiccated clean conditions, and constant control of humidity and temperature, these environments are inhospitable to microbial life and even considered "extreme" (Venkateswaran et al. 2001). Several procedures keep contamination from the outside as low as possible, but these conditions are also highly selective for indigenous extremophilic microbial communities (Crawford 2005). For space missions, it is crucial to control the contaminating bioburden as much as possible. On the other hand, for the development of novel cleaning/sterilization methods it is also important to identify and characterize (understand) the microbial community of spacecraft clean rooms.

The low biomass is generally problematic, since the sampling and recovery methods themselves are biased and characterized by significant losses during the procedure (Probst et al. 2010b). Furthermore, it is estimated that only 0.1–1% of all microbes present in any biotope can be cultivated using standard cultivation techniques (Amann et al. 1995), thus increasing the unseen microbial diversity in clean rooms significantly.

Information about the microbial diversity in clean rooms associated with space mission and spacecraft is quite sparse and only a few NASA and ESA supported reports have been published thus far (e.g., Puleo et al. 1977; Venkateswaran et al. 2001; La Duc et al. 2007; Stieglmeier et al. 2009). The first article about the microbial analyses of the two Viking spacecrafts reported that about 7000 samples were taken from both spacecraft surfaces during prelaunch activities in order to determine the cultivable microbial load (Puleo et al. 1977). Besides human-associated bacteria (pathogens and opportunistic pathogens), which were predominant among the microbes isolated from these samples, aerobic spore-forming microorganisms (*Bacillus*) were found frequently on spacecraft and within the facilities. The predominance of human-associated microorganisms and spore-formers has been confirmed in subsequent publications (e.g., Moissl et al. 2007), whereas the portion of *Bacillus* e.g., *Micrococcus* was reported to be significant. In general, 85% of all isolated microorganisms of the NASA JPL group were identified as Gram-positive bacteria (Newcombe et al. 2005).

All of these cultivated microorganisms were obtained from isolation attempts on heterotrophic rich media. However, the chances of surviving space flights are higher for organisms that can thrive under more extreme conditions (Stieglmeier et al.

Table 2. Sampling locations and specifics

	Friedrichshafen 1 (FR1)	**Friedrichshafen 2 (FR2)**	**ESTEC (ES)**	**Kourou (KO)**
Location[a] (site, city)	EADS[b], Friedrichshafen	EADS[b], Friedrichshafen	ESTEC[c], Noordwijk	CSG[d], Kourou
Country	Germany	Germany	The Netherlands	French Guiana
Sampling date	April 2007	November 2007	March 2008	April 2009
Clean room facility	Hall 6, room 6101-04	Hall 6, room 6101-04	Hall Hydra	BAF[e]
Clean room specifics	ISO 5[f]	ISO 5	ISO 8	ISO 8
Sampled surfaces[a]	Various clean room surfaces, e.g., floor, stairs, door knobs; spacecraft	Various clean room surfaces, e.g., floor, stairs; spacecraft	Various clean room surfaces, mainly floor; spacecraft	Various clean room surfaces, mainly floor; floor of Ariane5 container

[a]Further details are given in Stieglmeier et al. (2009) and Moissl-Eichinger (2010)
[b]European Aeronautic Defense and Space Company
[c]European Space Research and Technology Centre
[d]Centre spatial guyanais
[e]Final assembly building
[f]Clean room was nominally operated at ISO 5 but opened to ISO 8 section just before sampling

Table 3. Assortment of media used for the enrichment of microbes from clean rooms

Target microbes	Basic medium[a]	Supplements, modifications, and conditions	Gas phase[b]	Extreme conditions[c]
Oligotrophs	R2A	Diluted 1:10, 1:100	Ae	Low nutrients
Alkaliphiles	R2A	pH 9; pH 11	Ae	pH 11
Acidophiles	R2A	pH 5; pH 3	Ae	pH 3
Autotrophs	*MM* (methanogenic Archaea medium)[d]		H_2/CO_2	Carbon present as CO_2 only
	AHM (autotrophic homoacetogen medium)[d]		H_2/CO_2	
	ASR (autotrophic sulfate-reducer medium)[d]		H_2/CO_2	
	AAM (autotrophic all-rounder medium)[d]		N_2/CO_2	
Nitrogen fix	*N_2 fix* (Hino and Wilson N_2-free medium)[d]		N_2, N_2/O_2	N_2 as nitrogen source
Anaerobes	TGA (thioglycolate agar)[d]		N_2	Anaerobic medium
	TSA (trypticase soy agar)[d]			
	SRA (sulfate-reducer agar)[d]			
	TS (trypticase soy medium)[d]			
	TG (thioglycolate medium)[d]			
Thermophiles	R2A	Incubated at 50°C or 60°C	Ae	10°C
Psychrophiles	R2A	Incubated at 10°C or 4°C	Ae	4°C
Halophiles	R2A	Addition of 3.5 or 10% (w/v) NaCl	Ae	10% NaCl
Additional media				
Heterotrophs	R2A		ma	
Archaea	*MM* (methanogenic Archaea medium)[d]	Sodium acetate, methanol	N_2/CO_2	
	ASM (Archaea supporting medium)[d]	Antibiotics mixture; NH_4Cl or yeast extract	N_2, ma, ae	

[a]Liquid media are shown in italics
[b]Ae aerobic and ma microaerophilic (<3% O_2)
[c]The extreme conditions which were applied and are discussed in this chapter. If not given otherwise, the incubation temperature was 32°C
[d]Medium recipe and preparation provided in Stieglmeier et al. (2009)

2009). Nevertheless, the search for and successful isolation of microorganisms capable of growing also under more extreme conditions has been reported sparsely (La Duc et al. 2007; Stieglmeier et al. 2009).

In preparation for the recently approved ESA ExoMars mission and also for future lander and Mars Sample Return (MSR) missions, knowledge of the biological contamination in spacecraft assembly facilities, integration, testing, and launch facilities is absolutely necessary. In a very recent separate study of two European and one South American spacecraft assembly, clean rooms were analyzed regarding their microbial diversity, using standard procedures, alternative cultivation approaches, and molecular methods with the aim to shed light on the presence of microorganisms relevant for planetary protection. For this study, the Herschel Space Observatory (launched in May 2009) and the clean rooms used for housing it were sampled during ATLO activities at three different locations (Table 2). Although the Herschel Space Observatory did not demand planetary protection requirements, all clean rooms were under full operation (strict particulate and molecular contamination control) when sampled.

The cultivation procedures and media are summarized in Table 3 (see also Stieglmeier et al. 2009).

In the following chapters, results obtained by the author from cultivation attempts focusing on the extremophilic microbial community in spacecraft assembly clean rooms shall be compared to previously obtained results from US American studies.

1.5 Extremophiles and extremotolerants – definition

Generally, extremophiles are microorganisms that require extreme conditions for growth. For instance, psychrophiles are adapted to low-temperature environments and require temperatures lower than 15°C for optimal proliferation. A clean room itself is an extreme environment hosting mainly extremotolerant microorganisms that accept the extreme circumstances, but prefer moderate conditions for growth. In the following, the author will (for simplifying the terminology) not distinguish between real extremophiles and extremotolerants: for example, the term alkaliphiles concerning clean room isolates include also microorganisms that tolerate, but do not require, alkaline conditions.

2 Spore-forming microorganisms

2.1 Background

Spores are the resting states of bacteria and are usually formed when the organism recognizes a lack in nutrients (C or N source). Endospores of *Bacillus subtilis* are highly resistant to inactivation by environmental stresses, such as biocidal agents, toxic chemicals, desiccation, pressure, temperature extremes, higher doses of UV, and ionizing radiation (Nicholson et al. 2000; Nicholson and Schuerger 2005). They possess thick layers of coating proteins, and even their DNA is protected by small acid-soluble spore proteins (SASPs; Moeller et al. 2008). The gel-like core of a desiccated spore contains only 10–25% of the water available in a vegetative cell. Enzymes and therefore the metabolism of a spore are more or less inactive. Spores can survive hundreds or maybe even million years, when kept dry and protected against mechanical forces and lethal doses of radiation (Cano and Borucki 1995).

Germination of spores generally needs an activator, e.g., moderate heat. In culture, amino acids such as alanine seem to support the germination process. In total, almost 20 bacterial genera are able to form spores, but only *Bacillus* and *Clostridium* spores have been subjected to deeper characterization studies (Fig. 1).

The multiresistance properties of such spore-forming microorganisms make them ideal candidates for the survival of space flight. Additionally, commonly applied sterilization conditions such as dry heat or chemical disinfectants, that do not harm the spacecraft and its hardware, are not able to kill most bacterial spores (Crawford 2005). Since the microbial analysis of the Viking mission has proven the presence of a broad diversity of spore-forming microorganisms on spacecraft surfaces, they have become the main focus of attention in the past decades.

Although 99.9% of all *B. subtilis* spores were killed when exposed to a few minutes of simulated Martian surface conditions (in terms of UV irradiation, pressure, gas

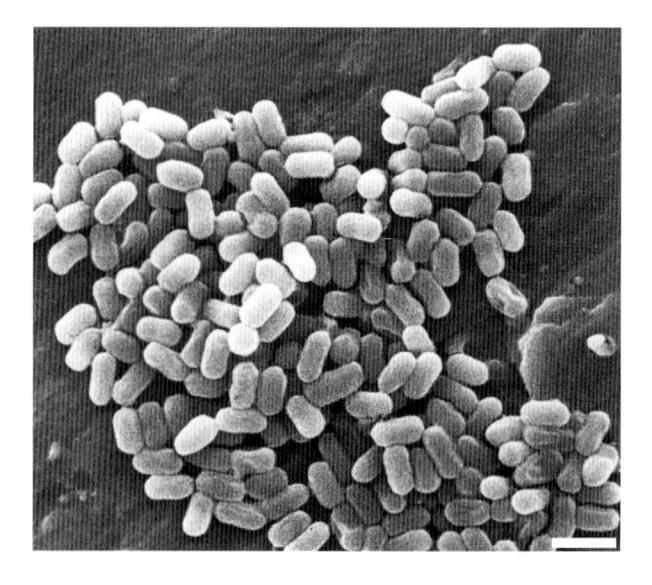

Fig. 1. Scanning electron micrograph of *Bacillus* spores on stainless steel. Bar, 2 μm

composition, and temperature), it has also been shown that dried spores were resistant to UV inactivation when mixed with Mars surrogate soil (Crawford et al. 2003; Schuerger et al. 2003; Osman et al. 2008). They were even resistant to sterilizing UV, as long as protected by a shallow layer of sand (Crawford et al. 2003). It can therefore be assumed, that highly resistant spores delivered to Mars could survive the travel to and the stay on Mars without further damage when located on lander parts which are not fully exposed to radiation or covered by a thin layer of dust (Osman et al. 2008).

For the selective enrichment of spore-forming microbes a heat shock at 80°C for 15 min is one important step within the procedure for measurement of biocontamination. Besides the effect that most vegetative cells are killed at 80°C, this heat step is also helpful in stimulating *Bacillus* spores to germinate. Newcombe et al. (2005) reported, that members of the genus *Bacillus* were the predominant microbes among the heat shock survivors, but the isolation of heat-shock-resistant *Staphylococcus*, *Planococcus*, and *Micrococcus* has also been described (Venkateswaran et al. 2001).

For the sake of completeness it shall be mentioned that some vegetative microbial cells can resist very harsh conditions such as extreme doses of (UV- and ionizing) radiation and desiccation (e.g., *Deinococcus radiodurans*, *Halobacterium* sp. NRC-1; Cox and Battista 2005; DeVeaux et al. 2007). Nevertheless, the information about vegetative, resistant bacteria from spacecraft assembly clean rooms is very limited.

It can be concluded that the current standard procedures of space agencies would not cover the broadest diversity of extremotolerant and spore-forming microbes but provide approximate numbers and estimations as a working basis.

2.2 Results

Bacillus is a typical spore-forming contaminant in spacecraft assembly clean rooms. Puleo et al. (1977) reported already the detection of more than 14 different *Bacillus* strains on the Viking spacecraft. Additionally, *Actinomycetes* and yeast have been detected, but were not characterized further.

Newer studies from spacecraft assembly clean rooms confirmed the presence or even predominance of spore-forming bacteria in cultivation assays based on rich heterotrophic media. Six different *Bacillus* strains have been detected on the Mars Odyssey spacecraft (La Duc et al. 2003), some of them revealing resistance against 0.5 Mrad γ-radiation, 5% H_2O_2 (60 min exposure) or higher doses of UV. In another study, further spore-forming organisms, such as *Sporosarcina*, *Paenibacillus*, *Actinomycetes*, and *Aureobasidium* have been detected (La Duc et al. 2004).

In our recent study of three clean rooms, 32 different culture media were used to target a wide range of different microorganisms (see Table 3). With this approach, the presence of a broader variety of spore-forming microorganisms in spacecraft assembly clean rooms was revealed. *Bacillus* and *Paenibacillus* were found in every facility. Overall 13 different *Bacillus* strains, 11 different paenibacilli, *Brevibacillus, Clostridium, Desulfotomaculum, Geobacillus, Micromonospora, Sporosarcina*, and two *Streptomyces* species were isolated. In general, spore-forming microorganisms accounted for about 5–25% of all microbes obtained via cultivation. The lowest percentage of spore-formers was found during the second sampling at EADS in Friedrichshafen (FR2). During that time, the clean room was operated at ISO 5 resulting in a higher percentage of human-associated microorganisms and a lower percentage of spore-formers. Most of the spore-forming bacteria observed are associated with environmental biotopes (such as soil) and therefore most likely introduced on items moved into the clean rooms or attached to humans and clothes. It can be concluded that the higher the operational cleanliness of a facility the less spore-forming microorganisms can be expected.

Since almost nothing is known about the resistance of spores obtained from spore-formers other than that of *Bacillus* and *Clostridium*, further analyses of our spore-forming isolates are definitely deemed appropriate.

Bacillus pumilus SAFR-032, an isolate originally obtained from a class 100K (ISO 8) clean room at the Jet Propulsion Laboratory spacecraft assembly facility (JPL-SAF) was reported to form spores with extraordinary UV resistance, outcompeting even a standard dosimetric strain of *B. subtilis* (Newcombe et al. 2005). Different strains of *B. pumilus* have very frequently been isolated from US American spacecraft assembly clean rooms (Puleo et al. 1977) and many of them were described to possess amazing resistances against H_2O_2 (Kempf et al. 2005) or UV light (Link et al. 2004; Newcombe et al. 2005). Generally, these isolates revealed a higher resistance to UV irradiation than the type strain *B. pumilus* (Newcombe et al. 2005).

Strains of *B. pumilus* have been isolated also during the second sampling at Friedrichshafen and the sampling at CSG Kourou, but were underrepresented (0.6% and 1.4%, respectively) among all other isolates. The most frequent *Bacillus* strains obtained were *Bacillus megaterium* or bacilli affiliated to the *B. thuringiensis/cereus* group. As reported by Newcombe et al. (2005), out of 125 aerobic strains isolated from US American spacecraft assembly facilities 65% were resistant against the heat shock implemented by the standard protocol of NASA. Among 15 different *Bacillus* sp. which were identified, *B. licheniformis* (25%) and *B. pumilus* (15%) were the most prevalent species.

3 Oligotrophic microorganisms

3.1 Background

Oligotrophic (or oligophilic microorganisms) are microbes that are adapted to low-nutrient conditions. Standard laboratory media are usually rich heterotrophic media providing a broad variety of carbon sources and other nutrients. In contrast, most of the microbes thriving in natural biotopes have to grapple with restriction of nutrients and competition with other organisms. Similarly, clean rooms are characterized by a significant lack of nutrients. Frequent cleaning and air filtering procedures remove particles that could provide nutritive substances, so that the microorganisms which are present either have to retreat into a resting state (such as spores) or have to adapt their metabolism to the extreme circumstances. To date, NASA's standard procedures recommend the usage of TSA medium for the cultivation of microbial biocontamination in spacecraft assembly facilities. However, a pharmaceutical clean room study revealed that the portion of cultivables from a clean room production unit could be increased by two orders of magnitude when a low-nutrient medium was applied instead of a rich medium (Nagarkar et al. 2001).

Additionally, when looking for possible hitchhikers to Mars, the search for microbes adapted to low-nutrient conditions is even more reasonable: So far, no complex organic molecules have been detected on the Martian surface or in its atmosphere.

3.2 Results

Until now, no data have been published with respect to oligotrophic microorganisms from spacecraft assembly clean rooms, and the afore mentioned study from a pharmaceutical clean room has not delivered data about the microbial strains which were detected (Nagarkar et al. 2001).

In our study of European and South American clean rooms, R2A medium was used for various cultivation attempts. This medium was originally developed to study microorganisms inhabiting potable water (Reasoner and Geldreich 1985); it is a low-nutrient medium that could stimulate the growth of stressed and slow growing microbes. For the detection of oligotrophs, R2A medium was applied even in a 1:10 and 1:100 dilution, respectively. Interestingly, a broad variety of bacteria was cultured on R2A medium diluted 1:100, including *Acinetobacter, Balneimonas, Brevundimonas, Citrobacter, Kocuria, Microbacterium, Micrococcus, Moraxella, Paenibacillus, Sanguibacter, Staphylococcus, Stenotrophomonas,* and *Streptomyces,* and also including the two spore-formers *Paenibacillus* and *Streptomyces*.

Interestingly, isolates from the first sampling at EADS in Friedrichshafen that were grown on R2A medium diluted 1:10 and 1:100, exceeded the number of cultivables on other (nutrient-rich) media (2.8×10^4 and 4.0×10^4 oligotrophic cultivables per m^2 clean room surface). In comparison, the samples from the other clean rooms revealed a high number of oligotrophs present. Since our approach of searching for oligotrophs is the first in this field, further studies in other clean rooms will be necessary and are highly recommended.

Based on own observations and results from other studies, the future ESA standard for measurement of biocontamination will rely on R2A medium instead of TSA for the cultivation of spacecraft assembly related microorganisms and will also recommend the usage of even more diluted medium for the growth of oligotrophs as an additional assay.

4 Alkaliphiles and acidophiles

4.1 Background

Although the pH of Mars' regolith was estimated to be about neutral (7.2 ± 0.1; Plumb et al. 1993), the presence of alkaliphiles and acidophiles in spacecraft assembly clean rooms could deliver valuable information for considerations in planetary protection and in particular for clean room maintenance: Most of the disinfectants and detergents used for (bio)cleaning in such clean rooms are either pH neutral or alkaline. It is unclear if thorough treatment with e.g. alkaline detergents could result in a positive selection effect. A preference of alkaline or acidic media by the microbial diversity in clean rooms was examined in two independent studies.

4.2 Results

Samples from diverse spacecraft assembly clean rooms were plated on R2A medium with a pH of 3, 9, and 10.6, respectively (La Duc et al. 2007), or with a pH of 3, 5, 9, and 11, respectively (this project). A broad variety of bacteria tolerating pH 10.6 was reported from US American clean rooms (La Duc et al. 2007) and our studies revealed also various alkaliphiles (growing at a pH of 9 or 11; Table 4). Interestingly, *Bacillus*, *Staphylococcus*, *Brevundimonas*, *Micrococcus*, and *Pseudomonas* were detected in both studies, growing in a medium of pH 10.6 and 11, respectively. Alkaliphiles were observed in every facility looked at, with numbers ranging from 1.6×10^2 (ISO 6, JSC-GCL) to 2.0×10^6 per m^2 (ISO 8, LMA-MTF). During the Herschel campaign (see Sect. 1.4) 4.6×10^2 (from the second sampling at Friedrichshafen, ISO 5; see Table 2) up to 8.4×10^3 (at ESTEC, ISO 8; Table 2) alkaliphilic microorganisms per m^2 were measured, whereas the number of alkaliphiles in the samples from Kourou (see Table 2) could not be determined, due to overgrowth of the agar plates.

The detection of acidotolerants was much more difficult. The colony counts on agar plates with medium of pH 5 were very low; no isolate was obtained during the Herschel campaign tolerating pH 3, whereas an acidotolerant colony

Table 4. Extremophilic isolates[a] from global spacecraft assembly clean rooms

Genus	Oligotrohs[b]	Psychrophiles[c]	Alkaliphiles[d]	Anaerobes	Thermophiles[e]	Halophiles[f]	CO_2 fix	N_2 fix	Species; in brackets: location of isolate[g]; type of extremophile[h]
Acinetobacter	○	○							*A.* sp., *A. ursingii* (FR2)
Actinotalea			○						*A. fermentans* (KO)
Aerococcus				○		○			*A. urinaeequi* (KO)
Arsenicicoccus				○			♦		*A. bolidensis* (FR2)
Arthrobacter			○	○				♦	*A.* sp. (KSC/alk, KO/an, N_2)
Bacillus			○ ○ ○ ○	○ ○ ○ ○	○ ○ ○	○			*B. thermoamylovorans* (ES/alk, FR1/an), *B. gibsonii* (FR2/alk), *B. licheniformis* (FR1/alk, ES/an,therm), *B. pumilus* (KO/an, JSC/alk), *B.* tc (FR2, ES, KO/an), *B. badius* (KO/therm), *B. coagulans* (ES, KSC/therm), *B. megaterium* (KO/halo), *B.* sp. (JSC/alk)
Balneimonas	○		○						*B.* sp. (KO)
Brachybacterium			○						*B. paraconglomeratum* (LMA)
Brevibacillus			○		○				*B. agri* (FR1)
Brevibacterium			○						*B. frigoritolerans* (ES)
Brevundimonas	○		○ ○						*B. nasdae* (FR2, KO), *B. diminuta* (KSC)
Cellulomonas				○ ○			♦		*C. hominis* (FR1, KO/an, KO/aut)
Citrobacter	○								*C. werkmanii* (KO)
Clostridium				○					*C. perfringens* (ES)
Corynebacterium				○					*C. pseudogenitalium* (ES)
Cupriavidus				○					*C. gilardii* (KO)
Dermabacter				○					*D. hominis* (FR2)
Desemzia						○			*D. incerta* (KO)
Desulfotomaculum				○					*D. guttoideum* (KO)
Enterococcus				○ ○			♦		*E. casseliflavus* (KO/aut), *E. faecalis* (ES/an), *E. faecium* (ES/an)
Facklamia			○	○					*F.* sp. (KO)

(continued)

Table 4 (continued)

Genus	Oligotrohs[b]	Psychrophiles[c]	Alkaliphiles[d]	Anaerobes	Thermophiles[e]	Halophiles[f]	CO_2 fix	N_2 fix	Species; in brackets: location of isolate[g]; type of extremophile[h]
Geobacillus					○ ○ ○ ○				G. caldoxylosilyticus (ES, KSC), G. stearothermophilus (JPL), G. kaustophilus (JPL), G. thermodenitrificans (JSC)
Georgenia			○						G. muralis (KO)
Kocuria	○		○						K. rhizophila (ES), K. rosea (KSC)
Lysobacter			○						L. sp. (KO)
Massilia		○							M. brevitalaea (KO)
Microbacterium	○		○ ○ ○					♦	M. oleivorans (KO/oligo), M. paraoxydans (KO/oligo, N_2), M. schleiferi (LMA), M. aurum (JSC), M. arborescens (KSC)
Micrococcus	○		○ ○ ○ ○			○		♦	M. sp. (KO/oligo, alk), M. flavus (KO/alk), M. indicus (ES/aut) M. luteus (KO, FR2, FR1/alk; KO/halo), M. mucilaginosus (KSC/alk)
Moraxella	○		○	○					M. osloensis (FR2/oligo; FR1/alk,an)
Oceanobacillus		○							O. sp. (JPL)
Paenibacillus	○		○ ○ ○	○ ○ ○	○		♦ ♦	♦	P. pasadenensis (FR1/oligo,alk, N_2; ES/alk), P. telluris (ES/alk), P. sp. (KO/alk), P. amylolyticus (FR1/alk), P. glucanolyticus (FR1/alk), P. sp. (ES/an), P. ginsengisoli (ES/an,auto), P. barengoltzii (FR1/an), P. cookii (FR1/therm), P. wynii (LMA/an)
Paracoccus								♦	P. yeeii (ES)
Propionibacterium				○ ○ ○					P. acnes (FR1, FR2, ES), P. avidum (ES)
Pseudomonas		○	○ ○					♦	P. luteola (FR2/N_2), P. xanthomarina (KO/alk,psy), P. stutzeri (KSC/alk)
Roseomonas		○	○						R. aquatica (KO)
Sanguibacter	○			○			♦	♦	S. marinus (KO)
Sphingomonas			○						S. oligophenolica (JSC), S. trueperi (JSC)

(continued)

Table 4 (continued)

Genus	Oligotrophs[b]	Psychrophiles[c]	Alkaliphiles[d]	Anaerobes	Thermophiles[e]	Halophiles[f]	CO_2 fix	N_2 fix	Species; in brackets: location of isolate[g]; type of extremophile[h]
Staphylococcus	○ ○		○ ○ ○ ○	○ ○ ○ ○		○ ○ ○ ○	♦ ♦		*S. haemolyticus* (ES, FR1/oligo, FR1, FR2, ES/alk, ES,KO/an, FR1, FR2, ES/halo,auto), *S. warneri* (KO/an), *S. pasteuri* (FR2/an, KO,ES/halo), *S. lugdunensis* (FR2/an), *S. hominis* (FR1/halo), *S. epidermidis* (JSC/an)
Stenotrophomonas	○		○	○				♦	*S. maltophilia* (FR2)
Streptomyces	○		○					♦	*S. luteogriseus* (ES)
Tessaracoccus				○					*T. flavescens* (KO)

[a]*Circles* and *diamonds* indicate each locations, where microbes were detected; each symbol represents one sampling location. *Black*: Isolates from this study (see Table 2 for abbreviations of locations). *Red*: Isolates from US American studies (La Duc et al. 2007)
[b]Oligotrophs were grown on R2A medium diluted 1:100
[c]Psychrophiles were grown on R2A medium at 4°C
[d]Alkaliphiles were grown on R2A medium of pH 10.6 or 11, respectively
[e]Thermophiles were grown at 50°C, 60°C or 65°C, respectively
[f]Halophiles were grown on R2A medium containing 10% NaCl
[g]Locations: KSC (Kennedy Space Center), JSC (Johnson Space Center), LMA (Lockheed Martin Aeronautics), and JPL (Jet Propulsion Laboratory)
[h]*an* anaerobic, *alk* alkaliphilic, *oligo* oligotrophic, *therm* thermophilic, *halo* halophilic, *aut* autotrophic, N_2 N_2 fixing, *psy* psychrophilic, and *B. tc* Bacillus thuringiensis/cereus goup

count was reported only for the US Lockheed Martin Aeronautics Multiple Testing Facility (LMA-MTF) and Kennedy Space Center Payload Hazardous Servicing Facility (KSC-PHSF) facilities (La Duc et al. 2007). All other samplings were negative for acidotolerants. None of the isolates obtained during the US American study was analyzed further or revealed multiresistant properties (La Duc et al. 2007).

A significant preference of alkaline media by bacteria was found in all facilities analyzed so far. One isolate obtained by the group of Venkateswaran was a salt and alkalitolerant bacterium, which was described very recently as *Bacillus canaveralius* (Newcombe et al. 2009). To date, the reasons for the shifts toward alkaliphily are unclear, but a positive selection by the usage of (alkaline) cleaning detergents seems probable and could result in an outpacing of acidophiles or nonalkalitolerants. If so, the selectivity via the pH of cleaning agents could be circumvented by using detergents with alternating pH.

5 Autotrophic and nitrogen fixing microorganisms

5.1 Background

The capability to fix nitrogen from the gaseous atmosphere or to grow autotrophically on CO_2 are important properties of primary producers. The activities of these microbes are the prerequisites for other microorganisms to colonize new nutrient-poor environments (Thomas et al. 2006). Since clean rooms and also the Martian environment are depleted in organic materials, primary producers, if present, could pave the way for secondary settlers.

Previous US studies have not reported any attempt to cultivate primary producers. All studies looking for cultivables from spacecraft and associated clean rooms have used heterotrophic, solidified rich media (except Stieglmeier et al. 2009; see below). The assortment of special media used in this study also included media selective for chemolithoautotrophs, using CO_2 as the only carbon source. Media providing nitrogen only in the gas phase were also applied.

5.2 Results

Seven isolated bacterial genera were able to fix CO_2 and eight were able to fix N_2. Further details about these bacterial genera are given in Table 4.

An interesting observation was that solely species of the bacterial genus *Paenibacillus*, *Micrococcus*, and *Sanguibacter* were able to perform both reactions, whereas *Sanguibacter marinus* was the only species that was isolated on an autotrophic and N_2 fixer medium in parallel (Kourou sampling). The type strain of this species was originally obtained from coastal sediment (Fujian province of China), and none of the observed properties had been reported in the original strain description (Huang et al. 2005). It can be imagined that the clean room isolate shows distinct properties due to the adaptation to the extreme biotope, or that the type strain has not been tested concerning these metabolic capabilities. *Paenibacillus* and *Streptomyces* were the only spore-forming microorganisms that were able to fix N_2 (both) or CO_2 (*Paenibacillus*).

6 Anaerobes

6.1 Background

The atmospheres of most planets within the reach of space missions contain only traces of oxygen, most likely not enough to support aerobic life as we know it from terrestrial biotopes (Thomas et al. 2006). Since the Martian surface is exposed to radiation and the soil is very oxidizing, the Martian subsurface could be an anaerobic biotope for possible life (Boston et al. 1992; Schulze-Makuch and Grinspoon 2005).

On Earth (facultative) anaerobes are widespread in different environments and can be detected in e.g. oxic soils, aerobic desert soils, or other biotopes, such as the human body (Tally et al. 1975; Peters and Conrad 1995; Küsel et al. 1999). In the latter case, they may become potential contaminants of spacecraft assembly facilities by staff, who is in close contact with flight hardware.

Generally, there are different types of anaerobic organisms. Facultative anaerobes always prefer aerobic conditions, but are able to grow under conditions with or without oxygen; aerotolerant anaerobes do not require oxygen for their growth and show no preference. Strict anaerobes (e.g., methanogens) never require oxygen for their reproduction and metabolism and can even be inhibited or killed by oxygen.

So far, not much is known about the presence of anaerobically growing microorganisms in spacecraft clean rooms. The presence of anaerobic microorganisms, which were enriched using the BD GasPaK system, in surface samples from US clean rooms has rarely been reported. Members of the facultatively anaerobic genera *Paenibacillus* and *Staphylococcus* have been isolated in the course of a study of extremotolerant microorganisms (La Duc et al. 2007).

A proper anaerobic cultivation necessitates the application of methods like the Hungate technique (Hungate 1969). Although this method has undergone a few simplifications during past decades, it still requires specialized equipment and practical experience. During our research, samples from the Herschel campaign were – for the first time – subjected to growth experiments performed with the anaerobic cultivation technology and a broad variety of microbes capable of anaerobic growth was isolated (Stieglmeier et al. 2009).

6.2 Results

A variety of anaerobic microorganisms was successfully isolated from all four clean room samplings. In total, 30 strains were isolated on anaerobic media. The greatest number and diversity of bacteria were obtained from the Kourou sampling (13 species). The following chart shows the oxygen requirements of isolates obtained from our campaign (see also Table 4 and Fig. 2).

In most cases, anaerobically enriched species were identified as facultative anaerobes comprising 16–78% of the total counts (numbers were calculated based on own observations and from published data for microbes grown on aerobic plates only; for comparison see Stieglmeier et al. 2009). Colony counts obtained on anaerobic complex media showed the presence of up to 5.8×10^2 anaerobes per m^2 clean room surface (Kourou sampling).

Only a comparatively low percentage of microbes grew strictly anaerobic (*Propionibacterium*, *Corynebacterium*, *Desulfotomaculum*, and *Clostridium*; 0.7–4% of all isolates at each location). *Propionibacterium acnes*, typically found on human

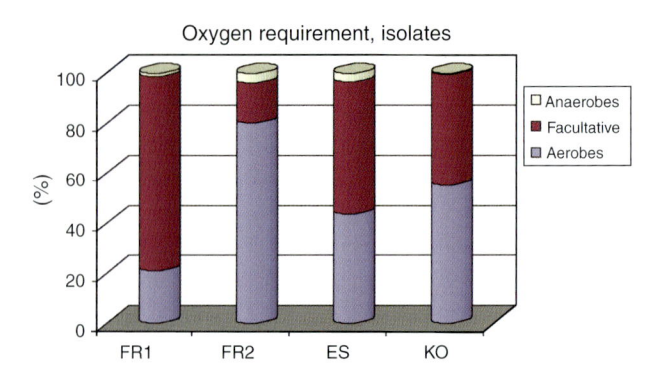

Fig. 2. Oxygen requirement of isolates. Numbers represent the percentage retrieved. Physiological capabilities are based on either own experiments or published data (for isolates grown on aerobic media only)

skin, was isolated from each European clean room and therefore the most prominent strict anaerobe detected. All strict anaerobes except *Desulfotomaculum* were opportunistic pathogens and isolated from rich, heterotrophic media. Interestingly, the *Corynebacterium* isolate (*C. pseudogenitalium*) could not be grown under aerobic conditions, although its type strain was described to be facultatively anaerobic (Stieglmeier et al. 2009).

Desulfotomaculum guttoideum was the only strict anaerobe isolated from the Kourou sampling. It was grown on sulfate-reducer-specific medium, but did not produce black colonies, which would have indicated a sulfate reducing activity. As it was clarified in a previous publication, *D. guttoideum* was misclassified and is actually affiliated to *Clostridium* cluster XIVa (Stackebrandt et al. 1997). This strain is therefore closely related to *Clostridium sphenoides*, a fermentative, saccharolytic, sulfite, and thiosulfate (but not sulfate) reducing spore-former. Other spore-forming microorganisms capable of anaerobic growth were *Bacillus*, *Paenibacillus*, and *Clostridium*. Further details are given in Table 4.

These data were confirmed by a very recent analysis of the anaerobic microbial diversity in NASA's clean rooms at the Jet Propulsion Laboratory. This study was based on a microbial enrichment of clean room samples under anaerobic conditions, which was subsequently analyzed via cultivation, 16S rRNA gene sequence analysis and microarrays (Probst et al. 2010a). *Clostridium* and *Propionibacterium* were the only strictly anaerobic microbes isolated, whereas additionally *Oerskovia*, *Dermabacter*, *Bacillus*, *Granulicatella*, *Sarcina*, *Leuconostoc*, *Paenibacillus*, *Staphylococcus*, and *Streptococcus* were detected during the molecular approach (Probst et al. 2010a).

Our results indicate that the facultatively and strictly anaerobic microbial community is quite diverse and may even be dominant in spacecraft assembly clean rooms.

7 Thermophiles and psychrophiles

7.1 Background

The Martian surface is very cold. Although the temperatures can reach up to 20°C in certain areas in the summer, the average temperatures are much below 0°C. Actually, also Earth's biosphere is quite cold – more than 70% of its freshwater occurs as ice, and the world's oceans reveal temperatures below 5°C (National Research Council 2006). A typical terrestrial biotope used for comparative studies is the Permafrost environment exhibiting a lively, highly diverse microbial community. It is assumed that Earth's psychrophiles could survive long-term in the Martian environment but would grow very slowly (National Research Council 2006).

Although (hyper-)thermophiles would probably not be able to proliferate on Mars, this group of microorganisms is often employed in studies concerning the origin and the evolution of life. Hot conditions prevailed on early Earth and many thermophiles exhibit "primordial" metabolic capabilities (for instance chemolithoautotrophy). Thermophiles are also generally considered more resistant than moderate or cold-loving microorganisms.

For these reasons, experiments were carried out searching for microorganisms that could grow under significantly higher or lower temperatures than the standard incubation temperature for all other experiments (32°C).

7.2 Results

A study of US American spacecraft assembly clean rooms did not report any growth of microorganisms on R2A medium following incubation for 10 days at 4°C (La Duc et al. 2007). Our study was selective for microorganisms capable of growing at 10°C and 4°C, respectively. The duration of incubation was prolonged up to 3 months. *Acinetobacter*, *Massilia*, *Pseudomonas*, and *Roseomonas* were isolated at 4°C, additionally, *Bacillus*, *Brevundimonas*, *Micrococcus*, *Moraxella*, *Paenibacillus*, *Sanguibacter*, *Sphingomonas*, *Sporosarcina*, *Staphylococcus*, and *Stenotrophomonas* were observed at 10°C. *Sporosarcina globispora*, a spore-forming bacterium, was the only isolate not able to grow at 32°C, the standard cultivation temperature.

Most of the "psychrophilic" isolates had also been obtained by using other cultivation methods at higher temperatures (32°C). It can be assumed that many (most?) of the present microorganisms in clean rooms are capable of growing at lower or very low temperatures, but that cell proliferation takes significantly longer than at higher, optimal temperatures.

The selective enrichment of thermophiles on R2A medium at 65°C (La Duc et al. 2007) or 60°C and 50°C, respectively (this study), allowed the isolation of three *Geobacillus* and one *Bacillus* strain (La Duc et al. 2007) and *Bacillus*,

Brevibacillus, Paenibacillus, and *Geobacillus* (this study). Interestingly, *Bacillus coagulans* and *Geobacillus caldoxylosilyticus* were isolated in both studies, whereas the latter was described as an obligate thermophile (*Saccharococcus caldoxylosilyticus*; see Ahmad et al. 2000), which is in accordance to the observations made in this study (no growth was observed at 32°C). It is unclear, how *Geobacillus* (spores) entered the clean room (although they would not be capable to proliferate under the thermal conditions of a clean room) and why this organism was detected in two independent studies in clean rooms on different continents. In general, *Geobacillus* spores have been described to be very resistant to environmental (thermal) stress (Head et al. 2008). Because of this high resistance the spores possibly survived strict cleanliness control conditions after being carried into the facilities by humans and materials.

8 Halophiles

8.1 Background

Halophiles have been discussed as possible survivors on Mars, since the Martian liquid water (if available) would contain high concentrations of different salts (Landis 2001). Additionally, a high resistance of salt-crystal-associated halophiles against UV radiation has been reported, making them potential survivors on the Martian surface (Fendrihan et al. 2009). Prevention of potential contamination of the Martian surface with halophiles is therefore highly important. Nevertheless, hardly any studies have been carried out thus far to investigate the potential presence of halotolerants and halophiles in spacecraft assembly clean rooms. In order to obtain insights into the distribution of these organisms in clean rooms samples from two studies were plated on R2A medium containing different concentrations of NaCl.

8.2 Results

Samples from US American clean rooms were plated onto R2A medium containing 25% (w/v) NaCl, but no growth was observed (La Duc et al. 2007). In contrast, samples obtained from the Herschel campaign (see Sect. 1.4) were plated on R2A medium containing 3.5% and 10% (w/v) NaCl, respectively. Growth on R2A medium containing 3.5% NaCl revealed, that most of the organisms isolated via other cultivation attempts were also capable of tolerating this comparatively low concentration of NaCl. The plates containing 10% NaCl revealed much lower cell counts: *Aerococcus, Bacillus, Desemzia, Micrococcus,* and *Staphylococcus* were observed on this medium (Table 4). The most prevalent species tolerating higher concentrations of NaCl were staphylococci, mainly originating from human skin, where they are exposed to higher levels of salt. Some staphylococci from this campaign were transferable to salt concentration up to 16% NaCl.

Despite the obvious presence of halophilic/halotolerant microbes, *B. megaterium* was the only spore-forming isolate that was detected on salty agar plates.

9 Archaea

9.1 Background

Archaea, the third domain of life, were considered for more than 20 years as extremophiles that were ecologically restricted and highly adapted to specific and often hostile biotopes. Meanwhile Archaea have been detected in almost any "normal" biotope such as marine and freshwater or soil (Bintrim et al. 1997; Karner et al. 2001; Rudolph et al. 2004). Many of the extremophilic Archaea have very interesting properties. In the eyes of many researchers they are "primitive" in their metabolism, which actually means that they can act as primary producers. This property could be an advantage when settling in new biotopes. Detailed experiments with vegetative (hyper-)thermophilic Archaea have revealed unexpectedly high tolerances against desiccation, vacuum, and UV or gamma radiation (Beblo et al. 2009). It is unclear however, whether these organisms could withstand the extremely harsh conditions during space travel, or lack of nutrients and low temperatures in extraterrestrial environments.

The main procedure to detect microorganisms in clean rooms is still directed toward Bacteria, but Archaea are more and more considered a possible source of biocontamination for spacecraft. The possibility that Archaea, such as methanogens or halobacteria, might be able to survive a space flight and survive or even to thrive on Mars, has already been discussed (Landis 2001; Kendrick and Kral 2006). Nevertheless, the existence of Archaea in human-controlled and rigorously cleaned environments has not been assessed before, so that it was unclear, if Archaea could even be found in spacecraft-associated clean rooms.

The vast majority of mesophilic and psychrophilic Archaea still resists cultivation, and also the attempt to cultivate Archaea from spacecraft assembly facilities failed (Moissl et al. 2008; Moissl-Eichinger 2010). For this reason, the detection of Archaea has to be based on molecular studies only, which are presented in the following section.

9.2 Results

In 2008, we reported the detection of archaeal 16S rRNA gene signatures in two US American spacecraft assembly clean rooms (Moissl et al. 2008). Using a very sensitive PCR approach 30 different cren- and euryarchaeal sequences were derived from NASA facilities. The omnipresence of Archaea in global clean rooms was confirmed in a very recent study (Moissl-Eichinger 2010). Archaea were detected in all clean rooms analyzed (second sampling Friedrichshafen, ESTEC and Kourou). As already reported in 2008 most of the gene sequences obtained clustered within the Crenarchaeota group 1b (now Thaumarchaeota) (Spang et al. 2010). The closest cultivated neighbor, candidatus *Nitrososphaera gargensis*, shows more than 4% difference in the 16S rRNA gene sequence; other close relatives from various natural biotopes are still uncultivated. Many representatives of this group were described to possess genes for ammonia oxidation (Hatzenpichler et al. 2008). Future studies looking for functional genes such as archaeal *amoA* genes could most likely enlighten the physiological capabilities of the Thaumarchaeota group detected in clean rooms.

The detection of these Thaumarchaeota in different spacecraft assembly clean rooms is significant: Different methods have been used and the studies were performed in two different laboratories. Overall, 48 different crenarchaeal sequences were obtained from five global clean rooms thus far (Moissl-Eichinger 2010). Additionally, molecular 16S rRNA gene sequence analysis data were supported by using fluorescence *in situ* hybridization (FISH) and real-time PCR (qPCR). Hybridization signals were obtained from samples taken in the clean room Hydra at ESTEC: rod shaped microbes reacted positively with Archaea-specific probes. QPCR revealed that the average number of Archaea per m^2 clean room surface amounted to around 2×10^4, approximately two to three logs lower than the estimated total number of Bacteria. Although the number of Archaea appears quite low, their presence was as consistent as typical bacterial clean room contaminants. Interestingly, most of the Bacteria detected are human commensals or opportunistic pathogens (e.g., *Staphylococcus*; Moissl et al. 2007; Stieglmeier et al. 2009). This finding could possibly hint toward a linkage of Archaea and humans, who might be carriers also for this group of microorganisms.

Besides Thaumarchaeota, methanogens and a halophilic archaeon were also (sporadically) detected (Moissl et al. 2008; Moissl-Eichinger 2010).

Our work has shown, that Archaea are an omnipresent, significant part of the microbial community in spacecraft assembly clean rooms all over the world. Since most of the data are based only on molecular analysis, it is unclear whether these Archaea could pose a threat with regard to planetary protection aspects. Further studies will be necessary and additional attempts for cultivation are recommended.

10 Other extremophiles

To complete the data presented here, other resistances of spacecraft assembly microbes shall be mentioned, although their presence has not been tested for the

Herschel campaign. The resistance against radiation (UV and gamma) as well as against hydrogen peroxide has been in the focus of interest, since these techniques are usually applied for the sterilization of spacecraft components (besides dry heat).

La Duc et al. (2007) reported the presence of microorganisms which were resistant to UV-C ($1000 \, J \, m^{-2}$) and 5% hydrogen peroxide. These resistances were found separately in spore-forming microbes only (*Bacillus*, *Nocardioides*, and *Paenibacillus*), whereas *B. pumilus* was resistant against both treatments. The detection of a hydrogen-peroxide-resistant *B. pumilus* has been reported even earlier (Kempf et al. 2005).

The multiresistant *B. pumilus* SAFR-032 was studied extensively at the Jet Propulsion Laboratory. The whole genome has been sequenced and annotated (Gioia et al. 2007). Although the sequence revealed differences to and, in addition, unknown genes compared to closely related species *B. subtilis* or *B. licheniformis*, the *B. pumilus* genome seems to lack genes functioning in resistance to UV or H_2O_2, which were found in other *Bacillus* strains. Further studies will certainly be necessary to understand the molecular basis of the extremotolerance in bacteria.

Interestingly, Eukarya have been detected only sparsely during the Herschel campaign. Solely one representative of *Coprinopsis* (fungi) and a few yeast strains have been isolated on R2A medium. The entire study focused on microorganisms, and the conditions would certainly have to be adapted for the cultivation of Eukarya. Although fungi can produce spores, their resistance properties, possible extremotolerance, and resulting impact on planetary protection considerations have not been studied yet and further research is highly recommended. Nevertheless, the detection of *Aureobasidium pullulans* from spacecraft assembly facilities was reported, a yeast-like fungus surviving 1 Mrad gamma radiation for 5.5 h (Bruckner et al. 2008).

11 Lessons learned from the Herschel campaign: extremophiles are everywhere

A broad variety of extremotolerant bacteria was successfully isolated from each facility analyzed so far. Microorganisms that were able to grow under extremely oligotrophic, cold, alkalic, anaerobic, warm, and high-salt conditions were detected (Table 4). Besides that, we have shown, that many microbes thriving or surviving in clean rooms are able to fix nitrogen and/or carbon dioxide, and could therefore serve as primary producers. The following list summarizes the main findings of our recent study:

- Spore-formers were present in each facility analyzed, but the total number did not exceed 25% of all isolates. The lowest percentage was seen in FR2, which was the "cleanest" environment sampled.

- The highest cell counts were obtained on media with lower nutrient contents and higher pH values, hinting toward a possible influence of environmental selective forces.
- Primary producers were found in unexpected diversity: autotrophs and N_2 fixing microbes were successully isolated.
- Strictly anaerobic bacteria were cultured from each facility, but are present in a low number only (up to 4%), although facultatively anaerobic microbes were found to add up to 78% of isolates.
- Thermophiles, psychrophiles, and halophiles were found in the clean rooms.
- Archaea were detected in each facility and their presence is significant. The properties of the (cren-)archaeal clean room community are unclear, but Archaea are able to persist and intact cells seem to be present.
- Cleanliness level of a clean room definitely influences the microbial diversity. The broadest diversity of cultivables was seen in Kourou and ESTEC samples (both clean rooms were operated at ISO 8). It can be assumed that also the environmental conditions have an influence on the microbial diversity within the clean rooms, since Kourou is located in a very humid environment.
- Most of the microbes detected in the overall study were human-associated, but the most resistant strains seem to be typical environmental organisms.

Furthermore, we have shown that not only *Bacillus* but also other (spore-forming) bacteria can play a significant role in clean room environments. For instance, several *Paenibacillus* strains have been detected and at least three of

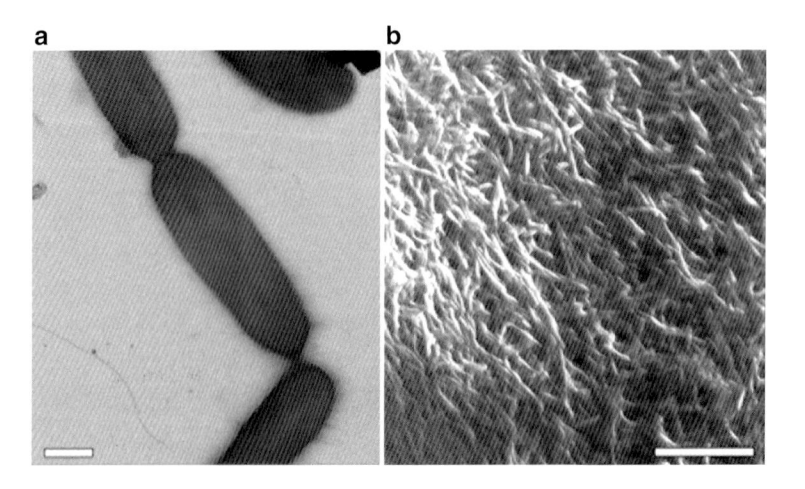

Fig. 3. *Paenibacillus cookii* FR1_23. (**a**) Electron micrograph of a dividing cell; bar, 600 nm. (**b**) Scanning electron micrograph of a colony; bar, 10 μm

them were identified to be novel isolates. Many bacilli are dependent on complex organic compounds for their metabolism. Interestingly, although closely related to bacilli, some paenibacilli seem to have the ability of nitrogen and carbon dioxide fixation. Initial resistance experiments with our isolates revealed a pronounced capability to survive desiccation, vacuum, heating, or Mars-cycle simulations (Fig. 3).

One novel isolate (ES_MS17, *Paenibacillus purispatii* sp. nov.) was shown to have profound metabolic capabilities for nitrogen conversion processes (Behrendt et al. 2010). Although paenibacilli have not explicitly been reported to be heat shock survivors, their capabilities hint toward multiresistance, combined with a high metabolic versatility. The study of paenibacilli from clean rooms could be beneficial for planetary protection considerations, and further research on this fascinating group of microbes is highly recommended.

In summary it can be stated that extremophilic and extremotolerant microorganisms are present in all spacecraft assembly facilities. Many of them reveal multiple resistances and capabilities of primary producers. Nevertheless, the information about indigenous microbial communities is still very limited: "How much is there?" and "What are they capable of?" are questions that will have to be answered. In order to preserve the integrity of future space travel, research in the field of planetary protection needs to be enforced further with the aim to understand the microbial communities in the spacecraft assembly facilities as much as possible.

12 The bacterial diversity beyond cultivation, or cultivation vs. molecular analyses

Cultivation as a singular procedure currently does not allow assessing the overall microbial diversity. Previous publications predicted a very low percentage (0.1–1%) of all microbes to be cultivable via standard laboratory techniques (Amann et al. 1995). Nevertheless, our own studies based on the usage of 32 different media and conditions led to the cultivation of approximately 0.3–5%, when compared to the qPCR results obtained (data unpublished).

Current molecular methodologies for assessing microbial diversity are mainly based on DNA extraction and subsequent PCR analysis (mostly 16S rRNA gene sequences), whereas LAL (limulus amoebocyte lysate) and ATP measurements have been reported as acceptable methods to obtain insights into the Gram-negative microbial diversity and the ATP content of clean room samples, respectively. LAL analysis is used for the estimation of the Gram-negative endotoxin-producing bacterial population and measures the presence of lipopolysaccharides. The ATP-based bioluminescence assay can help to obtain insights into the presence and the quantity of viable but nonculturable cells (La Duc et al. 2007; Bruckner et al.

2008). However, since cells do not contain the same amounts of ATP (which depends on the growth status or the size; Bruckner et al. 2008), quantitative measurements are strongly biased.

A strong bias has also been reported due to extraction methods for DNA and PCR. Up to now, no extraction method is able to fully extract DNA from spores, without disrupting the nucleic acid. For this reason, DNA-based molecular studies of clean room environments detect much more Gram negatives than Gram positives; the latter are generally harder to lyse or are spore-formers. It can be concluded that many microbes in spacecraft assembly clean rooms are present as spores, which are not detected by molecular methods, but are identified by cultivation attempts.

The bias of PCR has also been discussed in several publications and it is widely accepted, that no quantitative answers can be given based on standard PCR and subsequent cloning procedures (e.g., von Wintzingerode et al. 1997). Furthermore, selected primers are not universal for the entire microbial group in focus; mispairings can lead to lower PCR efficiency or even to a nonbinding of the primers to the target gene (Huber et al. 2002). This primer issue is also true for quantitative PCR approaches. Nevertheless, qPCR usually focuses on a specific microbial group and the entire methodology is designed for a very effective (up to 100%) amplification of the target gene. Since measurements are independent from cloning or other subsequent steps, qPCR allows at least quantitative, comparative predictions.

Compared to cultivation, molecular methods also detect dead cells that have influence on planetary protection issues only as a possible source of contamination with biomolecules. On the other hand, the sensitivity of PCR allows observing signatures of outnumbered and uncultivated inhabitants. Like previous studies based on standard 16S rRNA gene cloning, molecular studies of clean room environments revealed an unexpected broad diversity of microbes: The presence of diverse Bacteria and even Archaea has been shown only via molecular analyses (Moissl et al. 2007, 2008).

The knowledge of the microbial diversity in spacecraft assembly clean rooms has significantly increased with the recent studies (Bruckner et al. 2008). The methodologies to obtain insights into the qualitative (who is there?) microbial diversity are proceeding fast, but the overall (quantitative) microbial diversity of clean rooms (how much is there?) is still quite unclear. The next crucial step is the preparation of a typical model community, containing all main (and most resistant) microbes that were detected in spacecraft assembly clean rooms. With this artificial community the detection methods can be tested and more insights into quantitative information from molecular studies be gained (Kwan et al. 2011).

Although cultivation-independent methodologies are advancing fast, cultivation of microbes is still essential to investigate their abilities and resistances. Molecular techniques allow obtaining much more data in a much shorter time, and high-throughput methods such as microarrays and whole genome sequencing (454 sequencing and other technologies) are fascinating. A very recent study has reported the successful usage of microarrays (PhyloChips) also in the field of planetary protection (La Duc et al. 2009). Nevertheless, none of these techniques allows insights into the metabolic capabilities and resistance properties of microbes. Novel microarray techniques, such as the "GeoChip," searching for specific metabolic genes (He et al. 2007) could give stronger insights, but will not completely answer the open questions about resistance and properties. This information, however, is crucial for future planetary protection aspects. It is therefore advisable to put much effort into novel cultivation strategies (which are different from standard procedures) in order to increase the percentage of cultivables from spacecraft assembly clean rooms. Microbial analyses of biocontamination risks have to be as broad as possible and different technologies will be necessary to obtain the most complete picture on the specific microbial community in spacecraft assembly clean rooms.

Acknowledgments

I would like to thank the European Space Agency ESA for funding this project (contract no. 20508/07/NL/EK). Furthermore, Michaela Stieglmeier and Petra Schwendner for providing data, Alexander Probst, Annett Bellack, and Ruth Henneberger for critically reading the manuscript, and Gerhard Kminek (ESA) for wonderful discussions and valuable input. The preparation of graphical illustrations by Petra Schwendner and Alexander Probst is gratefully acknowledged.

References

Ahmad S, Scopes RK, Rees GN, Patel BK (2000) *Saccharococcus caldoxylosilyticus* sp. nov., an obligately thermophilic, xylose-utilizing, endospore-forming bacterium. Int J Syst Evol Microbiol 50:517–523

Amann R, Ludwig W, Schleifer KH (1995) Phylogenetic identification and *in situ* detection of individual microbial cells without cultivation. Microbiol Rev 59:143–169

Anonymous (1967) UN Outer Space Treaty. London/Washington

Anonymous (1999) NASA standard procedures for the microbiological examination of space hardware, NPG 5340.1D. Jet Propulsion Laboratory Communication, National Aeronautics and Space Administration

Anonymous (2002, amended 2005) In: Commitee on Space Research (COSPAR) (ed) COSPAR/IAU Workshop on Planetary Protection: Planetary Protection Policy. International Council for Science, Paris, France. http://cosparhq.cnes.fr/Scistr/Pppolicy.htm

Anonymous (2008) Microbial examination of flight hardware and clean rooms. ECSS-Q-ST-70-55C. European Cooperation for Space Standardization, ESA-ESTEC, The Netherlands

Beblo K, Rabbow E, Rachel R, Huber H, Rettberg P (2009) Tolerance of thermophilic and hyperthermophilic microorganisms to desiccation. Extremophiles 13:521–531

Behrendt U, Schumann P, Stieglmeier M, Pukall R, Augustin J, Spröer C, Schwendner P, Moissl-Eichinger C, Ulrich A (2010) Characterization of heterotrophic nitrifying bacteria with respiratory ammonification and denitrification activity – description of *Paenibacillus uliginis* sp. nov., an inhabitant of fen peat soil and *Paenibacillus purispatii* sp. nov., isolated from a spacecraft assembly clean room. Syst Appl Microbiol 33:328–336

Bintrim SB, Donohue TJ, Handelsman J, Roberts GP, Goodman RM (1997) Molecular phylogeny of Archaea from soil. Proc Natl Acad Sci USA 94:277–282

Boston P, Ivanov MV, McKay CP (1992) On the possibility of chemosynthetic ecosystems in subsurface habitats on Mars. Icarus 95:300–308

Bruckner JC, Osman S, Venkateswaran K, Conley C (2008) Space microbiology: planetary protection, burden, diversity and significance of spacecraft associated microbes. In: Schaechter M (ed) Encyclopedia of microbiology. Elsevier, Oxford, pp 52–66

Cano RJ, Borucki MK (1995) Revival and identification of bacterial spores in 25- to 40-million-year-old Dominican amber. Science 268:1060–1064

Cox MM, Battista JR (2005) *Deinococcus radiodurans* – the consummate survivor. Nat Rev Microbiol 3:882–892

Crawford RL (2005) Microbial diversity and its relationship to planetary protection. Appl Environ Microbiol 71:4163–4168

Crawford RL, Paszczynski A, Allenbach L (2003) Potassium ferrate [Fe(VI)] does not mediate self-sterilization of a surrogate Mars soil. BMC Microbiol 3. DOI: 10.1186/1471-2180-3-4

Debus A (2006) The European standard on planetary protection requirements. Res Microbiol 157:13–18

DeVeaux LC, Muller JA, Smith J, Petrisko J, Wells DP, DasSarma S (2007) Extremely radiation-resistant mutants of a halophilic archaeon with increased single-stranded DNA-binding protein (RPA) gene expression. Radiat Res 168:507–514

Fendrihan S, Berces A, Lammer H, Musso M, Ronto G, Polacsek TK, Holzinger A, Kolb C, Stan-Lotter H (2009) Investigating the effects of simulated Martian ultraviolet radiation on *Halococcus dombrowskii* and other extremely halophilic archaebacteria. Astrobiology 9:104–112

Gioia J, Yerrapragada S, Qin X, Jiang H, Igboeli OC, Muzny D, Dugan-Rocha S, Ding Y, Hawes A, Liu W, Perez L, Kovar C, Dinh H, Lee S, Nazareth L, Blyth P, Holder M, Buhay C, Tirumalai MR, Liu Y, Dasgupta I, Bokhetache L, Fujita M, Karouia F, Eswara Moorthy P, Siefert J, Uzman A, Buzumbo P, Verma A, Zwiya H, McWilliams BD, Olowu A, Clinkenbeard KD, Newcombe D, Golebiewski L, Petrosino JF, Nicholson WL, Fox GE, Venkateswaran K, Highlander SK, Weinstock GM (2007) Paradoxical DNA repair and peroxide resistance gene conservation in *Bacillus pumilus* SAFR-032. PLoS One 2(9):e928

Hatzenpichler R, Lebedeva EV, Spieck E, Stoecker K, Richter A, Daims H, Wagner M (2008) A moderately thermophilic ammonia-oxidizing crenarchaeote from a hot spring. Proc Natl Acad Sci USA 105:2134–2139

He Z, Gentry TJ, Schadt CW, Wu L, Liebich J, Chong SC, et al. (2007) GeoChip: a comprehensive microarray for investigating biogeochemical, ecological and environmental processes. ISME J 1:67–77

Head DS, Cenkowski S, Holley R, Blank G (2008) Effects of superheated steam on *Geobacillus stearothermophilus* spore viability. J Appl Microbiol 104:1213–1220

Huang Y, Dai X, He L, Wang YN, Wang BJ, Liu Z, Liu SJ (2005) *Sanguibacter marinus* sp. nov., isolated from coastal sediment. Int J Syst Evol Microbiol 55:1755–1758

Huber H, Hohn MJ, Rachel R, Fuchs T, Wimmer VC, Stetter KO (2002) A new phylum of Archaea represented by a nanosized hyperthermophilic symbiont. Nature 417:63–67

Hungate RE (1969) A roll tube method for cultivation of strict anaerobes. In: Norris JR, Ribbons DW (eds) Methods in microbiology. Academic Press, New York, pp 117–132

Karner MB, DeLong EF, Karl DM (2001) Archaeal dominance in the mesopelagic zone of the Pacific Ocean. Nature 409:507–510

Kempf MJ, Chen F, Kern R, Venkateswaran K (2005) Recurrent isolation of hydrogen peroxide-resistant spores of *Bacillus pumilus* from a spacecraft assembly facility. Astrobiology 5:391–405

Kendrick MG, Kral TA (2006) Survival of methanogens during desiccation: implications for life on Mars. Astrobiology 6:546–551

Küsel K, Wagner C, Drake HL (1999) Enumeration and metabolic product profiles of the anaerobic microflora in the mineral soil and litter of a beech forest. FEMS Microbiol Ecol 29:91–103

Kwan K, Cooper M, La Duc MT, Vaishampayan P, Stam C, Benardini JN, Scalzi G, Moissl-Eichinger C, Venkateswaran K (2011) Evaluation of procedures for the collection, processing, and analysis of biomolecules from low-biomass surfaces. Appl Environ Microbiol. 77(9):2943–53. Epub 2011 Mar 11

La Duc MT, Kern R, Venkateswaran K (2004) Microbial monitoring of spacecraft and associated environments. Microb Ecol 47:150–158

La Duc MT, Nicholson W, Kern R, Venkateswaran K (2003) Microbial characterization of the Mars Odyssey spacecraft and its encapsulation facility. Environ Microbiol 5:977–985

La Duc MT, Dekas A, Osman S, Moissl C, Newcombe D, Venkateswaran K (2007) Isolation and characterization of bacteria capable of tolerating the extreme. Appl Environ Microbiol 73:2600–2611

La Duc MT, Osman S, Vaishampayan P, Piceno Y, Andersen G, Spry JA, Venkateswaran K (2009) Comprehensive census of bacteria in clean rooms by using DNA microarray and cloning methods. Appl Environ Microbiol 75:6559–6567

Landis GA (2001) Martian water: are there extant halobacteria on Mars? Astrobiology 1:161–164

Link L, Sawyer J, Venkateswaran K, Nicholson W (2004) Extreme spore UV resistance of *Bacillus pumilus* isolates obtained from an ultraclean spacecraft assembly facility. Microb Ecol 47:159–163

Moeller R, Setlow P, Horneck G, Berger T, Reitz G, Rettberg P, Doherty AJ, Okayasu R, Nicholson WL (2008) Roles of the major, small, acid-soluble spore proteins and spore-specific and universal DNA repair mechanisms in resistance of *Bacillus subtilis* spores to ionizing radiation from X-rays and high-energy charged (HZE) particle bombardment. J Bacteriol 190:1134–1140

Moissl C, Osman S, La Duc MT, Dekas A, Brodie E, DeSantis T, Venkateswaran K (2007) Molecular bacterial community analysis of clean rooms where spacecraft are assembled. FEMS Microbiol Ecol 61:509–521

Moissl C, Bruckner JC, Venkateswaran K (2008) Archaeal diversity analysis of spacecraft assembly clean rooms. ISME J 2:115–119

Moissl-Eichinger C (2010) Archaea in artificial environments: their presence in global spacecraft clean rooms and impact on planetary protection. ISME J 2011 5(2):209–219

Nagarkar PP, Ravetkar SD, Watve MG (2001) Oligophilic bacteria as tools to monitor aseptic pharmaceutical production units. Appl Environ Microbiol 67:1371–1374

National Research Council (2006) Expanding our knowledge of the limits of life on Earth. In: Preventing the forward contamination of Mars. National Academies Press, Washington, pp 69–90

Newcombe DA, Schuerger AC, Benardini JN, Dickinson D, Tanner R, Venkateswaran K (2005) Survival of spacecraft-associated microorganisms under simulated martian UV irradiation. Appl Environ Microbiol 71:8147–8156

Newcombe D, Dekas A, Mayilraj S, Venkateswaran K (2009) *Bacillus canaveralius* sp. nov., an alkali-tolerant bacterium isolated from a spacecraft assembly facility. Int J Syst Evol Microbiol 59:2015–2019

Nicholson WL, Schuerger AC (2005) *Bacillus subtilis* spore survival and expression of germination-induced bioluminescence after prolonged incubation under simulated Mars atmospheric pressure and composition: implications for planetary protection and lithopanspermia. Astrobiology 5:536–544

Nicholson WL, Munakata N, Horneck G, Melosh HJ, Setlow P (2000) Resistance of *Bacillus* endospores to extreme terrestrial and extraterrestrial environments. Microbiol Mol Biol Rev 64:548–572

Osman S, Peeters Z, La Duc MT, Mancinelli R, Ehrenfreund P, Venkateswaran K (2008) Effect of shadowing on survival of bacteria under conditions simulating the Martian atmosphere and UV radiation. Appl Environ Microbiol 74:959–970

Peters V, Conrad R (1995) Methanogenic and other strictly anaerobic bacteria in desert soil and other oxic soils. Appl Environ Microbiol 61:1673–1676

Pillinger JM, Pillinger CT, Sancisi-Frey S, Spry JA (2006) The microbiology of spacecraft hardware: lessons learned from the planetary protection activities on the Beagle 2 spacecraft. Res Microbiol 157:19–24

Plumb RC, Bishop JL, Edwards JO (1993) The pH of Mars. In: Lunar and Planetary Inst., Mars: past, present, and future. Results from the MSATT program, Part 1:40–41

Probst A, Vaishampayan P, Osman S, Moissl-Eichinger C, Andersen GL, Venkateswaran K (2010a) Diversity of anaerobic microbes in spacecraft assembly clean rooms. Appl Environ Microbiol 76:2837–2845

Probst A, Facius R, Wirth R, Moissl-Eichinger C (2010b) Validation of a nylon-flocked-swab protocol for efficient recovery of bacterial spores from smooth and rough surfaces. Appl Environ Microbiol 76:5148–5158

Puleo JR, Fields ND, Bergstrom SL, Oxborrow GS, Stabekis PD, Koukol R (1977) Microbiological profiles of the Viking spacecraft. Appl Environ Microbiol 33:379–384

Reasoner DJ, Geldreich EE (1985) A new medium for the enumeration and subculture of bacteria from potable water. Appl Environ Microbiol 49:1–7

Rudolph C, Moissl C, Henneberger R, Huber R (2004) Ecology and microbial structures of archaeal/bacterial strings-of-pearls communities and archaeal relatives thriving in cold sulfidic springs. FEMS Microbiol Ecol 50:1–11

Rummel JD (2009) Special regions in Mars exploration: problems and potential. Acta Astronaut 64:1293–1297

Schuerger AC, Mancinelli RL, Kern RG, Rothschild LJ, McKay CP (2003) Survival of endospores of *Bacillus subtilis* on spacecraft surfaces under simulated martian environments: implications for the forward contamination of Mars. Icarus 165:253–276

Schulze-Makuch D, Grinspoon DH (2005) Biologically enhanced energy and carbon cycling on Titan? Astrobiology 5:560–567

Spang A, Hatzenpichler R, Brochier-Armanet C, Rattei T, Tischler P, Spieck E, Streit W, Stahl DA, Wagner M, Schleper C (2010) Distinct gene set in two different lineages of ammonia-oxidizing archaea supports the phylum Thaumarchaeota. Trends Microbiol. 18(8):331–340. Epub 2010 Jul 2

Stackebrandt E, Sproer C, Rainey FA, Burghardt J, Pauker O, Hippe H (1997) Phylogenetic analysis of the genus *Desulfotomaculum*: evidence for the misclassification of *Desulfotomaculum guttoideum* and description of *Desulfotomaculum orientis* as *Desulfosporosinus orientis* gen. nov., comb. nov. Int J Syst Bacteriol 47:1134–1139

Stieglmeier M, Wirth R, Kminek G, Moissl-Eichinger C (2009) Cultivation of anaerobic and facultatively anaerobic bacteria from spacecraft-associated clean rooms. Appl Environ Microbiol 75:3484–3491

Tally FP, Stewart PR, Sutter VL, Rosenblatt JE (1975) Oxygen tolerance of fresh clinical anaerobic bacteria. J Clin Microbiol 1:161–164

Thomas DJ, Boling J, Boston PJ, Campbell KA, McSpadden T, McWilliams L, Todd P (2006) Extremophiles or ecopoiesis: desirable traits for and survivability of pioneer Martian organisms. Grav Space Biol 19:91–104

Venkateswaran K, Satomi M, Chung S, Kern R, Koukol R, Basic C, White D (2001) Molecular microbial diversity of a spacecraft assembly facility. Syst Appl Microbiol 24:311–320

von Wintzingerode F, Goebel UB, Stackebrandt E (1997) Determination of microbial diversity in environmental samples: pitfalls of PCR-based rRNA analysis. FEMS Microbiol Rev 21: 213–229

Subject index

A

Acidophile   2, 9, 69, 99, 201, 236, 242, 245
Air-dried pharmaceutical   127
Alaska   55
Algae   11, 51, 93, 95, 100, 101, 104, 121, 133, 134, 136, 138, 141, 149, 157, 209, 210
Alkaliphile   2, 9, 99, 201, 236, 237, 242–245
Alkalithermophilic bacteria   201
Alkaline hot spring   44, 46–49, 51, 56
Alkaline phosphatase   145, 146, 151
Alpine environment   134
 – lake   8–10, 13, 21–23, 25, 26, 31, 88, 89, 92, 95–101, 103, 105, 108–111, 134, 136, 137, 139, 157, 173, 175–186, 188, 189, 191, 192, 208
Alvord desert   58
Amber   14, 207
Ammonium oxidizer   33
Amylase   141, 143–146, 148–150, 152, 153, 188
Anaerobe   4, 56, 135, 155, 236, 243–248
Anaerobic brines   22, 24, 28
 – condition   1, 3, 4, 6, 8, 9, 11–14, 21, 24, 26, 28, 33, 34, 37, 39, 46, 48, 50, 56–58, 62, 63, 69–73, 78, 80–82, 87–89, 92, 94, 99, 101, 107, 119, 122, 124–126, 128, 133, 139, 143, 150, 151, 154, 156–158, 185, 189, 191, 199–208, 212, 218, 234–241, 247–251, 253–255
 – cultivation   27, 28, 74, 81, 110, 144, 190, 234, 237, 239–242, 247–253, 255–257
Anhydrobiosis   123, 125, 127
Antarctic continent   87–90, 92, 93, 99, 107, 121, 140
 – environment   1, 3, 4, 6–11, 13, 14, 21, 23, 26, 27, 33, 37, 39, 43, 44, 48–50, 53, 56–63, 69–73, 77–82, 87–93, 97–100, 102, 103, 105, 107, 108, 110, 111, 119–121, 123–126, 133–136, 138–141, 143, 145, 148, 150–152, 154, 156, 157, 173–182, 185, 186, 188, 191, 192, 199–216, 218, 219, 221, 231, 234, 237, 238, 240, 246, 247, 249, 251, 253, 254, 256
 – extremophile   1, 3, 7, 9, 21, 23, 87, 88, 90, 97, 99, 102, 103, 109–111, 119, 199–201, 203, 204, 206, 212, 213, 215, 218, 219, 222, 231, 237, 251–253
 – ice core   101, 103, 110, 142
 – lake   136, 139
 – marine sediment   96, 98, 150

Subject index

– Peninsula   41, 89, 94, 102
– soils   9, 13, 41, 70, 72, 74–76, 82, 89, 91, 92, 96, 98, 101, 109, 134–137, 140, 145, 149, 151, 154, 155, 157, 185, 208, 210, 211, 247
Antibiotic   144–150, 189, 209, 236
Antibody microarray   213–216
Antifreeze protein   107, 142, 154, 158
Antinucleating protein   142
Atacama desert   119, 120, 122, 201–203, 208, 211, 217, 219
Atmosphere   3, 4, 88, 89, 122, 127, 204, 217, 220, 222, 241, 246
Autotrophs   51, 52, 218, 236, 254

B

Baeocyte   121
Balta Albă lake   182
Barberton Greenstone Belt   3
Beacon Valley   125
Bioburden   232–234
Biodegradation   148, 155, 186
Biofilm   9, 58, 110, 120, 154, 208–210, 216, 220, 221
Biogeography   39, 52, 60, 93
Biomarker   103, 201, 202, 204–216, 219–222
Biomining   155, 158
Biopolymer   152, 186, 189, 191, 201, 211
Bioremediation   79, 145–148, 151, 152, 155, 156, 158, 186
Bioscreening   144
Biosignature   11, 211, 219, 221, 222
Biotechnological application   33, 110, 186
– potential   14, 23, 26, 33, 34, 72, 79, 80, 88, 92, 96–99, 101, 102, 110, 111, 119, 122, 125, 126, 128, 140, 154, 157, 173, 178, 186, 187, 199, 201–204, 215, 220, 232, 247, 250
– process   3, 4, 9, 24, 26, 29, 31, 33, 59, 79, 91, 98, 103, 106, 108, 110, 124, 127, 133, 142, 144, 146, 153, 155, 156, 185, 187, 189, 203, 210, 211, 221, 222, 232, 233, 238, 255
Bratina island   100
Bride cave   183–185, 188
Brine basin   21, 22
– pool   21, 22, 25, 27, 29, 30, 41, 43–45, 49, 57, 58, 107, 175, 176

C

Cape Russell   100
CAREX   88, 223
Carotenoids   143, 209, 219
Carpathian   183, 184, 191
Cellulase   108, 141, 146, 148, 150, 154
Champagne pool   57, 58
Chasmoendolith   120
Chasmolith   96
Chefren mud volcano   23, 27, 30, 31
Chemocline   21, 24, 26, 31–33
Chemolithoautotroph   50, 246
Chitinase   150
Chlorophyll   92, 95, 209, 219
Cold active enzyme   104, 152
– acclimation protein   106, 142
– cave   98, 135, 138, 139, 183–185, 188, 191
– desert   9, 10, 58, 69–82, 91, 96, 119–126, 128, 136, 157, 201–203, 205, 208, 211, 217, 219, 247
– inducible protein   106
– lakes   139
– marine waters   139, 156
– seep   30, 35, 42, 90, 98
– shock   104, 106–110, 126, 127, 140, 143, 233, 239, 240, 255

264

Copiotroph   10
Cosmopolitan   10, 53, 61, 102, 109
COSPAR   14, 231, 232
Crenarchaeota   28, 78, 81, 252
Cryoconite hole   97, 138
Cryoprotection   143
Crystallizer pond   173, 185
Culture Collection   13, 121
Cyst   72, 77, 79, 80, 123

D

Dead Sea   8, 10, 26, 173, 176, 180, 192
Deception island   93, 94, 217
Dehydration   81, 125, 207
Desiccation tolerance   12, 71, 72, 79, 81, 82, 124, 125, 127, 128
Dextranase   149, 153
DHAB (Deep hypersaline anoxic basin)   21, 23–28, 31–34
Diatoms   107, 134, 137, 138, 141
DNA damage   70, 72, 80, 81, 124
– repair   14, 70, 72, 79–82, 110, 124, 125
Dormancy   70
Dormant state   13
Dronning Maud Land   103
Dry heat   238, 253
Dry Valley   89, 98–100, 103, 105, 119–121, 134, 136, 205, 216

E

Electron acceptor   21, 25, 26, 29, 63, 156, 205
Electron donor   31, 50, 63
Endolithic communities   122, 136
Endoliths   96, 134
Endospore   12–14, 71, 237
Epifluorescence microscopy   103

EPS (extracellular polysaccharide substances)   107, 125, 128, 143, 154, 189–191, 208–210, 215, 221
European Space Agency   212, 220, 257
Europa   110, 111, 199, 200, 203, 205, 232
Euryarchaeota   28, 32
ExoMars   212, 214, 217, 220, 237
Exopolysaccharide   100, 146, 189, 208
Extant life   120, 199, 203, 204, 207, 209, 217, 220, 221
Extinct life   209, 210, 220, 221
Extracellular enzyme   150, 188
– polymer   72, 128, 146, 153, 156, 190, 207, 209, 213–215
– polysaccharide   125, 146, 150, 189, 207, 209
Extraterrestrial life   204

F

Fatty acids   107, 140, 141, 148, 152, 157, 209, 211, 212
FISH (fluorescent in situ hybridization)   27, 30, 31, 102, 252
Fluid inclusion   8, 178
FOTON   12, 127
Fumarole   41, 43–46, 49

G

β-Galactosidase   145–151
Genospecies   38, 54, 61, 63
Geographical distribution   38, 60
Geothermal activity   58, 97, 99
– area   22, 23, 27, 37–39, 41–46, 48, 49, 52, 53, 55–59, 61, 69, 79, 88, 89, 93, 122, 133–137, 144, 152, 155, 156, 182–186, 200, 203, 205, 223, 249

– ecosystem   9, 22, 23, 33, 37, 50, 53, 59, 87–89, 92, 93, 98, 99, 101, 111, 120, 134, 135, 157, 187, 200, 202–205, 213
– field   9, 39, 41, 42, 44–46, 48, 51, 57, 98, 107, 111, 113, 127, 153, 158, 177, 181, 192, 200, 214, 215, 217, 231, 242, 255, 257
– habitat   1, 2, 6, 7, 10, 11, 22, 23, 28–30, 33, 37, 39, 48, 50, 53, 57, 60, 62, 63, 88, 94, 96, 98–100, 109–111, 122, 126, 130, 201, 203, 209, 220
– region   10, 21, 29, 40, 41, 44, 48, 50, 53, 55–58, 60, 61, 63, 71, 73, 77, 78, 87, 88, 90, 93–95, 99, 100, 110, 121, 133, 141, 175, 176, 179, 200, 205, 212, 213, 232
– vent   7, 9, 11, 48, 49, 56, 90, 93, 98, 200, 201, 203
– water   1–3, 7–13, 21–27, 31–33, 37, 41–49, 56–58, 60, 69–72, 78, 81, 82, 87–91, 95–97, 101, 106, 108, 110, 111, 119, 122, 123, 125, 126, 128, 133–136, 138, 139, 142, 150, 154–157, 173, 174, 180, 181, 183, 185, 187, 199–206, 208–211, 214, 217, 218, 232, 233, 238, 241, 250
Geyser   41
Great Salt Lake   8, 173, 177, 181, 192
Green Bath   183, 184, 188
Greenland   3, 101, 103, 138
Gulf of Mexico   21, 22, 24, 29, 180

H

Habitability   199, 202–204, 222
Halite   8, 25, 122, 178, 191
Haloarchaea   8, 14, 174, 178, 188
Halocin   187, 189

Halocline   23, 24, 26, 28, 32, 33
Halophilic archaea   175, 178, 186, 187, 190, 192
– bacteria   2–6, 8, 9, 11, 13, 14, 25, 27–33, 37, 38, 45, 47, 52, 53, 58, 59, 69, 70, 72–75, 77, 78, 80–82, 89, 90, 92–94, 96, 98–101, 103–105, 107, 109, 110, 121–125, 134–139, 141, 144, 145, 147, 150, 155–157, 173, 174, 179, 180, 181, 186, 187, 190, 192, 201, 202, 205, 208, 209, 235, 237, 239–242, 245, 247, 251–254, 256
Heat shock protein   108
– response   10, 21, 31, 55, 62, 69, 70, 72, 80, 82, 92, 104, 106–109, 123, 142, 209, 218
Heteropolysaccharide   128, 143, 153, 190
High pressure   10, 21, 23, 90, 97–99, 135, 154, 157, 201
High temperature field   41, 42, 44–46
Hot vent   200, 201, 203
Hydrogen peroxide   253
Hydrogenotrophs   30
Hydrothermal vent   11, 49, 56, 90, 93, 98, 203
Hypersaline environment   33, 173, 174, 181, 182, 185, 186, 191, 192, 209, 215
Hyperthermophile   2, 7, 60, 69, 109, 200

I

Iberian pyritic belt   200
Ice active substance   107, 110
– cave   98
Iceland   7, 11, 37, 41–50, 52, 53, 55–57, 59, 61, 63, 216

Subject index

International Space Station 119
Ionizing radiation 12, 75, 80, 81, 124–126, 128, 238, 239
Iron superoxide dismutase 124
Isua rocks 3
Italy 7, 21, 41, 119, 177

J

Japan 7, 11, 41, 53, 61, 152, 173, 177, 180, 181
Juan de Fuca ridge 60

K

Kamb Ice Stream 98
Kamchatka 41, 55
Kebrit Deep 22, 24, 28, 33
Kolyma Lowland 98, 105

L

Lake Fryxell 99, 105, 136
 – Magadi 177, 180
 – sediments 8–14, 23, 25, 27, 31, 41, 62, 96–98, 105, 109, 136, 139, 152, 156, 157, 208–210, 214
 – Vostok 87, 98, 110, 139
Lateral gene transfer 38, 53, 59
Lichen 11, 96, 100, 105, 106, 121, 123, 126, 135, 136, 240, 243, 253
Life detection 103, 199, 203–205, 212–215, 218, 219, 231, 233
Life Marker Chip 217
Lipase 135, 143, 145, 147–150, 152, 188
Lipid biomarkers 211
Lithobiont 120
Lithopanspermia 126
Low Earth Orbit 12, 126

M

Mars exploration 200
 – lander 202, 217, 218, 220, 232, 233, 237, 239
 – Odyssey 200, 201, 239
 – regolith 212, 218, 242
 – sample return 237
 – science laboratory 220
Marte project 214, 217
Martian analog 126, 205
 – atmosphere 3, 4, 88, 89, 122, 127, 204, 217, 220, 222, 241, 246
 – meteorite 78, 79, 103, 126, 127, 200, 207, 211, 212, 214, 215, 221
 – polar cap 95, 111
 – soil 9, 11, 13, 14, 41, 44, 45, 48, 49, 52, 70–72, 74–80, 82, 89, 91, 92, 96–98, 100, 101, 109, 120–122, 126, 134–137, 140, 145, 149, 151, 152, 154–157, 174–177, 179–181, 185, 202, 203, 208–211, 218, 239, 240, 246, 247, 251
 – subsurface 11, 14, 25, 41, 42, 56, 60, 200–205, 210, 212–214, 220–222, 246
 – surface 3, 4, 12–14, 27, 30, 41–45, 48, 56, 70–72, 74, 77–79, 81, 87, 90, 93, 95, 97, 98, 107, 110, 111, 120, 136, 141, 157, 183–188, 199–202, 207, 210–212, 217, 218, 232–235, 238, 241, 246, 247, 249, 250, 252
Mass spectrometer 211, 220
McKelvey Valley 121
McMurdo Dry Valley 89, 99, 103, 105, 119, 134, 136
Mediterranean Sea 21, 23, 25, 27, 28, 31, 173
Membrane fluidity 99, 106, 107, 140, 141

267

– lipids 29, 106, 123, 140, 141, 150, 152, 157, 178, 207, 209, 211, 213
Merzouga Dunes 73, 77
Metabolic activity 29, 99, 105, 122, 124, 204, 213, 217, 218
Metal tolerance 190
Methane oxidation 29, 30, 136
Methanogenesis 29, 32
Methanogenic archaea 103, 137, 236
Microarray 125, 213–216, 248, 257
Microbial ecosystem 98
– mat 27, 30, 31, 37, 45, 47–50, 53, 58, 62, 95, 96, 99, 105, 110, 135, 150, 233
Microfossils 3, 4
Microscopic eukaryote 133
Molecular biomarkers 207, 208, 210–212, 219
– methods 14, 28, 38, 48, 53, 57, 72, 78, 98, 110, 121, 136, 183, 187, 191, 219, 233, 234, 237, 247, 249, 252, 255–257
Mud pool 41, 43–45, 49
– volcano 27, 29–31, 40, 90
Mycosporine 219

N

Nanobiotechnology 144, 187
NASA 1, 12, 126, 214, 215, 220, 231, 233–235, 240, 241, 248, 252
Negev desert 121, 125
Nematodes 12
New Zealand 7, 41, 54, 56–58, 61, 106, 211

O

Ocnele Mari 182, 185
Oligotrophs 10, 236, 241, 242, 245

Oligotrophic environment 10, 82, 88, 97
Ornithogenic soil 91, 96
Osmoprotectant 72
Osmotic stress 88, 143, 187
OTU (operational taxonomic unit) 31, 74, 102
Outer space treaty 231

P

Pectinase 149, 152
Penguin guano 91, 96
Permafrost 12–14, 98, 99, 105, 135–137, 140, 157, 200–202, 205, 208, 210, 249
Photosynthesis 50, 95, 106, 119–121, 123, 127, 218, 222
Phylogenetic tree 5, 6, 178, 189
Phylogeography 54
Physico-chemical factors 1, 12
– limits 1, 3, 8, 10, 62, 87, 88, 103, 119, 120, 124, 133, 199, 203, 204, 222, 232, 234
Phytoplankton 92
Piezophiles 99
Pigment 95, 105, 106, 124, 174, 209, 219, 222
Pilbara formation 3
Planetary protection 231, 232, 237, 242, 252, 253, 255–257
Polar regions 10, 71, 87, 90, 100, 212
Polyextremophile 135, 201
Pressure 2, 10, 21, 23, 39, 47–50, 53, 69, 88, 90, 92, 93, 97–99, 111, 123, 126, 127, 135, 139, 154, 157, 185, 200, 201, 203, 206, 238
Protease 135, 143, 145, 147–152, 188
Protein damage 124
– microarray 213

Proteome   81, 128, 141
Psychrophiles   7, 8, 89, 95, 99–101, 105, 109, 133, 135, 140–144, 157, 158, 200, 236, 237, 243–245, 249, 254
Psychrotolerant   89, 100, 103, 133–136, 138, 140, 143–145, 150, 152, 154, 157, 158
PUFA   141, 146, 154

R

Radiation   4, 12, 13, 61, 63, 70, 72, 73, 75, 77, 80–82, 87, 88, 92, 93, 95, 105, 124–126, 128, 136, 157, 199–204, 207, 209, 211, 212, 219, 238, 239, 246, 250, 251, 253
Radiation tolerance   80, 81
Raman spectroscopy   178, 219
 – spectrometer   201, 211, 219–222
Red Bath   183–185, 188
 – Sea   2, 8, 10, 11, 21–28, 30–33, 42, 49, 57, 60, 89, 90, 93, 94, 98–100, 104, 107, 110, 134–135, 137, 138, 150, 152, 173, 175–177, 180–183, 185, 192
Reactive oxygen species   80, 123
Redox potential   26, 204
Rio Tinto   202, 206, 214, 220
Romania   133, 158, 173, 177, 182, 185, 190, 191
Rotifers   12, 123, 138
rRNA gene   28–32, 101, 103, 121, 122, 134, 178, 189, 248, 251, 252, 255, 256

S

S-layer   187, 188
Sabkha   173, 175, 179
Sahara   10, 70–73, 75, 81

Salinity   14, 21, 24–29, 33, 42, 87, 92, 173, 185, 189, 190, 200, 201
Salt cave   191
 – chamber   41, 42, 191, 218
 – deposit   8, 22, 122, 173, 184, 186, 200, 202, 214
 – mine   2, 9, 14, 82, 156, 175, 182, 184, 191, 202, 205, 216
 – sediment   2, 8–14, 23, 25, 27, 28, 30, 31, 41, 62, 96–98, 100, 104, 105, 109, 136, 139, 150, 152, 156, 157, 177, 208–210, 214, 246
Sand grain   74, 77, 81
Sequences   5, 6, 28, 30–33, 58, 72, 77–79, 81, 96, 103, 104, 107, 109, 122, 125, 136, 138, 139, 144, 178, 189, 251, 252, 255
Shaban Deep   22, 24, 28, 32, 177
Shepherd Bath   183, 184, 186, 188, 190
SIMCO   95, 138
Siple Station   103
Slănic Prahova   182–184
Soda lake   9, 10, 175, 176, 178, 179
Solar saltern   10, 173, 175–177, 179–181, 185
Solfatara   41, 44, 45, 49, 51, 57
SOLID (signs of life detector)   27, 74, 90, 111, 155, 182, 187, 208, 213, 215, 217, 220, 246
Solar radiation   12, 87, 92, 202
Solute   24, 69, 70, 107, 187, 208, 209
South Africa   205, 216
Space condition   13, 14, 125, 126
Spacecraft assembly clean room   231, 233, 237–243, 248–252, 256, 257
Speleotherapy   191
Spores   71, 78, 79, 123, 126, 135, 137, 209, 214, 234, 237–241, 250, 253, 256

Subject index

Sporulation 82
Sterility 232, 233
Sterilization 232, 234, 238, 253
Stress protein 127, 128
Subglacial lakes 88, 97, 98, 111
Subsurface ecosystem 202, 204
Sucrose 72, 127, 189
Sulfate reduction 136, 139, 157
Swaziland 3

T

Tardigrade 11, 12, 123, 126, 137, 138
Tataouine 73, 77, 78, 81
Taylor Dome 103
 – Valley 9, 10, 89, 98–100, 103, 105, 119–121, 125, 134, 136, 177, 181, 183, 205, 216
Techirghiol 182, 183, 185, 186, 188, 191
Telega 177, 183, 184, 186, 189
Terrestrial analog 120, 199, 200, 206, 211, 220
Thaumarchaeota 252
Thermal Emission Spectrometer 201
Tholin 212
Trehalose 72, 80, 125, 127, 143, 154, 209

U

Ultraviolet radiation 4, 95, 211
UREY instrument 212, 214
UV radiation 70, 80, 82, 105, 126, 136, 199, 201, 202, 207, 209, 212, 219, 250
UVB radiation 88, 93, 105

V

Vacuum 12, 125, 251, 255
VBNC (viable but not culturable) 110
Vestfold Hills 92, 108
Viking landers 217, 232
 – mission 12, 13, 121, 199, 200, 203, 204, 206, 211, 212, 214, 217–220, 231, 232–235, 237, 238, 246
Viruses 3–7, 10, 11, 14, 27, 55, 101, 134, 137, 138
Volcano 21–23, 25, 27, 29, 30, 31, 40, 90
Vonarskard 45, 46
Vostok station 87

W

Waiotapu 57
Water activity 2, 8, 63, 135, 136, 139, 157, 200–204, 206
Warrawoona 3

X

Xerophile 2, 8
Xylanase 149, 153, 156

Y

Yellowstone National Park 1, 7, 11, 41, 50, 53, 55, 60, 216

270

Organism index

A

Absidia psychrophila   135
Acacia   72
Acaryochloris marina   136
Acetobacterium bakii   136
  – *tundrae*   136
Achromobacter   145
Acidianus infernus   51
Acidithiobacillus ferrooxidans   157, 205
Acinetobacter   138, 142, 145, 151, 155, 241, 243, 249
  – *aceticus*   142
  – *ursingi*   243
Acremonium   135
  – *antarcticum*   99, 105, 135, 145, 156
  – *psychrophilum*   135
Actinoalloteichus spitiensis   75
Actinomadura namibiensis   75
Actinomycetes   149, 239
Actinopolyspora halophile   179
Actinotalea   243
  – *fermentans*   180, 243
Aerococcus   243, 250
  – *urinaeequi*   243
Aeromonas   145, 150, 151, 154
Agrococcus lahaulensis   75
Alkalibacillus   179–181
  – *salilacus*   179
  – *halophilus*   179
  – *flavidus*   180
  – *silvisoli*   181
Alkaliphilus transvaalensis   2
Alternaria   136
Alvinella pompejana   11
Amycolatopsis australiensis   75
Anabaena   78, 126, 128
  – *cylindrica*   126
Ancylonema   134
Ankistrodesmus   101
Aquisalibacillus elongatus   179
Archaeoglobus fulgidus   52
  – *profundus*   52
Arcobacter   30, 139
  – *sulfidicus*   30
Arhodomonas aquaeolei   181
Arsenicicoccus   243
  – *bolidensis*   243
Artemia salina   12
  – *franciscana*   12
Arthrobacter   72, 77, 137, 139, 140, 145, 149, 151, 155, 243
  – *psychrolactophilus*   145, 151, 156
  – *psychrophenolicus*   145, 155
Aspergillus   135, 151
Aureobasidium   135, 239, 253
  – *pullulans*   135, 253
Azotobacter   72, 79

Organism index

B

Bacillus   12–14, 51, 52, 72, 77, 137, 150, 201, 235, 238, 239, 240, 242, 243, 248, 249, 251, 253, 254
- *aidingensis*   179
- *badius*   243
- *cereus*   240, 245
- *chagannorensis*   179
- *canaveralius*   245
- *coagulans*   243, 250
- *firmus*   150
- *gibsonii*   243
- *halochares*   180
- *licheniformis*   240, 243, 253
- *megaterium*   240, 243, 251
- *persepolensis*   180
- *pumilus*   240, 243, 253
- *salarius*   179
- *subtilis*   13, 126, 145, 149, 150, 237, 238, 240, 253
  - *thermoamylovorans*   243
  - *thuringiensis*   240, 245

Bacteroides   74, 134
Balneimonas   241, 243
Bdellovibrio   96
Brachybacterium paraconglomeratum   243
Brevibacillus   240, 243, 250
- *agri*   243

Brevibacterium   243
- *antarcticum*   145, 156
- *frigotolerans*   243

Brevundimonas   241, 242, 243, 249
- *diminuta*   243
- *nasdae*   243

Burkholderia   155

C

Caldanaerovirga acetigignens   75
Candida   135
- *albicans*   149
- *antarctica*   149, 151

Cellulomonas   243
- *hominis*   243

Cellulophaga   139
Chamaesiphon   134
Chelatococcus   77
Chlamydomonas   104, 134
- *intermedia*   134
- *nivalis*   101
- *raudensis*   134

Chloroflexus   53
- *aurantiacus*   51

Chloromonas   101, 134
Chromobacterium   150
Chromohalobacter japonicus   180
Chromulina   134
Chroococcidiopsis   120–123, 125–129, 134
Chryseobacterium aquifrigidense   140
- *greenlandense*   138

Chrysosporium   135
Citricoccus alkalitolerans   75
Citrobacter   241, 243
- *werkmanii*   243

Cladosporium   135
Closterium   134
Clostridium   12, 13, 51, 135, 201, 238, 240, 243, 247, 248
- *algidixylanolyticum*   135
- *bowmannii*   135
- *frigidicarnis*   135
- *frigoris*   135
- *perfringens*   243
- *psychrophilum*   135
- *sphenoides*   248

Colwellia   135, 139, 1456, 146, 150, 155
- *demingiae*   145, 146
- *hornerae*   146
- *maris*   146
- *psychrerythraea*   133

– *psychrotropica* 146
– *rossensis* 146
Coprinopsis 253
Corynebacterium 243, 247, 248
– *pseudogenitalium* 243, 248
Cryptococcus 134
– *albidosimilis* 135
– *albidus* 149, 153
– *antarcticus* 135
– *capitatum* 149
– *cylindricus* 149, 152
– *laurentii* 149, 153
– *nyarrowii* 135
– *watticus* 135
Cupriavidus 243
– *gilardi* 243
Cyanidium 216
Cylindrocystis 134
Cystofilobasidium capitatum 149, 152
Cytophaga 74, 101, 136, 138, 139, 155, 157

D

Dactylsporangium roseum 146, 149
Deinococcus 13, 45, 72, 77, 79, 80, 81, 82
– *deserti* 75, 81
– *geothermalis* 81
– *gobiensis* 75
– *peraridilitoris* 75
– *radiodurans* 12, 13, 80, 81, 124, 125, 239
– *xinjiangensis* 75
Dermabacter 243, 248
– *hominis* 243
Desemzia 243, 250
– *incerta* 243
Desulfitobacterium 155
Desulfobacter 30
Desulfobulbus 30
Desulfocapsa 30

Desulfofaba 135, 157
Desulfofrigus 135, 157
Desulfomonile 155
Desulfosarcina 30
Desulfotalea 96, 135, 157
Desulfotomaculum 51, 240, 243, 247, 248
– *guttoideum* 243, 248
Desulfovibrio 157
– *thermophilus* 51
Desulfuromonas palmitatis 96
Desulphotalea psychrophila 143
Dietzia lutea 75
– *psychralkaliphila* 155
Dyadobacter alkalitolerans 75

E

Enterococcus 243
– *casseliflavus* 243
– *faecalis* 243
– *faecium* 243
Erythrobacter litoralis 146
Escherichia coli 10, 104, 127, 144
Exiguobacterium 98, 99, 105
– *antarcticum* 99, 105
– *sibiricum* 98, 99
– *undae* 99

F

Facklamia 243
Ferroplasma acidarmanus 2, 9
Fibrobacter succinogenes 146
Flavobacterium 101, 103, 138, 139, 146, 154, 156
– *frigidarium* 146
– *frigidimaris* 146, 152
– *hibernum* 146
– *limicola* 146, 156
Formosa 139
Fusarium 135, 136

Organism index

G

Galionella   138
Geobacillus   45, 56, 240, 244, 249, 250
  – *caldoxylosilyticus*   244, 250
  – *stearothermophilus*   244
  – *kaustophilus*   244
  – *thermodenitrificans*   244
Geodermatophilus   134
Geomicrobium halophilum   181
Georgenia   244
  – *muralis*   244
Geotrichum   151
Gigaspora   78
Glaciecola chatamensis   146
Gloeocapsa   121, 136
Gracilibacillus halophilus   179
Granulicatella   248

H

Haladaptatus litoreus   175
  – *cibarius*   177
  – *paucilophilus*   178
Halalkalicoccus   178
  – *tibetensis*   176
  – *jeotgali*   177
Halanaerobacter salinarius   179
  – *chitinivorans*   181
Halanaerobaculum tunisiensae   181
Halanaerobium congolense   179
Halarchaeum acidiphilum   177
Halarsenatibacter silvermanii   181
Haloarcula   137, 174, 178, 186
  – *amylolytica*   176
  – *argentinensis*   175
  – *hispanica*   177
  – *japonica*   177
  – *marismortui*   176
  – *quadrata*   176
  – *vallismortis*   177
Halobacillus mangrove   180

Halobacterium   174, 178, 239
  – *jilantaiense*   176
  – *noricense*   175
  – *piscisalsi*   177
  – *salinarum*   2, 174, 175
Halobaculum   137
  – *gomorrense*   176
Halobiforma   178
  – *haloterrestris*   176
  – *lacisalsi*   176
  – *nitratireducens*   175
Halocella cellulosilityca   181
Halococcus   174, 178
  – *dombrowskii*   175
  – *hamelinensis*   175
  – *morrhuae*   174, 176
  – *quingdaonensis*   175
  – *saccharolyticus*   177
  – *thailandesis*   177
  – *salifodinae*   12
Haloechinothrix alba   179
Haloferax   174, 178, 186, 188, 189, 190, 191
  – *alexandrinus*   176
  – *denitrificans*   178
  – *elongans*   175
  – *gibbonsi*   177
  – *larsenii*   175
  – *lucentense*   177
  – *mediterranei*   177, 190
  – *mucosum*   175
  – *prahovense*   177, 185, 189, 190, 191
  – *sulfurifontis*   178
  – *volcani*   176
Halogeometricum borinquense   177
  – *rufum*   175
Halogranum rubrum   175
Halomicrobium katesii   175
  – *mukohataei*   175
Halomonas   139, 155, 186
  – *alkaliantarctica*   100

- *almeriensis* 181
- *cerina* 181
- *fontilapidosi* 181
- *nitroreducens* 179
- *sabkhae* 179
- *variabilis* 181

Halonotius pteroides 175
Halopelagius inordinatus 175
Halopiger aswanensis 176
- *xanaduensis* 176

Haloplanus natans 176
- *vescus* 175

Haloplasma contractile 28
Haloquadratum walsbyi 175
Halorhabdus tiamatea 177
- *utahensis* 31, 177

Halorhodospira neutrifila 179
Halorubrum 134, 178, 186, 188
- *aidingense* 176
- *alkaliphilum* 176
- *arcis* 176
- *californiense* 178
- *chaoviator* 177
- *cibi* 177
- *coriense* 175
- *distributum* 177
- *ejinorense* 175
- *ezzemoulense* 175
- *kocurii* 175
- *lacusprofundi* 106, 175
- *lipolyticum* 176
- *litoreum* 175
- *luteum* 175
- *orientale* 175
- *sodomense* 176
- *tebequichense* 175
- *terrestre* 177
- *tibetense* 176
- *trapanicum* 177
- *saccharovorum* 178
- *vacuolatum* 177
- *xinjiangense* 176

Halosarcina limi 175
- *pallida* 178

Halosimplex carlsbadensis 178
Halospina denitrificans 180
Halospirulina tapeticola 180
Halostagnicola larsenii 176
Haloterrigena daquingensis 175
- *hispanica* 177
- *jeotgali* 177
- *limicola* 176
- *longa* 176
- *saccharevitans* 176
- *salina* 176
- *thermotolerans* 177
- *turkmenica* 177

Halothermothrix orenii 181
Halovibrio denitrificans 180
Halovivax asiaticus 176
- *ruber* 176

Helicobacter pylori 70
Hemichloris antarctica 134
Herminiimonas glaciei 138
Heterocephalum aurantiacum 135
Heteromita 101
Holosticha 101
Hydrogenobacter 38, 52, 216
- *thermophilus* 51

Hymenobacter deserti 75
- *xinjiangensis* 75

I

Idiomarina loihiensis 143
Intrasporangium 149

J

Janibacter 147, 149
- *melonis* 150

Jiangella gansuensis 76

K

Kineococcus xinjiangensis 76
Kocuria 241, 244
  – *aegyptia* 76
  – *rhizophila* 244
  – *rosea* 244
Kordiimonas gwangyangensis 146
Kushneria indalinina 181

L

Lactobacillus algidus 140
Lechevalieria atacamensis 76
  – *deserti* 76
  – *roselyniae* 76
Lentibacillus halophilus 181
  – *jeotgali* 180
  – *juripiscarius* 181
  – *kapialis* 181
  – *halodurans* 179
  – *lacisalsi* 179
  – *persicus* 180
  – *salinarum* 180
Leptospirillum ferrooxidans 205, 206
Leuconostoc 248
Leucosporidium watsoni 155
  – *antarcticum* 135
Lysobacter 244

M

Macrotrachela 12
Marinobacter 139, 155
Marinomonas 139, 150, 155
  – *prymoriensis* 150
Massilia 244, 249
  – *brevitalaea* 244
Mastigocladus laminosus 56
Mesorhizobium gobiense 76
  – *tarimense* 76
Mesotaenium 134

Methanobacterium 134
  – *thermoautotrophicum* 51
Methanobrevibacter 156
Methanococcoides burtonii 106, 143
Methanococcus igneus 52
Methanoculleus 139
Methanogenium frigidum 107, 134, 142, 143
Methanopyrus kandleri 2
Methanosaeta 156
Methanosarcina 134, 139, 156
Methanothermus fervidus 52
Methylococcus 136
Methylocystis rosea 136
Methylohalomonas lacus 180
Mesotaenium 134
Microbacterium 241, 244
Micrococcus 137, 144, 235, 239, 241, 242, 244, 246, 249, 250
  – *luteus* 244
  – *flavus* 244
  – *indicus* 244
  – *mucilaginosus* 244
Micromonospora 147, 149, 240
Mniobia 12
Modestobacter 134
Moraxella 147, 150, 151, 241, 249
  – *osloensis* 244
Moritella 135
  – *yayanosii* 10
Mortierella 135, 136
  – *alpina* 135
Mrakia 135
  – *blollopis* 134
  – *frigida* 104, 149, 152
  – *niccombsi* 134
  – *robertii* 134
Mucor 136
Mycanthococcus 101
Myxococcus 72

N

Natranaerobius thermophilus 179
— *trueperi* 179
Natrialba 178
— *aegyptiaca* 176
— *asiatica* 177
— *chahannaoensis* 175
— *hulunbeirensis* 176
— *magadii* 177
— *taiwanensis* 177
Natrinema altunense 176
— *ejinorense* 176
— *gari* 177
— *pallidum* 178
— *pellirubrum* 177
— *versiforme* 176
Natroniella acetigena 180
Natronoarchaeum mannanilyticum 177
Natronobacterium 9, 174, 178
— *gregoryi* 177
Natronococcus 9, 174, 178
— *amylolyticus* 177
— *jeotgali* 177
— *occultus* 177
Natronolimnobius 178
— *baerhuensis* 175
— *innermongolicus* 175
Natronomonas 9, 178
— *pharaonis* 176, 178
— *moolapensis* 175
Natronorubrum 178
— *aibiense* 176
— *bangense* 176
— *sediminis* 175
— *sulfidifaciens* 176
— *tibetense* 176
Natronovirga wadinatrunensis 179
Nesterenkonia halobia 180
Nitrososphaera gargensis 252
Nocardioides 134, 253

Nocardiopsis alkaliphila 76
— *kunsanensis* 180
— *salina* 179
Nostoc 78, 128
— *commune* 123

O

Oceanibulbus indolifex 147
Oceanobacillus 244
Oerskovia 248
Oleispira 155
— *antarctica* 147, 156
Orenia sivashensis 181
Oscillatoria 72, 78

P

Paecilomyces 135
Paenibacillus 72, 137, 239, 240, 241, 244, 246, 247, 248, 249, 250, 253, 254
— *amylolyticus* 244
— *barengoltzii* 244
— *cookii* 244, 254
— *gansuensis* 76
— *gingsengisoli* 244
— *glucanolyticus* 244
— *harenae* 76
— *pasadenensis* 244
— *purispatii* 255
— *tarimensis* 76
— *telluris* 244
— *wynii* 244
Pantoea anananatis 142
Paracoccus 244
— *yeeii* 244
Pelagibacter ubique 10
Penicillium 101, 135
— *antarcticum* 135
— *lanosum* 154
— *soppi* 154

Phoma exigua   135
Phormidium   72
Photobacterium   154
 - *frigidiphilum*   147, 152
Pichia lynferdii   149, 152
Picrophilus   45
 - *torridus*   9
Planobacterium taklimakanense   76
Planococcus   147, 150, 239
 - *antarcticus*   95
 - *maitriensis*   95
 - *psychrophilus*   95
 - *stackebrandtii*   76
Polaribacter   139
Polaromonas   147, 156
 - *hydrogenivorans*   140
 - *naphtalenivorans*   147, 156
Polypedium vanderplanki   12
Pontibacter akesuensis   76
 - *korlensis*   76
Porphyridium cruentum   149
Prauserella halophila   179
 - *sedimina*   179
Propionibacterium   244, 247, 248
 - *acnes*   244, 247
 - *avidum*   244
Proteus   144
 - *mirabilis*   149
Pseudoalteromonas   147, 149, 150, 151, 155
 - *elyakovii*   150
 - *haloplanktis*   107, 141, 143, 147
 - *tetraodonis*   150
Pseudomonas   101, 136, 138, 147, 149, 150, 152, 154, 155, 242, 244, 249
 - *duriflava*   77
 - *fluorescens*   139
 - *fragi*   140
 - *luteola*   244
 - *stutzeri*   244
 - *xanthomarina*   244

 - *xinjiangensis*   76
Pseudonocardia   134
Psychrobacter   139, 140, 148, 151
 - *adeliensis*   95
 - *arcticus*   143
 - *glacincola*   95
 - *okhotskensis*   147, 151
 - *salsus*   95
Psychromonas   135
 - *ingrahami*   2, 140, 143
Pyrobaculum islandicum   52
Pyrococcus   2
 - *furiosus*   52
Pyrodictium occultum   52
Pyrolobus fumarii   2

R

Ralstonia   155
Ramlibacter   78, 79
 - *henchirensis*   77, 78
 - *tataouinensis*   77, 78
Raphionema   101
Rhizobium   72
Rhodococcus   137, 148, 149, 155
 - *erythrococcus*   148, 155
 - *kroppenstedtii*   77
 - *ruber*   148, 155
Rhodomonas   134
Rhodospirillum   79
Rhodothalassium salexigens   181
Rhodothermus   52
 - *marinus*   54
Rhodotorula   134, 135
 - *psychrophenolica*   149, 155
Rhodovibrio sodomensis   180
 - *salinarum*   180
Roseobacter   139
Roseomonas   244, 249
 - *aquatica*   244
Rubrobacter   82

S

Saccharibacillus kuerlensis 77
Saccharococcus caldoxylosilyticus 250
Saccharomonospora 179
– *halophila* 180
Saccharomyces cerevisiae 10, 144
Saccharopolyspora halophila 179
Saccharothrix 77
Salicola marasensis 180
– *salis* 179
Salimicrobium halophilum 180
Salinibacillus aidingensis 179
Salinibacter rubber 181
Salinicoccus luteus 77
Salinicola halophilus 181
Salinisphaera dokdonensis 180
Salisaeta longa 180
Salmonella enterica 149
Salsuginibacillus halophilus 179
– *kocurii* 179
Sanguibacter 241, 244, 246, 249
– *marinus* 244, 246
Sclerotinia borealis 101
Selenihalanaerobacter shriftii 180
Serratia 151
– *proteamaculans* 148, 150, 151
Shewanella 104, 135, 136, 143, 148, 150, 151, 155, 156
– *donghaensis* 148
– *frigidimarina* 143, 148
– *gelidimarina* 148
– *livingstonensis* 106, 150
– *pacifica* 148
Skermanella xinjiangensis 77
Sphingomonas 244, 249
– *aerolata* 140
– *aurantiaca* 140
– *faeni* 140
– *oligophenolica* 244
– *paucimobilis* 148

– *trueperi* 244
Sphingopyxis alaskensis 109
Sporichthya 134
Sporobolomyces 134
Sporosarcina 239, 240, 249
– *globispora* 249
Staleya 139
Staphylococcus 239, 241, 242, 245, 247, 248, 249, 250, 252
– *aureus* 149
– *epidermidis* 89, 245
– *haemolyticus* 245
– *hominis* 245
– *lugdunensis* 245
– *pasteuri* 245
– *warneri* 245
Stenotrophomonas 241, 245, 249
– *maltophilia* 245
Streptococcus 150, 248
– *mitis* 10, 13
Streptomonospora alba 179
– *amylolytica* 179
– *flavalba* 179
– *halophila* 179, 180
Streptomyces 134, 148–150, 240, 241, 246
– *anulatus* 149, 153
– *fradiae* 149
– *luteogriseus* 245
Streptoverticillium 149
Sulfitobacter 139
Sulfolobus 45, 52, 55, 57
– *acidocaldarius* 51
– *islandicus* 55
– *solfataricus* 55
Sulfurihydrogenibium 45, 47, 50, 52, 53, 57, 58, 59, 60
Sulfurospirillum deleyianum 30
Symploca 78
Synechococcus 48, 53,
– *lividus* 51

T

Tenacibacter   139
Tessaracoccus   245
  – *flavescens*   245
Thermoanaerobacter   51
Thermobrachium   201
Thermococcus   38
Thermocrinis   50, 53, 58, 59
  – *alba*   47
  – *ruber*   47, 51
Thermofilum   51
Thermoplasma   45, 52
  – *volcanium*   51
Thermoproteus tenax   52
Thermotoga maritima   52
  – *neapolitana*   52
Thermus   38, 39, 51, 52, 53, 54, 56, 58, 59, 60, 61, 62, 63
  – *aquaticus*   52, 59, 60, 61
  – *antranikianii*   61
  – *arciformis*   61
  – *brockianus*   60, 61, 62, 63
  – *filiformis*   60, 61
  – *igniterrae*   60, 61, 62, 63
  – *islandicus*   61
  – *kawarayensis*   61
  – *oshimai*   52
  – *rehai*   61
  – *scotoductus*   52, 60, 62
  – *thermophilus*   54, 60, 61, 63
  – *yunnanensis*   61

Thiobacillus   45
Thiohalocapsa halophila   179
Thiohalorhabdus denitrificans   180
Thiohalospira halophila   180
Thiomicrospira halophila   180
Thiomonas   45
Trichoderma viride   135
Trichosporon dulcitum   155

U

Ulvibacter   139

V

Venenivibrio stagnispumantis   58
Vibrio   150
Virgibacillus salinus   179
  – *marismortui*   180
  – *salexingens*   181
  – *halodenitrificans*   187

X

Xanthoria elegans   126
Xeromyces bisporus   8

Z

Zygorrhinchus   136

List of contributors

Wafa Achouak
LEMIRE, UMR 6191
CNRS-CEA-Aix-Marseille Univ., Ibeb
CEA/Cadarache
13108 St-Paul-lez-Durance, France

Mohamed Barakat
LEMIRE, UMR 6191
CNRS-CEA-Aix-Marseille Univ., Ibeb
CEA/Cadarache
13108 St-Paul-lez-Durance, France

Daniela Billi
Department of Biology
University of Rome "Tor Vergata"
Via della Ricerca Scientifica
00133-Rome, Italy
E-mail: billi@uniroma2.it

Snaedis H. Björnsdottir
Matís, Vínlandsleið 12
113 Reykjavík, Iceland

Laurence Blanchard
LEMIRE, UMR 6191
CNRS-CEA-Aix-Marseille Univ., Ibeb
CEA/Cadarache
13108 St-Paul-lez-Durance, France

Sara Borin
Università degli Studi di Milano
DiSTAM, Via Celoria 2
20133 Milan, Italy

Daniele Daffonchio
Università degli Studi di Milano
DiSTAM, Via Celoria 2
20133 Milan, Italy
E-mail: daniele.daffonchio@unimi.it

Arjan de Groot
LEMIRE, UMR 6191
CNRS-CEA-Aix-Marseille Univ., Ibeb
CEA/Cadarache
13108 St-Paul-lez-Durance, France

Gilles De Luca
LEMIRE, UMR 6191
CNRS-CEA-Aix-Marseille Univ., Ibeb
CEA/Cadarache
13108 St-Paul-lez-Durance, France

Madalin Enache
Institute of Biology of the Romanian Academy
Splaiul Independentei 296
P.O. Box 56-53
Bucharest, Romania
E-mail: madalin.enache@ibiol.ro

Sergiu Fendrihan
Romanian Bioresource Centre
and Advanced Research
Aleea Istru nr. 2 C, bl. A14B, sc.8,
et.2 apt.113, sect.6
Bucharest, Romania
E-mail: ecologos23@yahoo.com

List of contributors

Olafur H. Fridjonsson
Matís, Vínlandsleið 12
113 Reykjavík, Iceland

Felipe Gómez Gómez
Centro de Astrobiología (INTA-CSIC)
Instituto Nacional de Técnica Aeroespacial
Carretera de Ajalvir, Km 4
Torrejón de Ardoz
28850 Madrid, Spain
E-mail: gomezgf@inta.es

Thierry Heulin
LEMIRE, UMR 6191
CNRS-CEA-Aix-Marseille Univ., ibeb
Bât 177, CEA/Cadarache
13108 Saint-Paul-lez-Durance, France
E-mail: thierry.heulin@cea.fr

Gudmundur Oli Hreggvidsson
Matís, Vínlandsleið 12
113 Reykjavík, Iceland
E-mail: gudmundur.o.hreggvidsson@matis.is

Takashi Itoh
Japan Collection of Microorganisms
RIKEN BioResource Center
Saitama 351-0198, Japan

Masahiro Kamekura
Halophiles Research Institute
677-1 Shimizu
Noda 278-0043, Japan

Francesca Mapelli
Università degli Studi di Milano
DiSTAM, Via Celoria 2
20133 Milan, Italy

Christine Moissl-Eichinger
University of Regensburg
Institute for Microbiology and Archaea Center
Universitaetsstr. 31
93053 Regensburg, Germany
E-mail: christine.moissl-eichinger@biologie.
uni-regensburg.de

Teodor G. Negoiţă
Romanian Institute of Polar Research
Bucharest, Romania

Philippe Ortet
LEMIRE, UMR 6191
CNRS-CEA-Aix-Marseille Univ., Ibeb
CEA/Cadarache
13108 St-Paul-lez-Durance, France

Víctor Parro
Centro de Astrobiología (INTA-CSIC)
Instituto Nacional de Técnica Aeroespacial
Carretera de Torrejón a Ajalvir, Km 4
Torrejón de Ardoz
28850 Madrid, Spain

David A. Pearce
British Antarctic Survey
Natural Environment Research Council
High Cross, Madingley Road
Cambridge CB3 OET, UK
E-mail: dpearce@bas.ac.uk

Solveig K. Petursdottir
Matís, Vínlandsleið 12
113 Reykjavík
Iceland

Gabriela Popescu
Institute of Biology of the Romanian Academy
Splaiul Independentei 296
P.O. Box 56-53
Bucharest, Romania

Helga Stan-Lotter
Department of Molecular Biology
University of Salzburg
Billrothstr. 11
5020 Salzburg, Austria
E-mail: helga.stan-lotter@sbg.ac.at